Natural Resources and the Environment Series

Series Directors, *Margaret R. Biswas and Asit K. Biswas*

Volume 23 Hazardous Waste Management

1 Environmental Impacts of Production and Use of Energy. *Essam El-Hinnawi,* United Nations Environment Programme.
3 Assessing Tropical Forest Lands: Their Suitability for Sustainable Uses. *Richard Carpenter,* editor, East-West Center.
5 Economic Approaches to Natural Resources and Environmental Quality Analysis. *Maynard M. Hufschmidt and E. Hyman,* editors, East-West Center.
7 Global Environmental Issues. *Essam El-Hinnawi and Manzur-Ul-Haque Hashmi,* editors, United Nations Environment Programme.
8 The World Environment 1971–82. *Martin Holdgate, Mohammed Kassas and Gilbert White,* editors; *Essam El-Hinnawi,* Study Director, United Nations Environment Programme.
9 Energy Alternatives in Latin America. *Francisco Szekely,* editor, United Nations Environment Programme.
10 Integrated Physical Socio-Economic and Environmental Planning. *Yusuf J. Ahmad and Frank G. Müller,* editors, United Nations Environment Programme.
11 Third World and the Environment. *Essam El-Hinnawi,* editor.
12 Development without Destruction: Evolving Environmental Perceptions. *Mostafa K. Tolba.*
13 Climate and Development. *Asit K. Biswas,* editor.
15 The Legal Regime of the Protection of the Mediterranean Against Pollution from Land-Based Sources. *Sachiko Kuwabara,* United Nations Environment Programme.
17 The Evolution of the Law of the Sea. *Bo Johnson Theutenberg.*
18 Photosynthesis in Relation to Plant Productivity in Terrestrial Environments. *C.L. Beadle, S.P. Long, S.K. Imbamba, D.O. Hall and R.J. Olembo,* United Nations Environment Programme.
19 Environmental Impact Assessment in Developing Countries. *Asit K. Biswas,* editor, United Nations University.
20 Global Aspects of Food Production. *M.S. Swaminathan and S.K. Sinha,* International Rice Research Institute.
21 Energy Storage Systems in Developing Countries. *S. Karkkainen and Y. ElMahgary,* United Nations Environment Programme.
22 Genetic Manipulation in Crops. International Rice Research Institute and Academia Sinica.
23 Hazardous Waste Management. *S. Maltezou, A.K. Biswas and H. Sutter,* editors, United Nations Industrial Development Organization and the International Association for Clean Technology.
24 Marine Environment Law in the United Nations Environment Programme. *Peter H. Sand.*

Hazardous Waste Management

Selected papers from an International Expert Workshop convened by UNIDO in Vienna, 22–26 June 1987

Edited by *Sonia P. Maltezou, Asit K. Biswas* and *Hans Sutter*

Published on behalf of
the United Nations Industrial Development Organization UNIDO
and the International Association for Clean Technology I.A.C.T.

TYCOOLY

London and New York

First Published 1989 by Tycooly Publishing, *A Cassell Imprint*
Artillery House, Artillery Row, London SW1P 1RT, England
125 East 23rd Street, Suite 300, New York 10010, U.S.A.

The designations employed and the presentation of material in this publication do not imply the expression of any opinion whatsoever on the part of the Secretariat of the United Nations Industrial Development Organization concerning the legal status of any country, territory, city or area or of its authorities, or concerning the delimitation of its frontiers or boundaries. Where the designation country or area appears in the heading of tables or figures, it covers countries, territories, cities or areas. The designations "developed", "developing" and "less developed" economies are intended for statistical convenience and do not necessarily express a judgement about the stage reached by a particular country or area in the development process. Mention of firm names or commercial products does not imply endorsement by UNIDO. This publication has not been edited by UNIDO.

British Library Cataloguing in Publication Data
Hazardous waste management : Selected papers from an international expert workshop convened by UNIDO in Vienna, Austria, 22–26 June 1987 — (Natural resources and the environment series ; v.23).
1. Dangerous waste materials. Disposal. Management
I. Maltezou, Sonia P. II. Biswas, Asit K. III. Sutter, Hans IV. United Nations, *Industrial Development Organization* V. International Association for Clean Technology VI. Series
363.7′068
ISBN 1-85148-027-7

Library of Congress Cataloging in Publication Data
Hazardous waste management.
(Natural resources and the environment series; v. 23)
"Published on behalf of United Nations Industrial Development Organization and International Association for Clean Technology."
1. Hazardous wastes—Management—Congresses. 2. Hazardous wastes—Congresses. 3. Hazardous wastes—Developing countries—Congresses.
I. Maltezou, Sonia P. II. Biswas, Asit K. III. Sutter, Hans. IV. United Nations Industrial Development Organization. V. International Association for Clean Technology. VI. Series.
TD1020.H39 1988 363.7′28 88-20175
ISBN 1-85148-027-7

This book has been printed and bound in Great Britain: typeset in Baskerville by Area Graphics Ltd, printed on Pegasus Wove paper by Redwood Burn Ltd. and bound by WBC Bookbinders.

Contents

Foreword

The organization of the Workshop on the Management of Hazardous Waste reflected the concern of the United Nations Industrial Development Organization (UNIDO) about the extraordinary complexity and importance of the topic. The task of UNIDO to assist in the promotion and acceleration of industrial development in the developing countries is closely related to the problems and issues associated with hazardous waste, clean technologies, environmental safety and risk management. UNIDO is aware of the role it can play in solving or alleviating these serious problems.

I hope that this book will demonstrate to the reader the enormous breadth of technical, methodological and managerial issues related to the disposal of hazardous waste and underscore the need for optimal choice.

Clearly, any rational effort to manage hazardous waste must concern itself with the technologies available for reducing the generation of waste and the associated costs. Endeavouring to improve water quality, one may unintentionally increase the pollution burdens on land and/or air. Environmental media and alternatives to managing hazardous waste should be viewed as an integrated whole, adapting an approach that blends engineering facts, economic theory and social circumstances. An understanding of these complex interrelationships was to be observed in the deliberations of the 170 experts during the workshop organized by UNIDO. These deliberations, contained in this book, point out that sustainable industrialization, environmental quality and quality of life are compatible goals.

Domingo L. Siazon, Jr.
Director-General
United Nations Industrial
Development Organization
Vienna
Austria

Acknowledgement

The organizers of the workshop on behalf of UNIDO express deep gratitude to the Federal Republic of Germany for giving the financial assistance that made this publication possible.

Note of Special Appreciation

The organizers of the Workshop express their appreciation for the valuable contributions of His Excellency Dr. Klaus Zeller, Extraordinary and Permanent Representative of the Federal Republic of Germany to UNIDO for his inspiring address; Professor Federico Mayor for his tireless effort to promote international co-operation and for his brilliant speech on "Societal Responsibility for Managing Hazardous Waste in the Environment;" Judge Will Irwin for his outstanding contribution as the Workshop Rapporteur.

The meaningful contributions of Dr. Hans Sutter and Professor Niels Lind for chairing the Workshop and Professor Sergio Almeida and Dr. Filemon Uriarte, Jr. for co-chairing the Workshop are greatly appreciated as well as the contributions of Professor Hanns Abele, Dr. Asit Biswas, Mr. Jack Brady, Dr. R.N. Chakrabarty, Professor J.M. Dave, Dipl.-Ing. Franz Defregger, Dr. A. Metry, Dr. Krishna Murti and Mr. Joop van Stratum for serving as Session Chairpersons and Members of the Panel of Experts and for making the Workshop a great success.

Dr Metry's creative and meaningful assistance in the overall editorial task is greatly appreciated. We are also grateful to Dr. Barbara H. Abele and Ms Helen Canada for providing valuable editorial insights and giving generously of their time.

Preface

During the brief period of about a decade, management of hazardous waste has become one of the most critical environmental problems facing mankind. While safe disposal of hazardous waste has always been a problem, the increase in the varieties and total quantities of such wastes generated in recent years, and the enhancement of our knowledge and understanding of the adverse impacts of hazardous wastes on human health and the environment, have meant that hazardous waste management has progressed towards the top of any environmental agenda of the present time. Furthermore, a decade earlier it was often assumed that only developed countries had to be concerned with hazardous waste management since they generated much of such wastes: developing countries had the luxury of waiting for several years before the problem became serious enough to warrant special attention. With increased industrialization, hazardous waste management has at present become an equally serious problem in many developing countries. Even for those developing countries for which it has not yet become a serious problem, it can be argued that it has to be taken seriously before the problem becomes so acute that it would take an extraordinary share of national resources and expertise for solution, which the countries can ill afford at their current state of development. Accordingly hazardous waste management is now a major environmental problem that is being faced by both developed and developing countries.

While there is general agreement that hazardous waste now constitutes a global problem, there is less unanimity on an acceptable definition of hazardous wastes. Generally it is accepted that such waste materials are generated during the production or use of hazardous materials, and include a compound or compounds that could be dangerous for human health and/or the environment. Such wastes may also be considered hazardous even though the concentration(s) of the harmful compound(s) could be below legal limits.

For the purpose of the present book, any waste can be considered to be hazardous if it meets with one of the following four criteria: *ignitability, corrosivity, reactivity, toxicity*.

Although hazardous wastes are produced in many different segments of economic activities, industry – both large and small-scale – is by far the most important source in nearly all countries. Generally chemical and allied products industry accounts for nearly 50% to 70% of all industrial hazardous wastes produced. Current estimates for the United States indicate that some 60% of all hazardous wastes is produced by the chemical and allied industry.

Latest available estimates of the quantities of hazardous wastes generated by industry in OECD countries indicate that the total production was some 300 million tonnes annually in 1985, out of which the United States accounted for some 268 million tonnes, and the European OECD countries produced another 24 million tonnes. Very little, if any, detailed and reliable information is available at present on the extent of hazardous wastes produced in most developing countries, as well as on their overall disposal and management practices. In 1984, a UNEP–WHO *Ad Hoc* Working Group of Experts on Environmentally-Sound Management of Hazardous Wastes concluded that "it is safe to say that virtually all developing countries have yet to develop a comprehensive hazardous waste management scheme, as compared to those already established in most industrialized countries". Regrettably the situation has not undergone much change in recent years.

For effective hazardous waste management, it is necessary to encourage all countries to adopt a holistic approach that can put into practice a reliable and integrated cradle-to-grave management system. Sound management of hazardous wastes must not only consider more than the current approach of careful storage and disposal practices, but also should consider the various other alternatives available, individually or in combination, to formulate an appropriate strategy. Among the options that are currently available and should be considered are the following:

1. Minimization of wastes generated by changing or modifying the industrial process
2. Reprocessing of the wastes produced in order to recover energy and materials
3. Transfer of the wastes to another industry which can use them as inputs
4. Separation of hazardous from non-hazardous wastes at the source, thus reducing the total volume to be disposed of in an environmentally-safe manner with a corresponding reduction in the overall costs of handling, transportation and disposal
5. Processing of the wastes physically or biologically to render them less hazardous or non-hazardous
6. Incineration of hazardous wastes
7. Disposal of the wastes in a properly designed and managed landfill which is regularly monitored to ensure that no threat is being posed to any form of life and the environment on a long-term basis

While any disposal practice for waste, hazardous or otherwise, has technical, economic, social, environmental, health, institutional, legal and aesthetic implications, the disposal of hazardous wastes has more serious, long-term, far-reaching and sometimes irreversible consequences than disposal of non-hazardous wastes. Because of the immediate potential adverse impact of hazardous wastes on the health of the people and the environment, most of the attention in recent years has only been focused on their eventual disposal practices, and an integrated approach to hazardous waste management has been generally neglected.

In order to control and manage effectively the rapidly growing hazardous waste disposal and industrial safety problems in various countries, the United Nations Industrial Development Organization (UNIDO), with the financial support from the Government of the Federal Republic of Germany, launched a major initiative on this critical issue by convening an Expert Workshop where some 150 leading international authorities were especially invited to discuss and review the situation. The Expert Workshop was held at the Vienna International Centre, Vienna, Austria, during 22–26 June 1987. Among the objectives of the Expert Workshop were:

1. Review and analysis of the present status of hazardous wastes management and industrial safety in different countries
2. Available current technologies and forthcoming new developments in this area
3. Effectiveness of various management practices from different parts of the world
4. Economic, environmental, social, legal and institutional implications of hazardous wastes management
5. North–south and south–south technology and information transfer

and

6. Formulation of an Action Plan for UNIDO

A. Vassiliev
Deputy Director-General
Department of Industrial Operations
United Nations Industrial Development Organization

A. Tcheknavorian-Asenbauer
Director
Industrial Operations Technology Division
Department of Industrial Operations
United Nations Industrial Development Organization

Editors' Overview of the Expert Workshop

Most of the invited experts presented papers during the Expert Workshop. On the basis of the discussions during the Workshop and the ensuing peer review, many of the papers presented were selected and are now published in this book. The papers are divided into three parts: management technology; economic, environmental and institutional issues; and experiences from developing countries.

Management Technology Technological alternatives available are an important component of any hazardous wastes management practice. Initially much of the remedial work, not surprisingly, was directed at how best to clean up the hazardous wastes of the past that had already been discharged to the environment, especially at dump sites where the probabilities of occurrence of catastrophic consequences were high. Later increasing emphasis was placed upon developing new treatment, containment and disposal technologies which could safely handle eventual disposal of hazardous wastes.

In terms of technological improvement, cleaner and more efficient technologies can make a significant contribution to hazardous wastes management. As the various papers in Part I of the book clearly indicate, technological advances occurring in recent years have resulted in the availability of cleaner and safer technologies, which have the potential of significantly reducing the total amount of wastes produced, including those which could be hazardous to human health and the environment. Among the alternatives currently available and referred to in Part I of the book are the following: new equipments and sensors to monitor accurately various levels of pollutants, development in microelectronics which provide greater control over production processes, advances in biotechnology which have improved the effectiveness of treating industrial wastes, and design changes in production processes as well as substituting of process operating conditions, which could allow more recycling and re-use of waste materials, and thus increase the production of useful products and reduce the amount of useless byproducts.

On the basis of recent experiences, as the various papers in Part I indicate, technological changes and increasing environmental awareness have significantly increased the re-use and recycling of wastes for material and energy recovery. This welcome development in terms of low-waste technology has decreased the amount of wastes that have to be disposed of in an environmentally sound fashion than otherwise may have been necessary earlier. Source reduction of hazardous wastes, or even elimination, is thus a very attractive management option whenever possible and applicable. There is no question that source reduction will be an economically sensible and environmentally feasible option in the years to come.

While the success of low- or non-waste technologies can now be seen world-wide, their use is still limited and must be increased significantly in the future. For example, in Bulgaria, the introduction of such technology has helped to reduce the production of industrial wastes by about 5.5 million tonnes annually. The re-use of wastes from the chemical, pharmaceutical, food processing and mining industries has increased markedly in recent years. Of about 22.5 million tonnes of industrial wastes generated in Hungary in 1985, some 6.5 million tonnes were recycled. Approximately 30 million tonnes of industrial wastes are recycled annually in the German Democratic Republic. Recycling and re-use of wastes have increased significantly in the OECD countries as well. For example, recycling of aluminium cans in the United States increased more than 21-fold, from 24,000 tonnes to 510,000 tonnes, within the 10-year period of 1972 to 1982. Glass recycling has doubled in Europe from 1.3 million tonnes in 1979 to 2.7 million tonnes in 1984.

It is heartening to note that the introduction of cleaner technologies has not only resulted in considerable reductions in production of wastes but also has often contributed to reduce the consumption of raw materials and energy. For example, the introduction of a new ferrosilicon furnace in Norway has contributed to higher yields, less raw materials and energy consumption and reduced air pollution emissions. Similarly, the reduction of high operating temperatures in the chloromethane/ethane industry in the United States has resulted in a decrease in the probability of formation of heavier more chlorinated byproducts, which generally end up as hazardous wastes requiring special disposal practices. The introduction of such low-waste technology or changes in operating conditions could be more economical than the processes used earlier, and thus makes good sense in terms of both economic and environmental housekeeping. Such technologies also enable industry to comply with the mandatory environmental regulations more fully and effectively.

Economic, Environmental and Institutional Issues While technology is an important consideration for hazardous waste management, technology by itself cannot provide the complete solution: other factors like economics, environment, institutions and legal framework are also equally necessary considerations. It is not enough to ask only technological questions such as what are the technological alternatives currently available to reduce the generation of hazardous wastes? What is the extent of source reduction possible with the current technology? What are the technical options to dispose of such wastes? Such technological questions have to be complemented with other economic, environmental and legal questions in order to ensure an integrated approach to the problem. For example, it would be essential to know, what are the costs of the various alternative technological solutions proposed? What are the environmental and social impacts of the various alternatives? How do these solutions fit in with the existing legal framework? What package of incentives and disincentives can be proposed to ensure industry follows socially acceptable solutions? Only by asking a complete set of hard questions can an optimal solution be obtained. The papers in Part II of the book deal specifically with the economic, environmental and legal issues associated with the management of hazardous wastes.

It is clear that an important consideration for hazardous wastes management has to be proper environmental impact assessment (EIA) of existing and planned industrial projects. While the EIA process was made mandatory in the United States in 1969 through the National Environmental Policy Act, its acceptance and implementation so far in most developing countries has been somewhat slow. The reason for this slow development is not because of the consideration as to whether the EIA methodology is desirable for industrial projects and programmes but because of the lack of an operational methodology that can be successfully applied in developing countries with limited expertise, resources and time. As discussed in Part II, the EIA methodologies that are being used in industrialized countries are not directly transferable to developing countries for various socio-economic and institutional reasons. Even in those developing countries where multilateral and bilateral aid agencies have carried out fairly comprehensive environmental impact analyses of development projects, primarily with foreign experts and consultants, their overall impact on industrial project formulation and operation in developing countries has appeared to have been generally minor.

Complex, lengthy, expensive and time-consuming EIAs as practised in developed countries, are not the right tool to assess the impacts of industrial projects in developing countries. Under certain conditions, complex EIAs may even prove to be detrimental, and may hinder rather than enlighten the overall process of industrial development. What is urgently necessary is to develop operationally-sound EIA methodology that is flexible and at the same time can be carried out within the cost, timeframe and expertise limits of developing countries. This is an issue that has to be addressed squarely in the future in order to ensure that EIA becomes an integral component of the industrial development process, including proper management of hazardous wastes from various industrial projects.

In addition to what has been discussed earlier, proper management of hazardous wastes requires regulations and legislations covering standards for generators of hazardous wastes, standards for transporters as well as standards and permit requirements for owners and operators of facilities that treat, store or dispose of hazardous wastes. It is not enough to have appropriate legislations and regulations: their enforceability must be an important consideration. If legislations and regulations cannot be enforced for whatever reasons, their usefulness becomes very limited. For those developing countries which have established regulatory mechanisms and have promulgated environmental standards, or those countries that are in the process of taking these steps, practical enforcement requirements must be taken into account.

Export of hazardous wastes for eventual disposal is a factor that requires more attention than it has received thus far. The transport of hazardous wastes from one country to another generally reflects the absence of adequate disposal facilities in the originating countries. Transborder shipments are now quite significant. In the European member countries of OECD alone, such transborder shipments amounted to 2.2 million tonnes of hazardous wastes in 1983, requiring some 100,000 border crossings. It clearly indicates that hazardous wastes management can no longer be resolved only at the national level: a regional or even international approach is becoming an increasing necessity.

Experiences from Developing Countries Various papers in Part III of the book clearly indicate that developing countries have some special requirements in terms of hazardous waste management.

One important problem stems from the preponderance of small-scale industry in developing countries. It is much easier to control the generation and the disposal of hazardous wastes from a large-scale industrial project which often has the financial resources and technical expertise to manage such wastes. Such projects can be brought under increasingly strict regulation, and for the most part, economically it makes sense to institute strict control and treatment procedures for their wastes.

Small-scale industry presents a different set of problems. Often these are "mom and pop" operations, with very little resources and equally deficient expertise. The volume of wastes generated is small and thus may not be economical to treat. Furthermore, the entire operation, which may be marginal to start with, may collapse if the owners are forced to spend additional resources on treatment and disposal. And yet, even though individually such small-scale industrial projects may generate little waste, collectively the impact on the environment of a series of such small projects could be significant, and in many instances could be worse than a single large plant. Hazardous wastes from small-scale industry have generally been neglected, and for the most part unregulated. This is a typical problem of most developing countries that needs immediate priority attention.

Export of hazardous wastes from developed to developing countries, when their safe disposal in industrialized countries could constitute a serious problem, is an increasing concern. This type of export of hazardous wastes to developing countries should not be permitted without the prior knowledge and consent of the importing governments. It is essential that export of hazardous wastes to developing countries not be looked on as a "solution" since it simply transfers the problem to countries which may be even less able to prevent or mitigate adverse health and environmental impacts. One possible solution could be that there should not be any export of hazardous wastes, until and unless it can be shown that the receiving country possesses the disposal, treatment and monitoring capabilities and facilities which will enable it to meet one of the two standards, either that set by an international agreement or those existing in the exporting country itself. More international attention is needed in this area.

Overall the papers indicate that there is an increasing awareness of the problems posed to human health and the environment by the generation and disposal of hazardous wastes. The earlier attitude of "après moi, le déluge" has to give way to a realistic attempt at a holistic approach which will not contribute to health and environmental problems for present as well as future generations. It is hoped that the papers included in this book will prove to be a useful contribution to the proper management of hazardous wastes in both developed and developing countries.

Part I

Management Technology

1

Review of Hazardous Waste Management Systems as Applied by the Government and Private Sectors

Hans Sutter

The Generation of Hazardous Wastes

Wastes are generated by almost all branches of modern industry, but a few major groups are most likely to produce hazardous toxic wastes which require special treatment. Industries with a high potential for generating hazardous wastes are mainly: inorganic and organic chemicals, petroleum refining, iron and steel, nonferrous metals (smelting and refining), leather tanning and finishing, paint and coatings, electroplating and metal finishing.

In general, the industrial production system generates products, wastes and interim products; it consumes raw materials, energy, manpower, etc. The waste products generated enter an industrial waste stream and are moved to the waste management sector for further processing, treatment and, ultimately, disposal. Figure 1.1 illustrates the various points at which the generation, storage, handling and transport of hazardous wastes may arise. It also shows the hierarchy of alternatives in waste management, which involves the following three options:

1. Avoidance at source
2. Recycling of the hazardous waste, and
3. Disposal of the waste

The chemical process plant (shown inside the dashed area) will produce wastes that may be recycled in-house. If wastes, either from the plant itself or from the recycling unit, are hazardous, then storage, handling and transport problems will arise when it is transferred to the waste disposal industry as shown on the diagram. Hazardous waste will also arrive from industries which use toxic chemicals. (Such a plant is shown inside the outlined area.) For

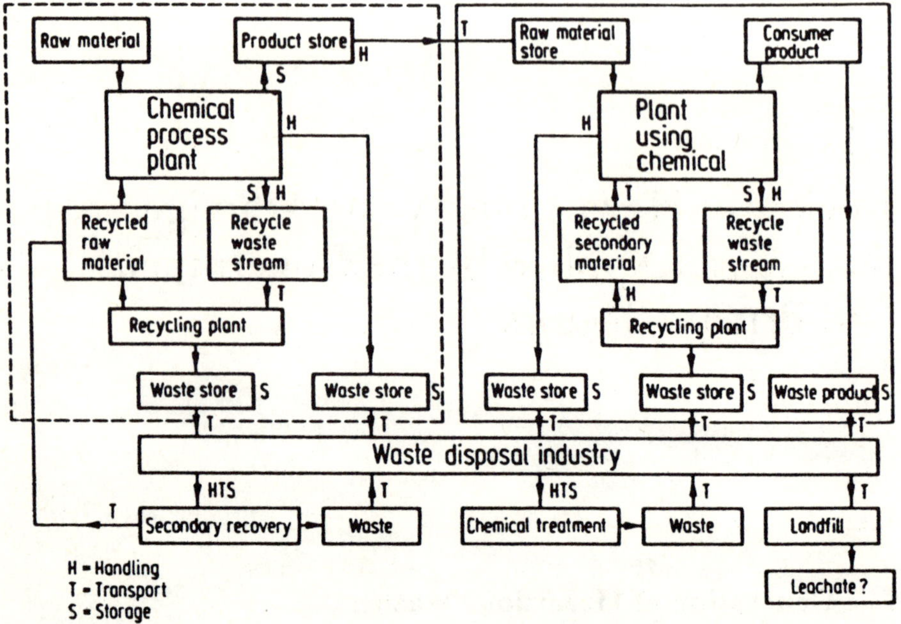

Figure 1.1 Generation and Handling of Hazardous Waste

example, the metal finishing industry uses a wide range of chemicals which may subsequently appear in liquid, sludge or solid wastes.

Ecological Aspects

In general, hazardous waste is a material that has physical, chemical or biological characteristics which require disposal procedures to avoid risk to health and other adverse environmental effects.

A hazardous waste can present either a short-term acute hazard or a long-term environmental hazard:

1. Short-term acute hazards are, for example, acute toxicity by ingestion, inhalation, or skin absorption; corrosivity or other skin or eye contact hazards; or the risk of fire or explosion
2. Long-term environmental hazards include chronic toxicity upon repeated exposure, carcinogenicity, resistance to detoxification processes such as biodegradation, or the potential to pollute underground or surface waters.

Many wastes that offer no significant short-term hazard may cause severe long-term hazards due to their physical or chemical properties. For example, the short-term hazard of certain halogenated hydrocarbon solvents is low because they are nonflammable and of low order of acute toxicity. They can cause problems in landfills, however, because their slow rate of breakdown may consequently bring the risk of surface run-off or damage to groundwater.

The long-term hazard posed by the waste will depend on the chosen disposal route because a waste containing a contaminant can only be a hazard if there exists an exposure channel by which this contaminant can be made available to the environment. Hazardous wastes have been and are still being handled and disposed of by various methods, many of which are not environmentally safe. A few of the environmentally unsatisfactory practices that have been employed include the following:

1. Disposal on-site without proper precautions to avoid groundwater and surface water contamination
2. Disposal of sludges and slurries in landfills where leachate may contaminate groundwater
3. Incineration without controlled emission of toxic or corrosive gases to the atmosphere
4. Discharge to waterways or sewer systems without understanding the potential danger and harmful environmental effects

Many cases document the immediate and long-term danger to man or his environment from improper disposal of hazardous wastes.

Legal Aspects

Definition and classification of hazardous wastes are crucial points of the whole problem. In the Federal Republic of Germany, the "Waste Disposal Act" has been in existence since 1972. "Hazardous Wastes" are defined in article 2 of this Act as such wastes from commercial or trade companies which, due to their nature, composition or quantities, are especially hazardous to human health, air or water, or are explosive, flammable or may cause diseases. Their disposal must be subject to additional requirements according to the Act. Since this definition falls short in practice, in 1977 the Federal Government established a list defining hazardous wastes (Table 1.1). It should be noted that the wastes are characterized by type as well as indication of the "active" component and, for each of them, the generator type is also presented. Each type of hazardous waste is assigned a five-figure number in a reference system. The present list will be expanded in the future. The Waste Disposal Act of 1972 and its amendments have provided a relatively comprehensive instrument for the supervision of waste collection and transport and the supervision of waste disposal plants.

An important task in a hazardous waste disposal system is, and will continue to be, to ensure that a hazardous waste is directed along the right disposal route from waste generator to waste transporter and to a treatment facility where the waste may be properly treated. The surveillance of this waste transfer is a difficult task to manage in the overall waste disposal control system as far as the manpower to do it efficiently and the great number of individual checks, which have to be performed, are concerned.

Table 1.1 Classification of Hazardous Waste

Waste Type	*Characteristics*	*Waste Code No.*	*Origin*
Liming sludge		14401	Leather production
Tanning sludge		14402	Tanneries
Kiln excavations/ from metal process	Containing arsenic, lead, cadmium, cyanide, mercury		Nonferrous metal production
		31108	Blast-furnace works
Light metal slag	Aluminium containing	31205	Light metal producing industry
	Magnesium containing	31206	Light metal-finishing industry (mills, foundries, melting works)
Salt slag	Aluminium containing	31211	Light metal melting works and parts of such
Used filter and absorption mass (diatomite, activated earth, activated coal)	Containing halogenous organic solvents or mixtures of solvents (waste code No. 55201 to 55213, 55220).		Chemical industries
Asbestos dust		31437	Asbestos mining
Beryllium containing dust		35318	Beryllium production
			Beryllium processing industry
Electroplating sludge	Cyanide	51101	Electroplating factory and parts of such (surface finishing, surface treatment)
	Chrome +6	51102	
	Cadmium	51106	
Arsenic lime		51513	Non-ferrous metal production
Hardening salts	Cyanide	51533	Hardening salts production and parts of such
	Nitrate Nitrite	51534	
Acids acid mixtures pickling (acids)		52102	Chemical industry, surface metal treatment factories, and parts of such
			Surface metal treatment
			Surface metal-finishing factories and parts of such (Surface metal-finishing)
Lye Lye mixtures Pickling (basic)		52402	Same as above
Concentrate, baths	Sulphur containing	52711	Metal and synthetic
	Chrome +6 containing	52712	Surface treatment factories and parts of such (Metal and synthetic surface finishing factories and parts of such)
	Cyanide containing	52713	
	Metal salts containing	52716	

Table 1.1 (cont.)

Waste Type	*Characteristics*	*Waste Code No.*	*Origin*
Semi-concentrate	Chrome +6	52717	Same as above
	Cyanide	52718	
	Metal salts	52719	
Waste from herbicide and pesticides (insecticides, herbicides, fungicides, rodenticide, acaricide, nematicide, molluscizide)		53104	Chemical industry production of vegetation protection and pest control agents
Waste from the production of pharmaceutical products		53502	Industrial production
Synthetic cooling agent and lubricants		54401	Metal processing industry
			Metal processing industry
Drilling and grinding oil emulsion, emulsion mixtures	Mineral oil containing	54801	Waste oil refinery
			Lubricating oil refinery
Fullers earth	Mineral oil containing	54802	Same as above
Sludge from mineral oil refinery		54803	Mineral oil refinery
Sludge from the treatment of crude oil and coal processing	Phenol containing	54903	Petro-chemical industry
	Mercaptan containing	54904	Cokeries
	Cyanide containing	54923	Gas works
Halogenous organic solvents	Containing:		
	Ethyl chloride	55201	
	Chlorobenzene	55202	
	Chloroform	55203	
	Dichlorophenol	55204	
	Cooling agent	55205	Chemical industry
	Methyl chloride	55206	Production of dyes, varnish, varnish colour, coating agents
	Monochloracetic phenol	55207	Gas works
			Production of cooling agents
	Paraffin (chlorinated)	55208	Cokeries
			Plastic production industry
	Perchloro ethylene	55209	Plastic finishing factories
	PVC-softener	52210	Production of solvents
			Metal works industry
	Carbon tetra chloride (tetra)	52211	Surface treatment industry
			Petrochemical industry
	Trichlorethane	55212	Pharmaceutical industry
	Trichlorethylene (tri)		

Table 1.1 (cont.)

Waste Type	*Characteristics*	*Waste Code No.*	*Origin*
Halogenous organic solvents	Containing: Halogenous organic solvents (waste code No. 55201 to 55213)	55220	
Halogen free organic solvents	Containing:		
	Acetone	55301	
	Ethyl acetate	55302	
	Ethylene glycol	55303	
	Ethyl glycol	55304	
	Ethyl phenol	55305	
	Benzene	55306	
	Butyl acetate	55307	
	Cyclohexanone	55308	
	Decahydronaphthalen	55309	
	Diethyl ether	55310	
	Dimethyl formamide	55311	Chemical industry Production of dye and varnish, varnish colour, coating agents
	Dimethyl sulphide	55312	
	Dimethyl sulphoxide	55313	Gas works Cooling agent production
	Dioxane	55314	Cokeries
	Methyl alcohol	55315	Plastic industry Plastic finishing
	Methyl acetate	55316	Solvent production
	Methyl ethyl ketone	55317	Metal-finishing industry Surface treatment industry
	Methyl isobutyl ketone	55318	Petrochemical industry
	Methyl phenol	55319	Pharmaceutical industry
	Pyridine	55320	
	Carbon disulphide	55321	
	Tetrahydrofurane	55322	
	Tetrahydronaphthaline	55323	
	Terpentine	55324	
	Toluene	55325	
	Benzene (cleansing)	55326	
	Xylene	55327	
Halogen free solvents	Containing: Halogen free organic solvents (waste code No. 55301 vis 55327)	55370	
Solvents containing sludge	Containing: Halogen containing organic solvents or other solvent mixtures (waste code No. 55201 vis. 55213, 55220)	55401	Chemical industry Production of dyes and varnish, varnish colour, coating agents, Plastic manuf. industry Plastic finishing industry Distillation of solvents

Table 1.1 (cont.)

Waste Type	*Characteristics*	*Waste Code No.*	*Origin*
	Containing: Halogen free organic solvents or solvent mixtures (waste code No. 55301 vis. 55327, 55370)	55402	Refining of solvents Metal manuf. industry Metal processing industry Surface treatment industry
Varnish sludge Paint sludge	Containing: Halogenous or halogen free organic solvents or solvent mixtures (waste code No. 55201 vis. 55213, 55220 or 55327, 55370) and/or heavy metal	55503	
Paints		5507	Industrial production of paints and coating agents
Coating agents	Containing: Same as above	55508	Industrial production of paints and coating agents
Cautchouch solvents		55704	Rubber producing industry Rubber processing industry
Catalysts	Heavy metal containing	59507	Chemical industry Mineral oil processing
Polychlorinated Biphenyls and terphenyls		59901	Industrial production and processing

Processors of hazardous waste are required by art. 4 sec. 3 of the Waste Disposal Act to act with particular care, especially when selecting and checking the suitability of waste disposal contractors. They must at least check whether the contractor possesses the necessary permits and whether the means for the ultimate disposal of the substances have been secured. Producers, collectors and carriers of hazardous waste, as well as those who dispose of such waste, must notify the appropriate authorities when they come into possession of and when they pass on the waste (art. 11).

The supervision of collection and transportation is ensured by the requirement to obtain approval from the appropriate authorities (art. 12). A system of accompanying documents (trip-ticket-system) enables the responsible authority to trace the path of notifiable wastes from their source to ultimate disposal. Copies of the notification forms enter the files of the authority as certificates on the orderly transport and disposal of the wastes.

As a result of the Third Amendment to the Waste Disposal Act, the export and import of hazardous waste are subject to strict approval requirements. So-called "waste tourism" is only permissible in a few exceptional cases.

The supervision of waste disposal plants is divided into:

1. Planning of waste disposal: according to art. 6, the Federal Republic's states are responsible for the preparation of waste disposal plans in which particular attention is given to hazardous waste. These plans are currently in various stages of development
2. The granting of permission to plants for disposal by the authority responsible: only approved plants may dispose of hazardous waste (art. 4).
3. The monitoring of plants by the authority responsible: art. 11 sect. 4 grants the supervisory authorities the right of access to waste disposal plants for the purposes of inspection

The organization of hazardous waste disposal may be divided into disposal by the waste producer or disposal by third parties. The legal basis is art. 3, sect. 4, according to which the processor is responsible for the disposal of waste which may not be disposed of with domestic waste. This is generally so in the case of hazardous waste. A third party may, however, act on behalf of the producer responsible for disposal. Big companies, for example in the chemical industry, often operate their own disposal facilities. In contrast, small- and medium-sized industrial firms often use centralized disposal plants owned by third parties.

The Fourth Amendment to the Waste Disposal Act covers the aspects of waste avoidance and waste utilization in addition to waste disposal. This means that the principle of low-waste technology will be included in future German waste legislation. In this respect, the amendment introduces the following requirements:

1. The production of wastes must be reduced to the minimum as far as is technologically feasible and economically acceptable
2. Wastes must be re-used or recycled unless it is economically unreasonable to do so
3. The avoidance and utilization of wastes becomes compulsory. Only wastes that cannot be avoided or used with reasonable costs shall be disposed of in an environmentally sound way
4. The avoidance and utilization of waste take precedence over customary waste disposal provided that it is technologically feasible and economic

The objective of the new regulation in waste legislation is primarily to give priority to the avoidance and recycling of waste over disposal.

Waste Management Technology

Depending on waste composition, volume, economics and other factors, proper hazardous waste disposal procedures may involve one or more of the following methods:

1. Waste reduction by applying low- and non-waste technologies
2. Conversion of hazardous waste to less- or non-hazardous forms by chemical, thermal or other means of treatment
3. Ultimate disposal in specially designed hazardous waste disposal sites

Low- and Non-Waste Technologies

Until recently, hazardous wastes were looked upon as unavoidable byproducts of production requiring further treatment to render them harmless. However, with the increasing cost of raw materials and energy, the problem of avoiding wastes has gained in importance. Non- and low-waste technologies provide a technical solution to the elimination of wastes at their source. There are two main approaches to low- and non-waste processes: the recycling of wastes generated and the development of new technologies which produce less or no waste.

Because of the growing importance of low-waste technologies in solving hazardous waste problems, research and development projects focussing on the avoidance and utilization of hazardous waste in different industrial areas have been initiated in recent years. The new developments in low-waste technology have led to a pronounced reduction in hazardous waste for disposal in certain key areas such as the chemical and metal-finishing industries.

A brief review of the research and development of low-waste technologies in these industrial areas is given below.

Chemical Industry

Large amounts of chlorinated hydrocarbon residues are generated in the chemical industry, for example in the production of vinyl chloride, propylene oxide, per- and trichlorethylene, chlorinated benzene, hydrocarbons and certain pesticides. In 1981, the Federal Republic of Germany manufactured 2.8 million tons of chlororganic products, which generated about 185,000 tons of chlorinated hydrocarbon residues. Of this amount, 45,000 tons had to be disposed of by incineration at sea. The major portion of 140,000 tons was treated using low-waste technologies to allow re-use of the residues as raw materials. Sixty thousand tons were used as input for perchlorination processes, and 26,000 tons for the production of hydrogen chloride. In the coming years the amount presently incinerated at sea will be reduced further by avoidance and recycling processes.

Chlorinated hydrocarbon residues may be utilized by burning to obtain hydrochloric acid. The two main types of procedures are those with either:

1. Extraction of a 30% concentration of hydrochloric acid (Figure 2) or
2. The direct introduction of flue gas into chemical reactors (Figure 3)

A procedure for treating chlorinated hydrocarbon residues is currently being developed for their utilization in chemical processes. Residues of solvents containing chlorinated hydrocarbons are separated into a water-based phase and an organic phase. The organic phase is burned in a special oven using heat, and the hydrochloric acid which results is extracted as technical acid. The water-based phase is purified by extraction.

Avoidance may also be promoted by process substitution. The substitution of chlorine-based processes is primarily of importance in the production of propylene oxide according to the chlorohydrin procedure and of isocyanide.

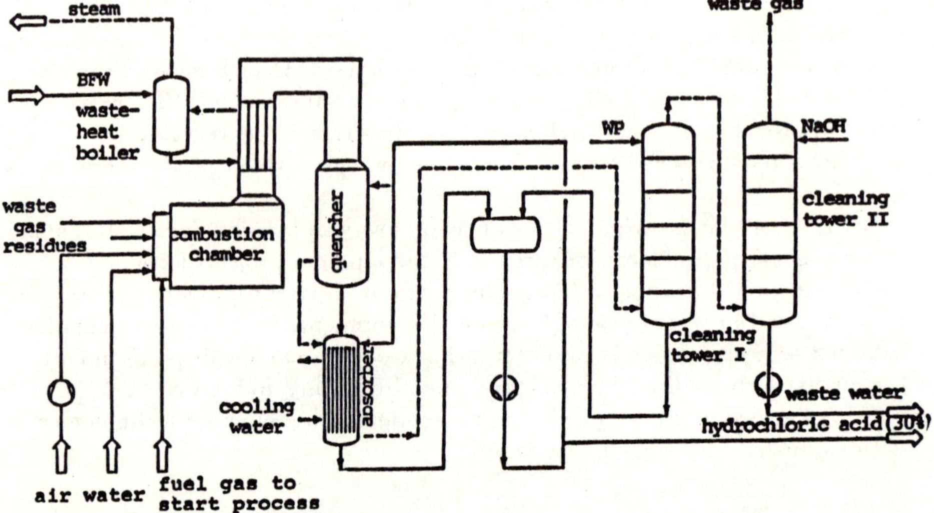

Figure 1.2 Production of 30% Concentrated HC1

Another example is the substitution of the chlorine production by the amalgam process to the membrane process which results in avoiding the mercury-contaminated brine sludges.

Numerous waste acids arise from the use of sulphuric acid in the organic chemical industry and the production of pesticides. Often the wastes are heavily contaminated by organic substances. A broad spectrum of waste acid sludges and tars containing sulphur may be treated in a recently developed process.

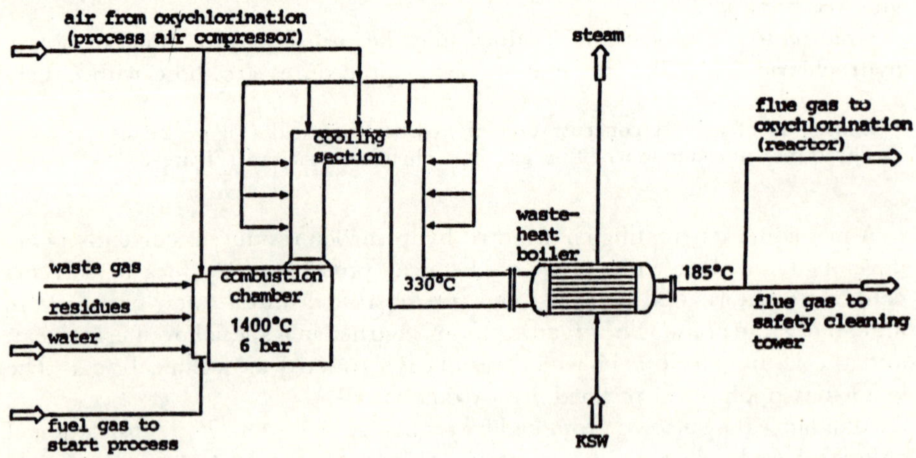

Figure 1.3 Direct Introduction of Flue Gas

The process involves producing SO_2 of sufficient purity to be used in the manufacturing of further products (Figure 1.4).

Both desirable and undesirable reactions take place concurrently during thermal splitting. The desired reaction is the complete combustion of organic compounds; this requires high temperatures and an excess of oxygen. The undesirable reactions arising under these conditions result in the formation of nitric oxide and sulphur trioxide. Separating the overall reaction into several stages enables not only the organic compounds to be completely oxidized, but also the formation of nitrogen oxides and sulphur trioxide to be largely suppressed. As a further consequence, the procedure can also be well regulated and, therefore, is not so sensitive to variations in the input materials.

The titanium-dioxide industry in the Federal Republic of Germany creates about 2 million tons of diluted acid waste per year which is dumped at sea. This amount of waste cannot be realistically disposed of on land in the traditional way, i.e. neutralization and then storage in landfills. For that reason, all concepts for solution of the problem must be based on avoidance or re-utilization. Thus, the aim is always the recycling of the dilute acid under energetically advantageous conditions. Bearing this in mind, research and development projects for the solution of the waste problem of the TiO_2 industry have been carried out. The research projects showed different reactions:

1. How can the production process be changed so that lower concentrated sulphuric acid can be used?
2. A further goal was the development of specialized concentration methods for the diluted acid from the TiO_2 production
3. A third goal was to find more uses for ferrous sulphate

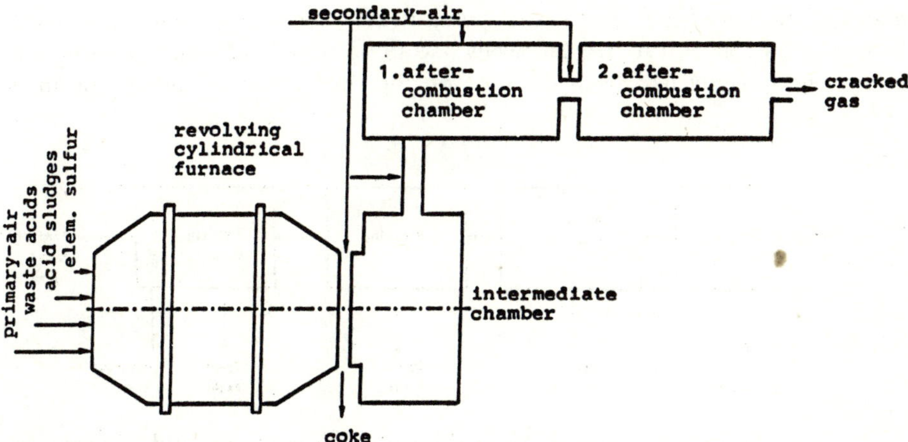

Figure 1.4 Scheme of the Installation to Utilize Waste Sulphur Acid

The research projects provided a basis for a national programme for step-wise reduction and termination of the dumping of wastes for all German TiO_2 production plants.

According to the programme, dumping will cease by the end of 1989. This means that 2 million tons of diluted acid per year will be utilized to produce about 700,000 tons of sulphuric acid (96%), which will be handled in a closed-loop production system.

Metal-finishing Industry

The metal-finishing industry uses a large number of industrial processes for machining and surface treatment of metals and other materials. Figure 1.5 shows typical operations in machine shops, i.e. in shops where metals or other materials like metal-plated plastics are formed and a surface treatment is carried out. The operation comprises machining to form or cut metal parts; the parts are then cleaned and degreased; and, after this, pickling and electroplating are done. There are shops where just machining or just electroplating is done. There are other shops where all the illustrated operations are used to produce the finished metal parts.

The industry undertakes work for a wide range of customers, typically in a large number of widely dispersed installations. Often small firms, with a maximum of ten employees, are performing these operations.

The processes shown in Figure 1.5 involve the use of a wide range of chemicals which may subsequently appear in liquid, sludge or solid waste arising from these operations. The main hazardous waste streams are shown in Figure 1.5. These are: spent solvents and lubricating oils in emulsion form, spent chlorinated solvents, spent acids and lost plating solutions in the form of spent baths and drag-out rinse containing toxic heavy metals like cadmium. This type of industry causes severe environmental problems. Investigations in the Federal Republic of Germany have shown that in the area surrounding many of the machine shops, the groundwater contained such chlorinated hydrocarbon concentrations that wells had to be closed. The main reason was that spent solvents had been dumped on the site. There are other examples

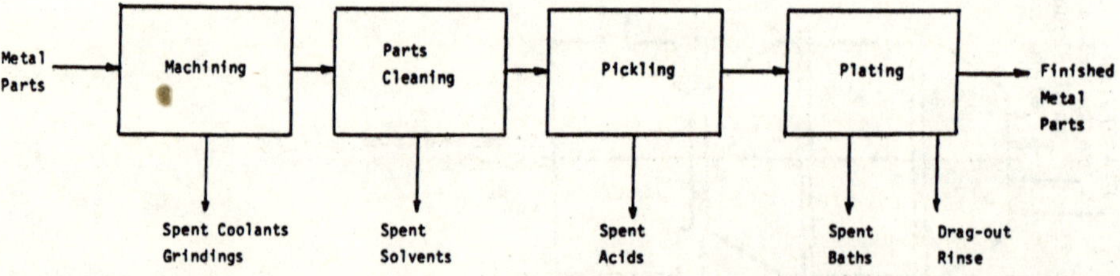

Figure 1.5 Simplified Typical Metal-Finishing Operations and Hazardous Waste Streams

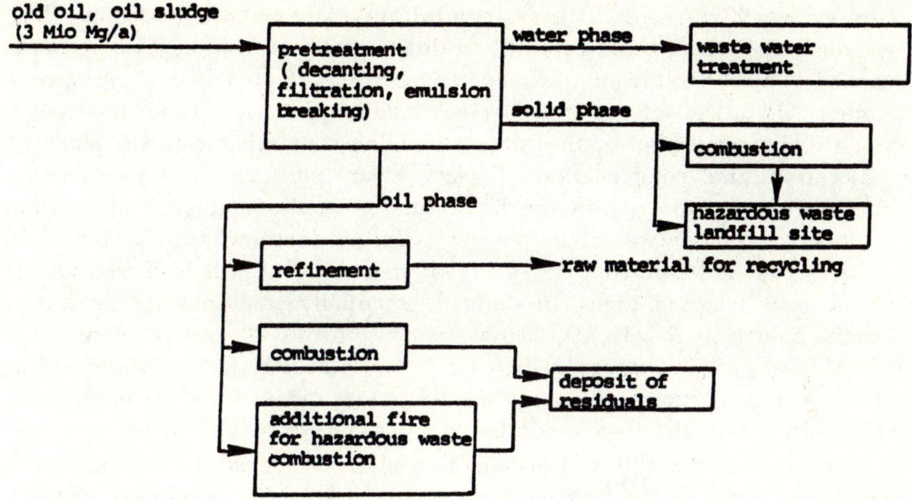

Figure 1.6 Treatment of Emulsions

where the improper handling of spent plating solutions has led to severe chromium and cadmium contamination.

In machine shops, emulsion oil is used for both cooling and lubricating. The waste emulsions may be essentially aqueous, contaminated with quantities of emulsion oils or spent lubricating oils. The degree and nature of the contaminants present will depend on the use from which the oil was derived. For example, water-soluble cutting oils will contain less than 10% mineral oils finely dispersed in water plus such substances as emulsifiers, biocides, and special lubricating oil additives.

The emulsions can be separated by different methods in a water phase and an oil phase which contain most of the contaminants. The type of contaminants affects whether the oil can be re-used in a refinery or whether it must be incinerated in specialized plants (Figure 1.6).

The problem of used-oil disposal is currently under discussion in the Federal Republic of Germany. Some 500,000 tons of used oil are produced annually. From this figure, approximately 270,000 tons are currently being reprocessed. Utilization of used oils by conventional methods is possible provided that mixtures containing oils with a high content of chlorine and PCB/PCT do not arise. An appropriate alteration to the current legislation should help ensure this. On the suggestion of the states of the Federal Republic of Germany, used oil will be treated after 1988 if it does not contain more than 10 ppm total PCB in the organic phase. A legal limit of 50 ppm is currently under discussion.

For parts cleaning, mainly halogenated solvents are used. The metal parts are degreased either by cold or hot dipping in a solvent bath or by contact with solvent vapour. Whichever method is employed, a sump residue accumulates with use which contains cutting oils, lubricants, fine metal particles etc. This residue is allowed to build up until an unacceptable level of contamination is reached. Then the solvent must be removed and treated in an appropriate manner. Often the solvent is recovered. Figure 1.7 shows a solvent-recovery system. The separation of the solvent and the contaminants takes place by evaporation and condensation. If very clean products are necessary, a fractionation (as indicated by the dotted line) must also be done. As a result of the recovery, one gets a solvent as the useful product and an oily sludge. In Figure 1.7, it is assumed that 100 kg of spent solvent will lead to 75 kg of product and 25 kg of sludge. The sludge is potentially hazardous because it may contain halogenated solvents, heavy metals and oils. Therefore, it must be treated in a special manner. In Figure 1.7, it is indicated that the waste will be incinerated to destroy its hazardous characteristics, and the incineration residues are brought to a landfill.

In the Federal Republic of Germany during the last decade, a large variety of closed-loop processes and recovery procedures have been developed for clean technology in electroplating and metal-finishing. In this way the consumption of chemicals has been effectively reduced. Simultaneously, smaller volumes of process solutions are lost, resulting in a lower impact on the environment by wastewater and sludges. It should be noted that this technical progress has not been achieved in one large step but as a result of many efforts. The technical development in this field can be characterized, in general, as a sequence of activities beginning with improved collecting and waste treatment systems,

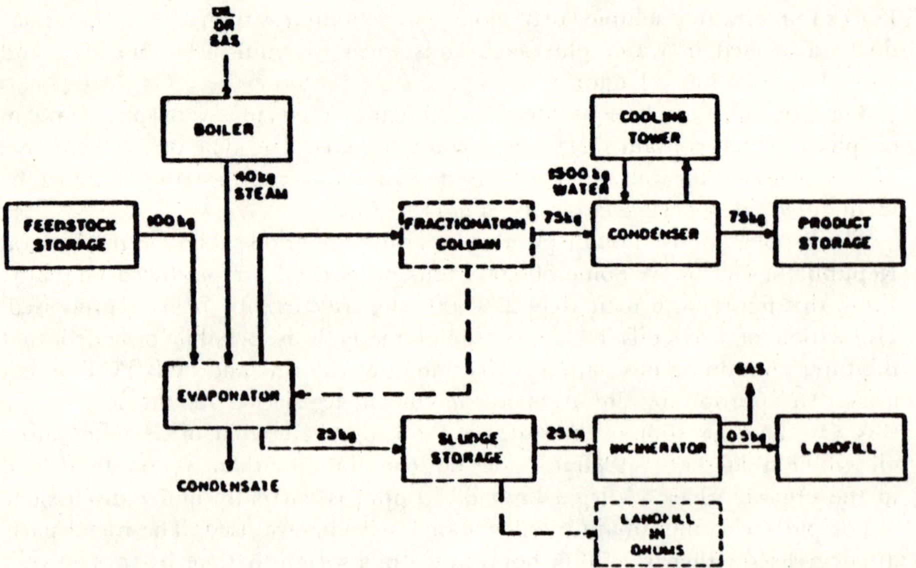

Figure 1.7 Solvent Recovery System

followed by the development and introduction of low-waste technologies; and the objective of further research work for water economy and waste management is focussed on a reduction of wastes by low-effluent and effluent-free production methods, mainly attainable by operating closed-loop processes.

In electroplating, pollution emissions mainly result from rinsing processes and draining of spent solutions. An improved rinsing technique has been successfully applied for low-effluent and, in some cases, effluent-free operation of plating systems. A lot of research work has been performed to regenerate process solutions or to recover heavy metals and other valuable materials. Summing up, we will find that the impact on the environment by electroplating processes can be reduced quite effectively.

Another important development for clean technology in metal-finishing should be also cited in this short review: recovery of overspray of wet paints in spray-painting chambers. When spray painting methods are used, it is unavoidable for some of the paint to be sprayed on other objects. The amount of such "overspray" is about 50%, which leads to paint sludges. Economic conditions for the implementation of reduction measures are favourable because the annual losses from overspray are in the order of several hundred million Marks in the Federal Republic of Germany. Figure 1.8 illustrates the principle of paint recycling in the spray booth by means of a recovery disc. The disc is arranged directly behind the workpiece and collects part of the overspray. A stripper removes the paint, almost in its original condition, from this rotating disc. In practice, up to 80% of the overspray has been recovered , made up by addition of solvents, filtrated and returned into the storage tank.

Disposal Technology

Wastes that cannot be avoided or re-used should be treated to convert them from hazardous into less- or non-hazardous materials. The residues that are produced in these processes must be disposed of permanently. Therefore, the waste disposal technology consists of the following systems:

1. Physical-chemical pre-treatment to detoxify, neutralize, de-emulsify and dehydrate
2. Thermal treatment to reduce the bulk of the waste and hazard presented by the waste
3. Depositing of pre-treated wastes at chemical landfills
4. Highly toxic wastes which cannot otherwise be treated are stored in an underground landfill

Physical–Chemical Pre-treatment

Chemical and physical treatments usually involve processes that either convert or separate the hazardous wastes. Physical treatments are largely concerned with separation of solids from liquids and liquids from liquids. Solid–liquids are usually based on simple settlement or on some form of filtration. Liquid–liquid separations are mainly applicable to immiscible liquids. On standing, these will separate into two (or more) layers which can be separated by decanting.

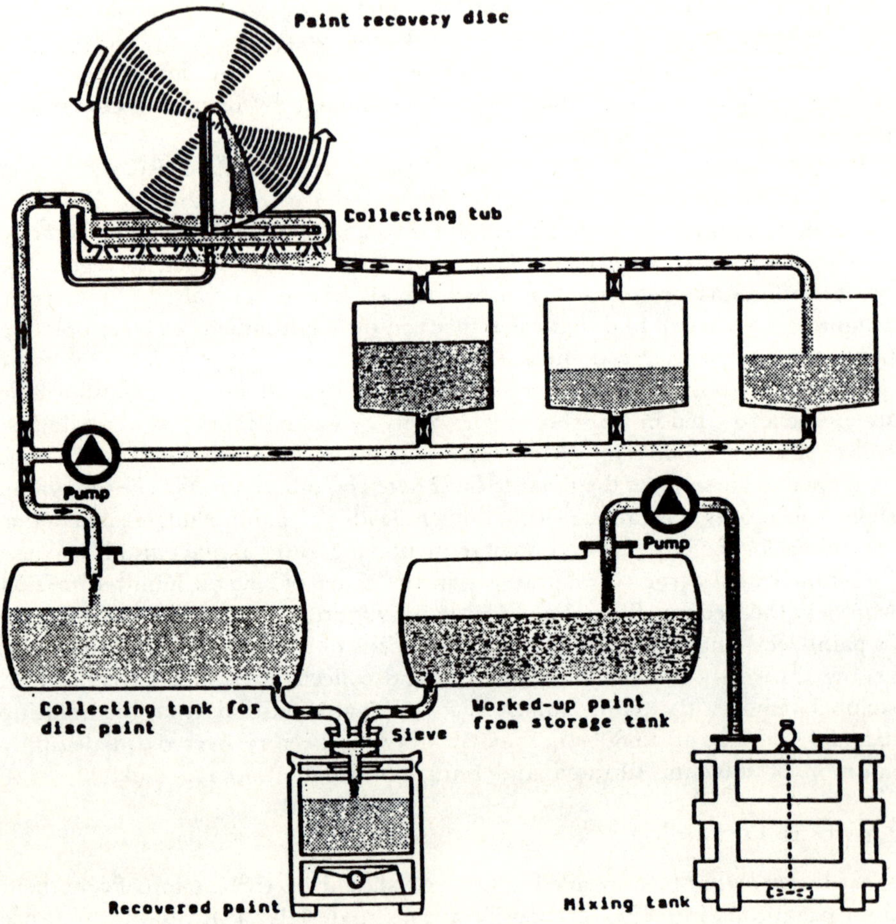

Figure 1.8 Paint Recovery in a Booth by Means of the Rotating Disc

Chemical treatments are almost always concerned with reducing the hazard of waste. There are a multitude of possible reactions that can be used to achieve this objective, but comparatively few have achieved much use in hazardous waste treatment. These methods are:

1. Neutralization of acids and alkalis
2. Heavy metal precipitation
3. Oxidation of cyanides and nitrates
4. Reduction of hexavalent chromium

Figure 1.9 presents an inorganic treatment facility for plating waste. Waste streams from the electroplating industry are considered potentially hazardous and need special treatment, because they contain chemicals such as cyanide,

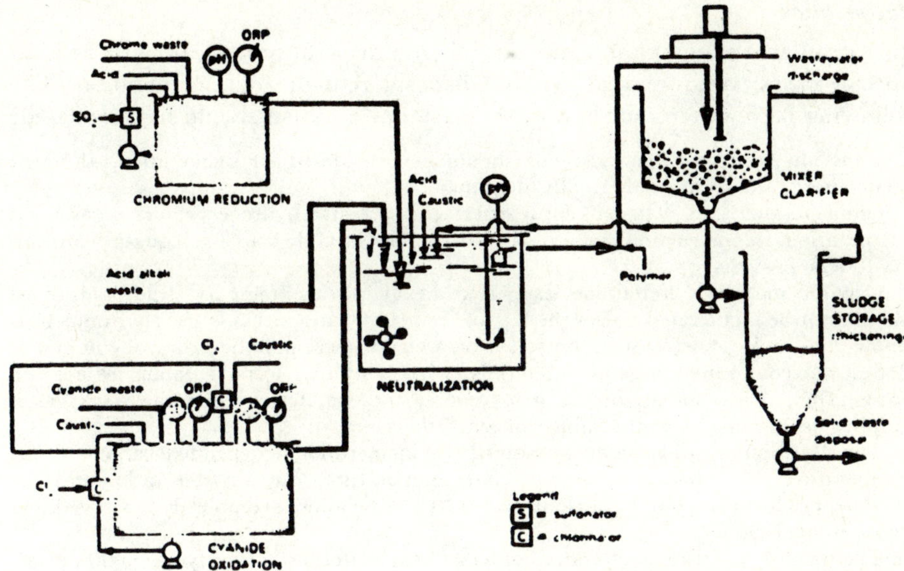

Figure 1.9 Electroplating Industry (Conventional Treatment)

chromium, nickel, cadmium and zinc which are classified as toxic substances. Figure 1.9 is a schematic diagram of a conventional treatment facility which is widely used in the metal-finishing industry for treating electroplating wastes containing chromium and cyanides in addition to other heavy metals, acids and alkalis.

Chromium complexes are usually present in electroplating waste streams as trivalent chromium or as hexavalent chromium. Hexavalent chromium must first be reduced to the less toxic trivalent chromium. Reduction usually is done by reaction with gaseous sulphur dioxide or a solution of sodium bisulphate.

Cyanide-containing waste streams must be treated separately to oxidize the highly toxic cyanide, first to less toxic cyanide, then to harmless bicarbonates and nitrogen. The oxidation reagent is usually chlorine, which can be introduced into the system by adding chlorine gas or sodium hypochlorite.

The mixed acid/alkali waste streams from the various operations are combined in the neutralizer with the effluent from the chromium reduction and cyanide oxidation steps. By pH adjustment most heavy metals are precipitated readily as insoluble hydroxides in the neutralization.

The neutralization is then followed by gravity settling for separation of suspended solids and sludge thickening. The waste treatment will result in the discharge of two streams: overflow from the classifier and sludge from the thickener. This sludge should be deposited in a chemical landfill. There is the danger that the heavy metals in the sludge will dissolve in a sanitary landfill because of the low pH there.

Incineration

Incineration is also a treatment rather than a disposal process, because in most special cases, residues that require disposal remain after incineration. The following basic criteria for hazardous waste incineration should be considered:

1. Generally, only organic materials should be considered for incineration, although some inorganics can be thermally degraded
2. Chlorine and other halogens form extremely corrosive hydrogen chloride gas upon incineration. Construction materials should be suitable, and adequate emission scrubbers provided
3. Organic materials containing dangerous heavy metals (mercury, lead, cadmium) should not be incinerated unless the fate of the metal components in the environment is known or can be satisfactorily controlled by air and water pollution control equipment
4. Sulphur-containing organic materials will normally form sulphur oxides on incineration. Care should be taken to remove these materials from the stack gas if sulphur is present in appreciable concentrations
5. The destruction ratio of a given material by incineration is dependent largely on the temperature, the residence time at that temperature, the air/feed ratio, and the turbulence. Most organic hazardous materials can be almost completely destroyed in 2 seconds at 1200°C
6. The hazard posed by incineration of hazardous materials is largely a function of the amounts, nature, and distribution of the combustion products entering the environment. Care must be taken that these materials are not dispersed or disposed of in concentrations capable of affecting health or damaging the environment
7. Incineration can be employed for combustible solids, semi-solids and liquid wastes.

The following wastes are commonly treated in hazardous waste incinerators: solvent waste and sludges, waste mineral oils, varnish and paint wastes and sludges, plastics, rubber and latex waste sludges and emulsions, oil emulsions and oil/water mixtures, phenolic wastes, mineral oil sludges, resin wastes, grease and wax wastes, pesticide wastes, acid tar and spent clay and organic wastes containing halogen sulphur or phosphorus components.

The main technology for incineration of hazardous wastes is the rotary kiln. It has established itself as a universal incineration system. It is suitable for the simultaneous incineration of solid, liquid and semi-solid wastes of a wide range of calorific values. It is now, however, the best solution for the incineration of solely liquid wastes. There are numerous simple incinerators that can be used for liquid wastes which are much cheaper than solid waste incinerators.

Municipal solid waste incinerators, however, as currently installed and operated, are not designed to destroy hazardous material.

Incineration of hazardous waste as a pre-treatment prior to land disposal could be a beneficial means of protecting the land environment by destroying the organic fraction and leaving only inorganic ashes. Volume reduction is frequently very high, depending on the relative concentration of organic and inorganic materials.

Chemical Landfill

Treatment of hazardous wastes by various methods such as incineration or other chemical, physical, or biological processes should lessen the quantity and leave the waste in its least harmful form. Then, ultimate disposal in landfills may be a reasonable alternative. Landfill disposal is by far the most important method of disposal of waste in most countries.

The objectives of modern landfill practice are to dispose of waste, with minimal risk to ground- or surface-waters. The approach to modern landfill practice is, therefore, to distribute those waste types with a higher potential for adverse impact on the environment to those sites with a higher degree of security. In turn, waste categories with lesser or minor potential impact may be directed to those sites with a lower level of security. In this fashion, the highly toxic, persistent hazardous wastes can be directed to the highest security sites, and wastes with a lesser degree of potential impact can be sent to sites with a lower level of security. The net result of this approach is not only protection of environmental quality, but also provision of cost-effective disposal. For example, the large quantities of non-hazardous domestic refuse can be sent to the lower-cost landfills which do not provide the high security feature necessary for toxic wastes.

Table 1.2 depicts a typical matrix matching waste classifications with disposal site classifications.

Table 1.2 Typical Waste-Disposal Site Matrix

Site Classification	*Waste Classification*
I Secure Landfill	Group A: Hazardous Wastes (can accept group B and C wastes)
II Sanitary Landfill	Group B: Industrial Non-Hazardous Waste, Municipal Wastes (can accept group C waste)
III Inert Fill	Group C: Wastes only

Basically three types of landfill sites may be considered to incorporate both natural site conditions (geology and soils, surface and groundwater use potential) and design features which pertain to the level of leachate containment and collection. Chemical landfills provide the highest security. Leachate collection must be used if natural conditions are not highly impermeable. Complete protection must be provided for all time for the quality of ground- and surface-water. Chemical landfills are designed especially for the ultimate disposal of hazardous wastes. Liquids and semi-liquids must be treated prior to disposal.

Sanitary landfills provide a lower level of security than chemical landfills. Leachate control through natural attenuation or leachate collection is generally required. They accept wastes consisting of or containing chemically or biologically decomposable materials which do not include toxic substances. Generally these are industrial non-hazardous and municipal wastes. Class B includes wastes that are not hazardous and are generated from industrial sources. Municipal wastes, class C, include those wastes which are generated as a result of municipal associated activities. These typically include: domestic refuse, commercial waste, garbage, municipal dewatered stabilized sewage sludge.

In an inert fill, protection is provided from group C wastes by location, construction and operation which prevent erosion of deposited material. Only inert wastes, i.e. wastes consisting entirely of non-water soluble, non-decomposable inert solids are accepted. These may include such materials as rock, brick, concrete, demolition debris. The most common existing facilities are land disposal sites used for municipal wastes. Hazardous waste disposal in these landfills includes the following problems:

1. Percolation of toxic liquids to groundwaters
2. Dissolution of solids followed by leaching to groundwater, as rain or groundwater moves through the fill
3. Dissolution, with subsequent leaching of solid hazardous materials, such as heavy metals, by acidic leachates generated by decomposing municipal refuse
4. Potential for undesirable chemical reactions within the fill, creating explosive situations or releasing toxic gases
5. Volatilization of wastes, releasing toxic or explosive gases or vapours to the atmosphere
6. Corrosion of hazardous materials containers, thus releasing their contents.

Much has been learned about landfills because of their long use in the disposal of various kinds of substances. The improper use of sanitary landfills, like unregulated dumping areas, has resulted in unidentified and uncontrolled hazardous waste disposal sites that must now be reclaimed at a high cost.

Economic Considerations

The main factors affecting the cost of hazardous waste management for a waste generator are the amount of hazardous waste produced and the ultimate disposal method chosen.

There are several factors which can have a major impact on the quantity of waste requiring disposal. As already mentioned, there are examples where fundamental process changes and plant modifications might well be justified by the raw material savings and the cost implied for waste disposal. Waste separation and concentration can also reduce waste. Even with the minimum amount of waste, it is possible to isolate the more hazardous or toxic waste streams from the mixture in which they occur. Waste separation early in process-stream flows, as well as simple isolation of similar wastes into separate disposal containers, can reduce waste handling and disposal costs. Concentration of wastes by dewatering will reduce the amount of wastes requiring treatment or disposal.

The cheapest method for waste disposal is a sanitary landfill without predisposal treatment. The cost of physical, chemical and thermal pre-

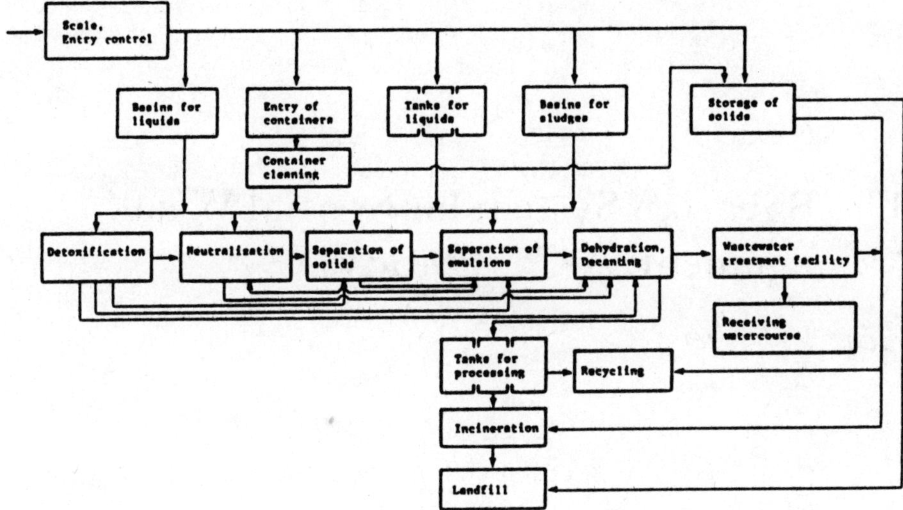

Figure 1.10 Centralized Hazardous Waste Disposal Facility

treatment is sensitive to waste characteristics. The most expensive hazardous waste management option for a waste generator is pre-treatment and ultimate disposal in a secure chemical landfill. However, it is essential to compare not only short-term but also long-term costs of different options. Specifically, cost comparisons of different technologies must be made on an environmentally equivalent basis. The cost must be considered from the time of waste generation until the hazardous characteristics have been rendered permanently harmless or can be assured of being immobilized over very long periods of time. As a result, landfills may, in fact, place a larger financial burden than other options that are not now considered to be cost-effective.

Smaller firms usually do not have sufficient resources of their own to properly dispose of their wastes at reasonable cost. For the small- and medium-sized firms, therefore, a solution must be found at a level above that of the individual firms. The major advantage of centralized waste treatment is that the investment required and the operating cost for a single, large facility is much less than that associated with installing a treatment plant at each company. In the Federal Republic of Germany, for example, hazardous waste streams from small- and medium-sized firms are treated off-site in centralized plants (Figure 1.10).

2

The Bavarian System for Special Waste Management: A Case Study

Franz Defregger

The Federal Republic of Germany was one of the first European countries to pay serious attention to the problem of hazardous waste. The Federal Waste Disposal Act was not enacted before January 1972. It has now been actualized by four amendments. The Federal Law provided only a framework and guidelines within which the problem could be addressed but it was soon complemented by the corresponding State legislation. The Federal Law together with the State's legislation makes the towns and counties (i.e. the municipal sector) responsible for the disposal of waste. There is, however, one important exception. Those types of waste that cannot be codisposed with domestic refuse (because of their amount or characteristics) can be excluded from the disposal obligation (hazardous or special waste). If the municipalities make use of their exclusive right – and that is with only few exceptions the regular case – the waste generator himself automatically becomes responsible for the safe disposal of his waste.

As a consequence of this regulation a legislative "grey area" remains concerning most of the waste material produced by industry. The Waste Disposal Plans that each State is obliged to create for its area according to article 6 of the Waste Disposal Act was intended to close this loophole.

These Plans should be comprehensive in so far as they should not only regulate the disposal of domestic but also of hazardous waste. But up to now few of the Federal States have published Waste Disposal Plans that uphold this idea. Among those States that have, Bavaria was the first to enact a Hazardous Waste Disposal Plan (in 1977) strictly regulating the safe disposal of hazardous waste statewide by offering industry a dense network of collection stations and central treatment plants.

Disposal Organizations

Within the system of waste treatment in the State of Bavaria an elaborate charge system has developed during its 20 years of operation. Being the largest of the federal states (7,500 sq. km, almost 11 million inhabitants), Bavaria has a relatively low density of population and industry with respect to central European conditions. Industry, however, does enjoy a comparatively high technological level combined with the near absence of classic heavy industry. These factors have led to a peculiar mix of advantages, problems and corresponding solutions.

With the foundation of a municipal co-operative (ZVSMM: Zweckverband Sondermüllplätze Mittelfranken) working on a strictly regional basis as early as 1966, and the foundation of a state-wide semi-public organization (GSB: Gesellschaft zur Beseitigung von Sondermüll in Bayern; Association for Disposal of Special Waste) in 1970, Bavaria represents the forerunner in hazardous waste treatment practices in the Federal Republic of Germany and beyond its frontiers, and was able, therefore, to shape federal legislation by presenting practicable solutions.

In operation at present there are 10 regional collecting stations, 3 central facilities (incineration, chemo-physical treatment, landfill), and 1 recovery plant for contaminated solvents (Figure 2.1).

With the exception of the treatment plant of Schwabach (ZVSMM) all other facilities and collecting stations are run by GSB. Its stock fund rose from DM1 million in 1970 to DM21 million today. The shareholders of the company are the Bavarian State (78%), 3 municipal organizations (8%) and 76 industrial firms (14%). The activity of the company is conducted on a non-profit basis but being a private company the GSB has to search continuously for new ways of treatment to obtain a proper disposal or recovery of raw materials from this waste.

As a consequence, waste treatment costs are relatively high, thus creating potentially a substantial incentive for waste producers to ship their wastes to surrounding states or countries with lower treatment quality and costs. For this reason, waste exports from Bavaria are generally forbidden; the use of the above-mentioned facilities is compulsory with only few exceptions granted for some already existing waste treatment plants or sites of large companies. In effect, GSB and ZSVMM have a monopoly in handling waste in Bavaria.

Collection and Transport

The collecting stations are fairly uniform in their service and in their equipment (Figure 2.2). They provide an intermediate holding function and have equipment which is fairly simple to run. They pretreat as much waste as possible to cut down the volume. They have a wastewater treatment plant, in which the separation of the usually large volumes of oilwater mixtures into

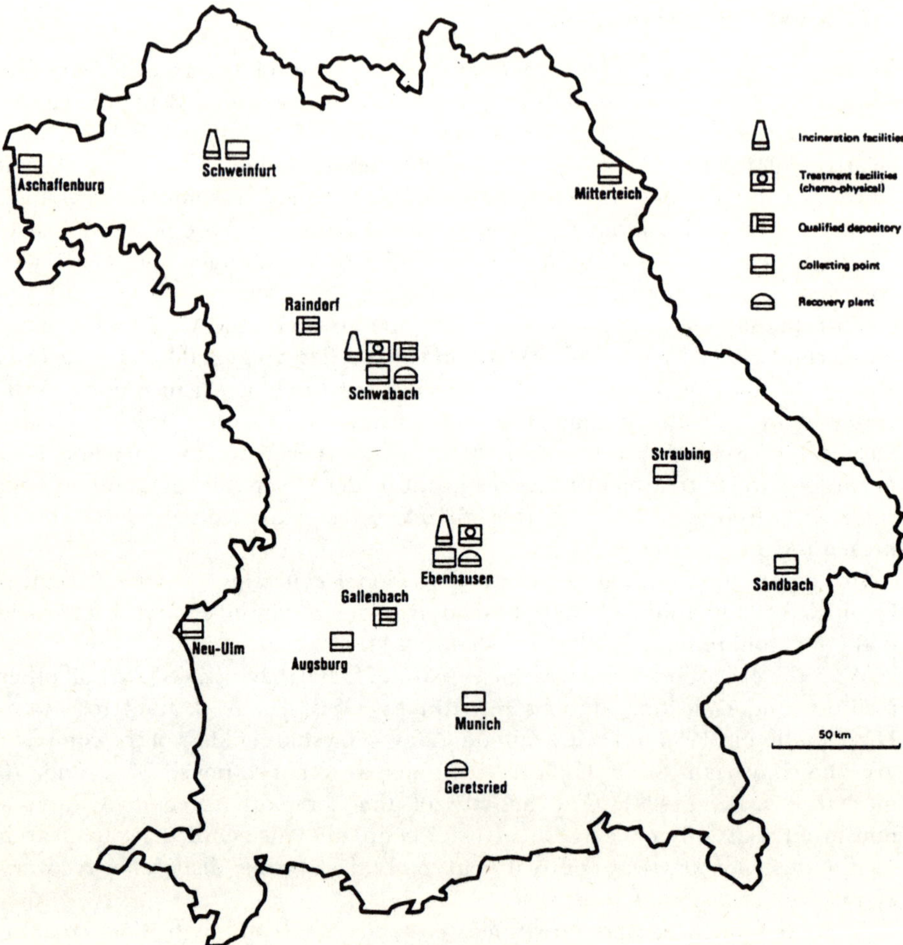

Figure 2.1 Special Waste Management System of Bavaria, 1986

water and solids takes place. The treated wastewater is discharged locally. The waste volume is cut back to 10% of the original. Where necessary, facilities for neutralizing and sludge thickening are in place. All collecting points have an administration building and test lab, vehicle scales and an area for oily soil and containers to receive and provide intermediate storage for industrial sludges of all types.

Central Treatment Facilities

Ebenhausen Central Treatment Facility

The largest and most modern of the three Bavarian hazardous waste facilities is

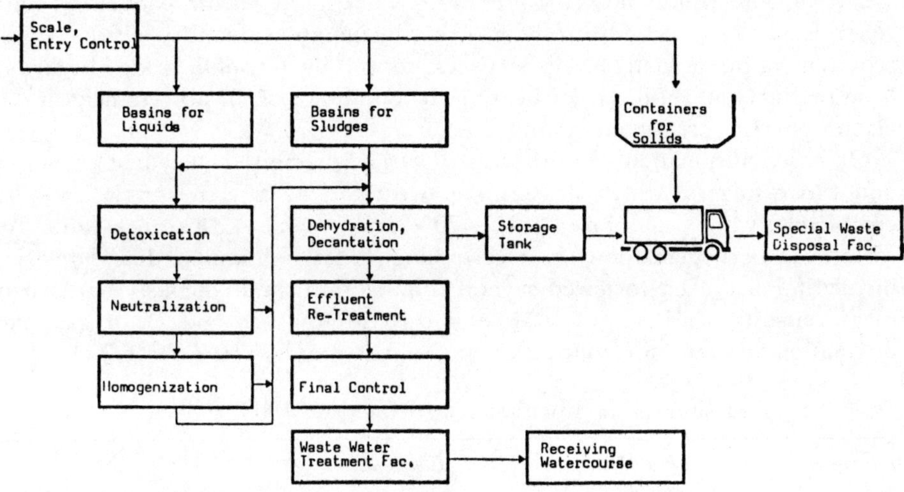

Figure 2.2 Collection Centre Processes

the Ebenhausen plant which became operational in 1976 on a 4-hectare site in Ebenhausen near Ingolstadt. The Ebenhausen plant comprises a laboratory, a chemo-physical treatment plant for organic and inorganic substances, a water purification plant, and the FRG's largest waste incineration plant with a capacity of 100×10^6 power Btu/h or 80,000 metric tons (Figure 2.3).

Incineration

The incineration plant has two parallel rotary kilns for solid and pasty wastes and a common burner-chamber with a set of six burners for liquid wastes. The

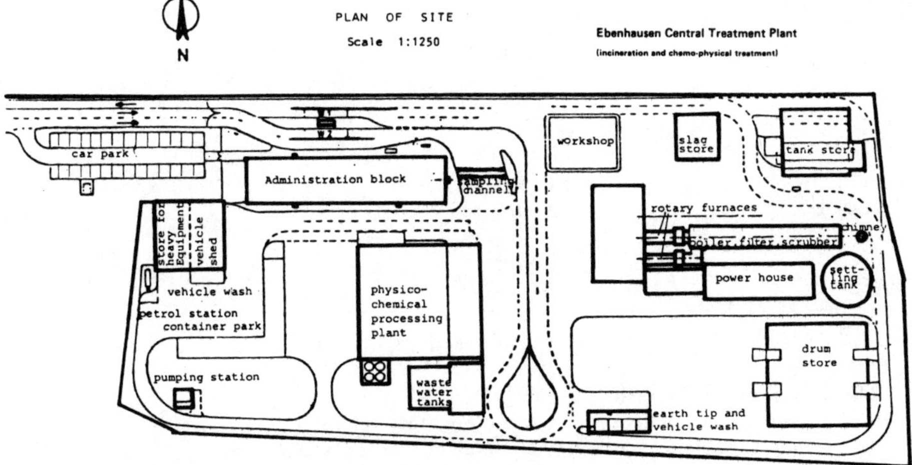

Figure 2.3 Ebenhausen Central Treatment Plant

heat of the off-gases from the after-burners is utilized in a steam boiler and in a steam turbine while the remainder is condensed in an air condenser; this electric energy not only provides the incineration plant's entire power requirement but actually leaves an excess supply for the public grid. The steam from the turbine is utilized for heating the building and for process heat in the chemo-physical treatment plant.

Off-gases purification comprises an electrical precipitator for dust retention and a two-stage venturi-type scrubber which removes HCl and HF almost completely, while retention of SO_2 is 70% and is enhanced by flocculants. As regards airborne emissions and scrubbing water pollutants, this industrial incinerator has been reviewed several times. From simultaneously performed measurements of HCl, SO_2, HF and dust in raw gas and clean gas, the fluctuation margins for routine operation can be derived (Table 2.1).

Table 2.1 Emissions of the Ebenhausen Incinerator, 1985

Parameter	*Raw Gas* mg/m^3	*Clean Gas* mg/m^3	*Limit* mg/m^3
HCl	1000–4000	22–90	100
SO_2	450–21,000	40–300	400
HF	62–260	1–3	5
Dust	1000–2000	10–20	50

A new legislative order for emission control was introduced on 1 March 1986, which sets new and still more stringent requirements as to the performance standards (Table 2.2).

Table 2.2 Emission limits in the FRG, 1986

Parameter	*Limit**†
CO†	100.0 mg/m^3
$C_{org.}$†	20.0 mg/m^3
NO_x†	500.0 mg/m^3
SO_2†	100.0 mg/m^3
HCl†	50.0 mg/m^3
HF†	2.0 mg/m^3
Particulate matter†	30.0 mg/m^3
class I‡	0.2 mg/m^3
class II‡	1.0 mg/m^3
class III‡	5.0 mg/m^3
PAH§	0.1 mg/m^3

* values calculated for normal gas conditions, dry gas with 11% O_2

† to be measured continuously

‡ special inorganic elements in the particulate matter: class I: Hg, Cd, Tl; class II: As, Cr(VI), Co, Ni, Se, Te; class III: Sb, Pb, Cr, Cu, Mn, V

§ various polycyclic aromatic hydrocarbons

In order to meet these more stringent emission limits or even to exceed them because of special orographic situations, the incineration plants at Ebenhausen and Schwabach are to be equipped with additive air pollution control technologies based on different processes. They range from a quasi-dry scrubbing system combined with highly efficient fabric filters over a wet, but wastewater free scrubber to an ionizing wet scrubber in the GSB central facility supplementary to the existing, quite efficient scrubbing system.

The installation of a computerized data gathering and processing unit for those parameters to be measured continuously (including temperature in the afterburner chamber) has been made a requirement for the waste incineration facilities in Ebenhausen. The resulting records are ultimately evaluated by the Bavarian Environmental Protection Agency which is generally responsible for the control and surveillance of hazardous waste treatment and disposal facilities.

Chemo-physical Treatment

The inorganically polluted liquids and thin sludges especially from the electroplating industry are pretreated in the chemo-physical plant. These wastes are contaminated by cyanide, chromate, nitrite, heavy metals, acid or basic solutions and normally need detoxification and filtration to eliminate the toxic substances and the solids. The detoxification is achieved by oxidation (using sodium-hypochlorite or hydrogendioxide), by reduction (using sodium-hydrogensulfite or sodium dithionite) and by metal precipitation as hydroxide (using lime) or sulphide (using sodium sulphide). For separation of solids a filter press has been approved. The filter cake is landfilled and the filtrates after being analysed are discharged to the municipal sewage plant. The annual capacity of the chemo-physical plant is about 50,000 tons.

The de-emulsification plant carries out the following treatment:

1. Separating of solids first by sedimentation in a four-chambered discharge and sedimentation tank and second by a decanter
2. Chemical treatment by adding iron (III) chloride as de-emulsification reagent, neutralizing the solution by sodium hydroxide and thus absorbing the oil particles at the surface of the precipitation of iron (III) hydroxide
3. Separation of sludge and water phase in vacuum drum-filters

The residues of the treatment processes (dewatered, neutralized sludges, fly-ash, slag) and other solid wastes are deposited at the two landfill sites in Schwabach and Gallenbach under special security conditions. A third landfill site, located in Raindorf, went into operation in 1985 (capacity: 800,000 m^3; basic clay-cover: 2.0 m; owner: ZVSMM). While the landfill site of Schwabach is located together with the treatment facilities, the Gallenbach site is dislocated (over 40 km) from the Ebenhausen treatment facilities because of the hydrogeological conditions.

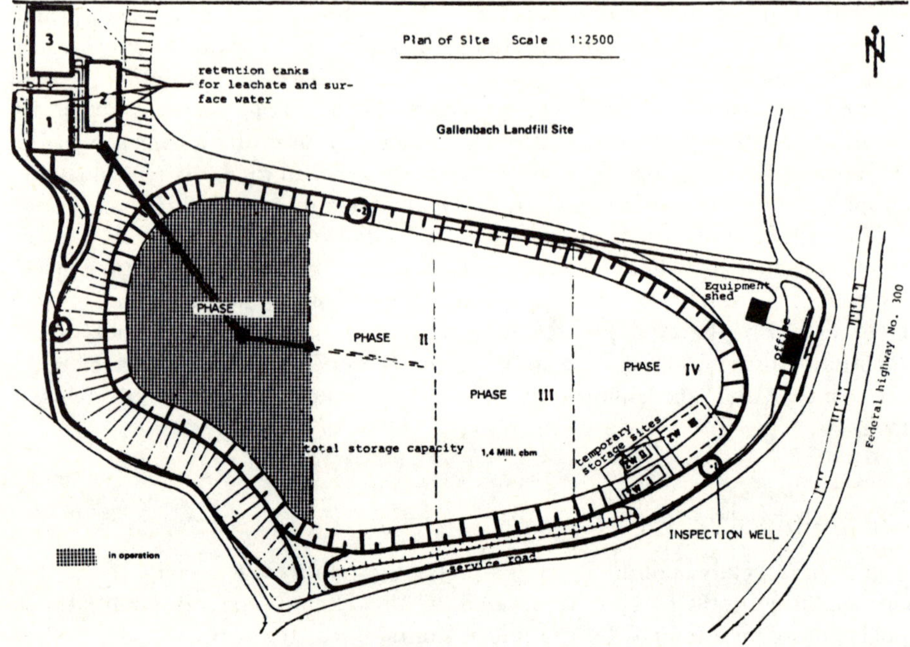

Figure 2.4 Gallenbach Landfill Site

Gallenbach Landfill

The Gallenbach Landfill, in operation since 1975, comprises the following installations (Figure 2.4):

1. Operational building with amenity rooms of various types for the staff
2. Laboratory for sampling the substances delivered
3. Vehicle weighbridge
4. Vehicle and instrument shop
5. Control systems (drainage, retaining basins) for collecting surface water and leachate

The site is operated by the "area method" of landfilling on a 100 cm thick, pre-constructed clay pad with a reported permeability of 10^{-9} m/s. Watersoluble solid wastes containing heavy metals are deposited in drums and covered with concrete to reduce the contamination in the leachate. Leachate is collected by a series of underdrains, located in the clay pad. It is then channelled to two closed-concrete storages from where it is taken by truck to the Ebenhausen plant for treatment. Average leachate characteristics are identified in Table 2.3.

Table 2.3 Average Leachate Concentrations in Gallenbach (mg/1)

Parameters	*1980*	*1981*	*1982*	*1983*	*1984*	*1985*	*1986*
pH	7.4	7.4	7.5	7.6	7.6	7.8	7.6
BOD_5	4526	4383	3520	2515	2853	1963	2075
COD	3850	3098	3900	3585	4343	4470	4640
TOC	3062	3530	3130	196	2109	1986	1940
Cr	0.3	0.7	0.5	0.3	0.1	0.2	0.1
Cu	0.4	0.4	0.3	0.1	0.1	0.2	0.2
Cn	<0.1	<0.1	<0.1	<0.1	<0.1	<0.1	<0.1
Ni	4.6	3.9	2.4	1.4	1.5	0.8	0.6
Zn	0.4	0.7	0.6	0.7	0.4	0.7	0.6
Cd	0.2	0.7	0.2	0.1	0.1	0.1	0.1
Fe	4.5	3.0	3.5	1.7	1.7	0.9	1.0
Sn	34	3	<3	<3	<3	0.4	<1
Pb	1.2	4.8	1.8	0.6	0.4	0.2	0.1
Phenole	11	14	9	12	17	17	16
Chloride	47,900	48,900	4300	28,600	33,600	36,100	43,200
Sulphate	3717	4060	1600	3814	2543	2841	2600
Ammonia	944	970	960	960	783	698	240

The leachate is characterized by high organic (BOD, COD, TOC) and inorganic concentrations, in particular salt contents between 30 g/l and 50 g/l and ammonia between 200 mg/l and 1000 mg/l, but heavy metals have low concentrations.

Advanced Techniques for Leachate Treatment

In view of the high contamination of the leachate, it has to be treated in the chemo-physical plant and sewage plant in Ebenhausen. The purpose of the treatment is to reduce the heavy-metal content by precipitation and changing pH values. It is evident from conventional treatment that several contaminants cannot be reduced in a biological way or eliminated in a common oxidation process (high salt content, COD).

Therefore, on behalf of the Bavarian Ministry of Environment a research programme was undertaken in 1983 to treat the leachate in a two-stage distillation pilot plant in Ebenhausen. In the first step (acid distillation), the contaminants were crystallized and reduced in a salt concentrate. In the second step (alkaline distillation), the acid contaminants were again reduced. The results made clear that TOC and COD could be reduced by over 90%, particularly by using discontinuous distillation.

Based on this experience, the first full-scale distillation unit in the world went into operation in 1986 in Schwabach to purify leachates and comparable, aqueous hazardous wastes. The treatment consists of purging the liquid (removal of ammonia and highly volatile organics), a two-stage distillation process and separation of organic and salt residues. The latter, together with $(NH_4)^2SO_4$ resulting in the purging step, is disposed of in the landfill under special conditions, thus closing the loop between waste and leachate. Volatile organics and organic residues are fed into the incinerator.

Charges System and Investments

Disposal costs in Bavaria are reasonable, probably reflecting the facts that initial capital investments are not being amortized and that later capital investments are subsidized through interest-free or low-interest loans. The pricing formulas are quite involved and are summarized in Table 2.4. Waste treatment charges are established for each of the processes used in the major waste categories. The charges are designed roughly to cover the transportation from the collection centres and the cost of waste treatment for each process.

Table 2.4 Average Treatment Charges per Ton (in DM)

	DM	*DM*
Landfilling	75	190*
Chemo-physical	120	300
Incineration	325	1500†

* deposition in concrete jacket
† PCB

The hazardous waste disposal facilities in Bavaria have required an investment of DM150 million in the last 20 years. Financing is handled through subsidies, government loans and favourable interest terms and the company's own resources obtained by economic activity.

Control

The central task in each hazardous waste disposal system is, and will continue to be, to ensure that a hazardous waste is directed along the right disposal route from waste generator to waste transporter and to a disposal facility, where the waste may be properly treated. Because of the manpower requirements and the many individual checks which have to be performed, surveillance of this waste transfer is perhaps the most difficult element to manage in the overall waste disposal control system.

The existing regulations in the FRG allow for thorough control of the wastes from their source to their disposal on the basis of a "trip-ticket-system". Therefore in the FRG there exists a control system in which hazardous waste generators can and usually are required to initiate a trip-ticket, cradle-to-grave recording system for special waste involving six separate copies. Waste haulers, waste disposers and responsible agencies receive and/or transmit appropriate copies. Thus, there are numerous places for cross-checking the kind and quantity of waste as it travels to its final destination (Figure 2.5).

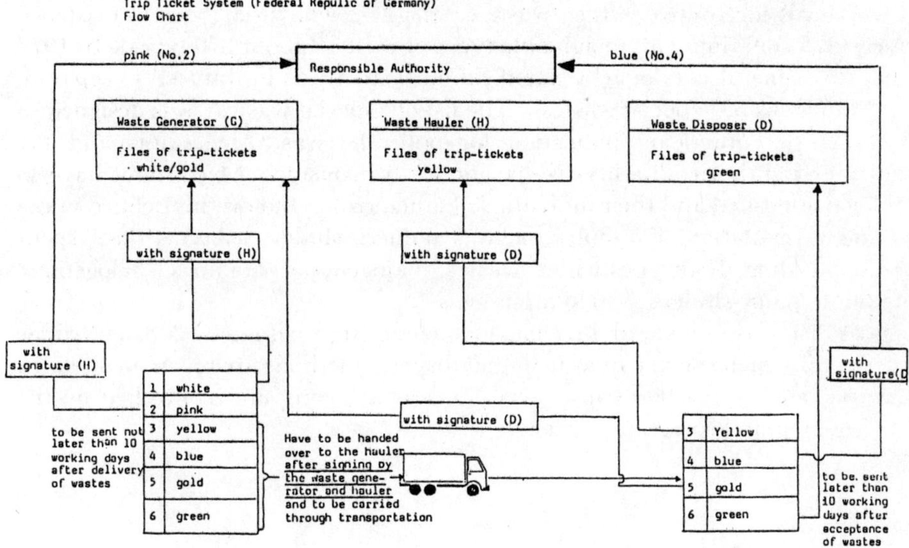

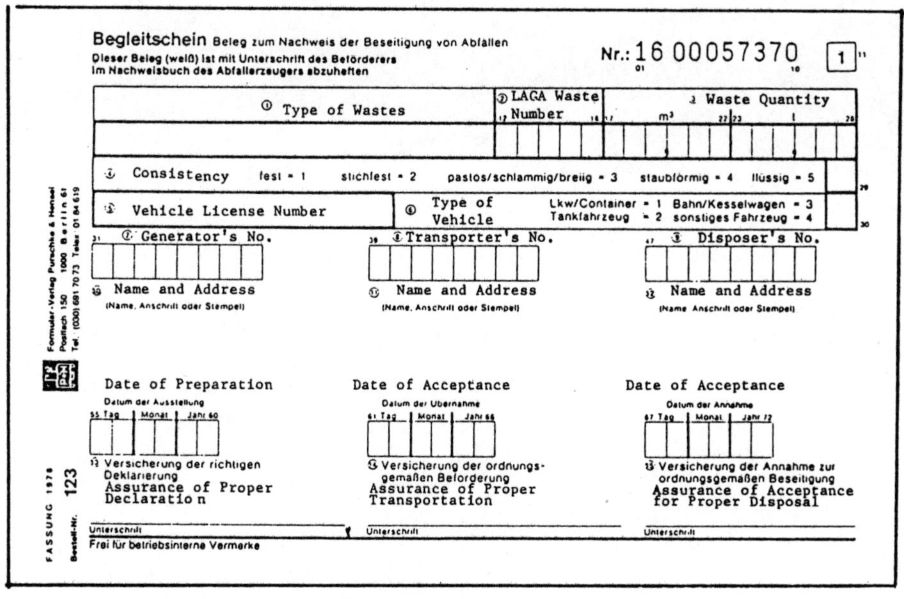

Begleitschein Beleg zum Nachweis der Beseitigung von Abfällen
Dieser Beleg (weiß) ist mit Unterschrift des Beförderers im Nachweisbuch des Abfallerzeugers abzuheften

Nr.: 16 00057370 [1]

① Type of Wastes | ② LAGA Waste Number | ③ Waste Quantity m³ t

④ Consistency fest = 1 stichfest = 2 pastos/schlammig/breiig = 3 staubförmig = 4 flüssig = 5

⑤ Vehicle License Number | ⑥ Type of Vehicle Lkw/Container = 1 Tankfahrzeug = 2 Bahn/Kesselwagen = 3 sonstiges Fahrzeug = 4

⑦ Generator's No. | ⑨ Transporter's No. | ⑩ Disposer's No.

Name and Address (Name, Anschrift oder Stempel) | Name and Address (Name, Anschrift oder Stempel) | Name and Address (Name, Anschrift oder Stempel)

Date of Preparation (Datum der Ausstellung: Tag, Monat, Jahr) | Date of Acceptance (Datum der Übernahme: Tag, Monat, Jahr) | Date of Acceptance (Datum der Annahme: Tag, Monat, Jahr)

Versicherung der richtigen Deklarierung — Assurance of Proper Declaration | Versicherung der ordnungsgemäßen Beförderung — Assurance of Proper Transportation | Versicherung der Annahme zur ordnungsgemäßen Beseitigung — Assurance of Acceptance for Proper Disposal

Unterschrift | Unterschrift | Unterschrift

Frei für betriebsinterne Vermerke

Formular-Verlag Purschke & Hensel, Postfach 150, 1000 Berlin 61, Tel. (030) 681 70 73, Telex 01 84 619

FASSUNG 1978 Bestell-Nr. 123

German Special Waste Manifest

Figure 2.5 Hazardous Waste Control System in the Federal Republic of Germany

The basis for control is the "waste catalogue", which has been in existence since 1975 containing all imaginable types of wastes (about 570 types). In 1977 from this general catalogue a classification system was introduced to separate out hazardous or "special wastes". The list of "special wastes" was designed to facilitate the practical application for both the waste generators and the controlling authority. Eighty-six specific wastes considered hazardous have so far been identified and their individual characteristics and origins defined in the statutory regulation. Examples include: tannery sludge, asbestos dust, spent acids, pickling bath, pesticides wastes, halogenated organics, halogenfree organics, paint-sludges and hardening salts.

This list was increased in 1986 to include approximately 200 hazardous wastes, for which specific treatment and disposal methods are recommended. In Bavaria, almost 100,000 trip-tickets per year are computer controlled by the Environmental Protection Agency.

3

Incineration of Toxic and Hazardous Wastes

Jack D. Brady and J.A.C. van Stratum

Incineration Equipment and Operating Characteristics

A number of different thermal oxidation systems are used for incineration of hazardous waste materials. Some of the more common systems are described below.

Rotary Kilns

A rotary kiln is a refractory lined steel cylinder which rotates at very low speed and is mounted on a slight incline so that solid materials introduced at one end will move through the kiln and be discharged at the other end. A burner is mounted at the same end of the kiln as the solids feed mechanism. The burner is fired with natural gas, oil or waste solvents. Combustion air is introduced at the burner end and combustion gases are exhausted at the opposite end of the kiln. Solids residence time in the kiln will vary from 10 minutes to 30 minutes. Combustion gas residence time will be less than one second. Exhaust gases exit the kiln at temperatures of between 815°C and 1100°C. The rotary kiln operates with as much as 100% excess air and oxygen. This results in relatively complete combustion of the waste products and the fuels. Some of the more difficult waste materials to burn are volatilized into the gaseous state. These are then fully combusted in an afterburner where supplemental fuel and additional air are added to elevate the exhaust gas temperature to approximately 1200°C. The afterburner insures complete dissociation of these compounds, including the halogenated organic compounds, to allow their complete destruction.

Figure 3.1 shows a typical rotary kiln incinerator with secondary combustor or afterburner. Because the kiln has seals at each end it operates at negative pressures sufficient to prevent fugitive emissions. Excess air rates are more difficult to control than in other types of incinerators. Rotary kiln systems typically operate at higher total exhaust gas flow rates than other systems operating on comparable waste materials. The major advantage of the rotary kiln is its ability to handle solid waste materials with widely varied sizes, liquid wastes using atomizing burners centrally located at the inlet end of the kiln, high moisture content wastes and sludge-like materials, and materials which form molten slags. The gas residence time in the secondary combustion section is typically one second. Relatively high particulate loadings are discharged from rotary kiln systems, since the ash material is constantly being disturbed by the rotation of the kiln and since much of this ash becomes airborne in the kiln section. Very efficient particulate collectors must then be used to allow compliance with air pollution control regulations in most areas.

Because these systems depend upon the refractory material lining the kiln and the secondary chamber to prevent thermal damage to the mechanical components, and because most refractory materials have temperature limitations of about 1650°C, rotary kiln systems are seldom operated at temperatures exceeding 1370°C to allow some safety factor over the ultimate temperature limitation of the refractory. The most conservative designs utilize 2 seconds of gas residence time in the rotary kiln and 2 seconds of gas residence time in the afterburner.

Pyrolytic or "Starved Air" Systems

This type of thermal oxidizer is equipped with a fixed hearth onto which the solids are introduced via a ram feeder. The same ram feeder also pushes the ash material towards the discharge end of the hearth. Combustion air is controlled to provide for essentially no excess air. Only enough air to burn the fuel and some of the organic waste material is introduced. Temperatures of approximately 815°C are produced in this primary chamber. In the primary chamber, the organic materials are vaporized and driven from the solid materials into gas phase. The exhaust gases enter a secondary combustion chamber where excess air is added and where supplemental fuel is sometimes added to elevate the temperature to about 1100°C to ensure complete combustion of hazardous components. Solids residence time is typically 30 minutes, while gaseous residence time is between one-half and 2 seconds. Because of the need for mechanically introducing the waste material and mechanically pushing the ash out of the system, the combustion chamber length is relatively short, compared with a rotary kiln system. This limits the capacity of the systems, and pyrolytic units are used for relatively small mass flow rates. Most of these systems handle 1.1 metric tons or less of waste material per hour. They cannot be used for waste materials which melt during the combustion process, nor can they handle large solid objects. Abrasion of the refractory material on the fixed hearth can be a serious problem with these systems. Because the combustion air flow rates can be very carefully controlled, the exhaust gas flows are relatively small compared with an equivalent capacity rotary kiln system. This reduces the size of any secondary emission control system. Because the waste material is not agitated in the combustion zone, particulate emissions are quite small relative to those produced by the rotary kiln system. Figure 3.2 shows one such system.

Liquid Injection Systems

The simplest of all hazardous waste incinerators are the liquid injection systems. If only combustible liquid wastes must be burned, they are introduced through atomizing burners where they are intimately mixed with combustion air, elevated in temperature to about 1100°C, maintained in contact with excess air for between 1.5 and 2 seconds, and then exhausted to secondary emission control systems. Frequently, only a single refractory lined chamber is used for the entire combustion process with no afterburner being required. Supplemental fuel can be mixed with the waste or can be introduced through a separate orifice in the burner to maintain the required temperature for complete destruction of the waste components.

The key to proper operation of these systems is the design of the burner. Most of the burners which have been designed especially for waste materials introduce air in a vortex section surrounding the liquid atomizing nozzle. This creates intimate mixing of the vaporized fuel and waste materials and the combustion air, yielding uniform flame temperatures and high destruction efficiencies. Figure 3.3 shows one such system in its simplest form.

Multiple Hearth Furnaces

For the very high moisture content solid materials, or for sludge-like materials which must be dried before they burn completely, multiple hearth furnaces are used. In these systems, the wet feed material is introduced at the top of the combustor. Rotating rabble arms distribute the wet waste material across a doughnut shaped hearth and simultaneously introduce combustion air. The hearth is perforated so that combustion gases can pass through the hearth counter to the flow of the solid materials. The rabble arm pushes the solids around until they fall through the openings in the hearth to an additional hearth mounted directly below the first. Combustion gases are discharged to secondary air pollution control systems while the solids are discharged through ash hoppers at the bottom of the furnace. Figure 3.4 shows one system of the multiple hearth type.

Fluid Bed Combustors

Two advantages are offered by fluid bed combustors over other hazardous waste incineration systems. Solid absorbent materials can be introduced simultaneously with the waste feed to absorb acid gases during the combustion process, rather than after the combustion process. In addition, solids residence time can be controlled by simply decreasing the feed rate and the ash removal

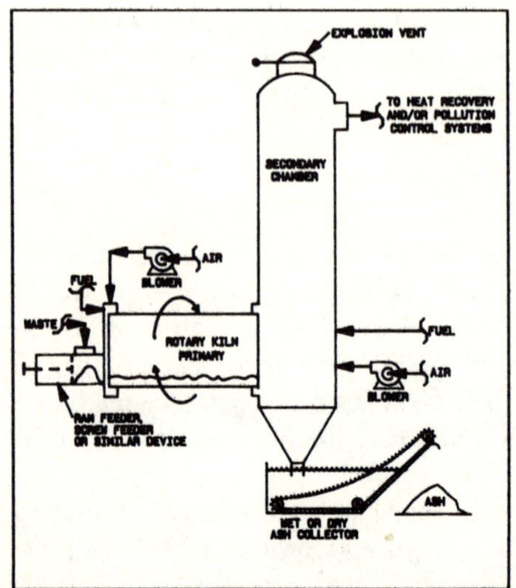

Figure 3.1 Rotary Kiln Incinerator

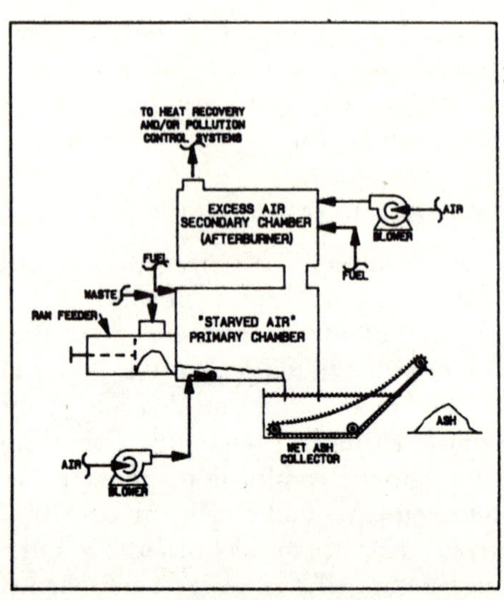

Figure 3.2 Pyrolytic Incinerator (Controlled Air Type)

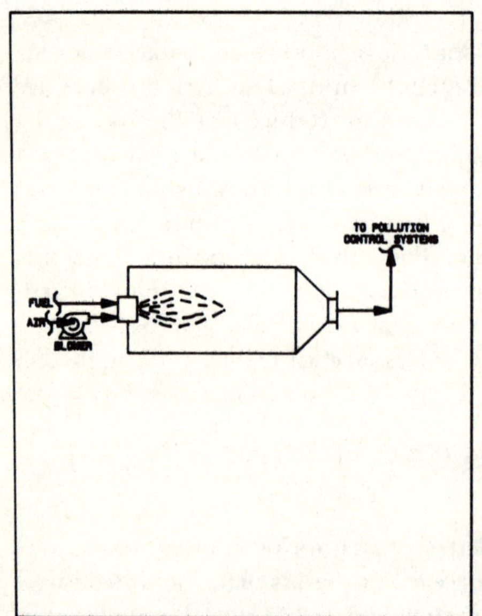

Figure 3.3 Liquid Injection Incinerator

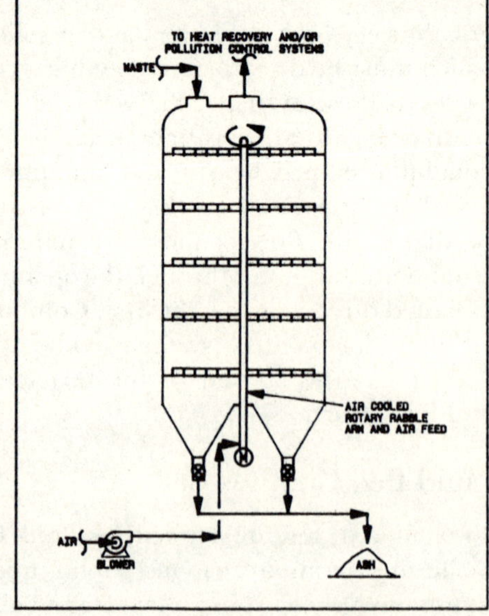

Figure 3.4 Multiple Hearth Incinerator

rate while maintaining a constant temperature. This allows control over organic "burn-out," which is normally not possible with the other systems. In these systems, the waste material is typically introduced by a screw feed mechanism into the lower portion of the fluid bed. Air, fuel, or a mixture of both, is introduced through a distributor plate below the solid bed to fluidize the solid material and provide for good fuel, air, and waste mixing. If acid gases are expected as products of combustion, lime or limestone can also be added to the feed material and will absorb the acid gases during the combustion process. Most of the solid material is then removed by a primary cyclone collector before secondary pollution control systems remove residual particulate matter. Normally, secondary acid gas absorbers are not needed since the absorbent materials in the bed accomplish this same acid gas removal. Figure 3.5 shows one type of fluid bed combustor.

The primary limitation on a fluid bed unit is that the waste material must be of relatively uniform particle size and no large or very fine particulate matter can be accommodated in the bed. Liquid waste can be introduced using atomizing burners directly above the fluid bed.

Heat Recovery

Exhaust gas temperatures from the incinerator systems are typically as high as 1100°C. At these temperatures, most of the operating costs are related to fuel costs. Waste heat recovery represents one method for reducing operating costs. We will discuss three different types of waste heat recovery systems that are available for these applications.

Waste Heat Boilers

Both fire-tube and water-tube boilers have been used for waste heat recovery on hazardous waste incinerators. There are some special considerations which must be taken into account. First,the gas temperatures are typically higher than in conventional boilers so that the tubes in the boiler must be protected from the excessive temperature levels. For fire-tube boilers, it is common to use ceramic inserts in the tube ends to minimize thermal shock at the hot gas entry point into the tubes. On water-tube units, it is common to use bare rather than finned tubes for the first few rows of tubes to prevent high temperature gradients in the tubes and the heat transfer surfaces. It is of critical importance that the temperature of the exhaust gas from the waste heat boiler be maintained at greater than any of the acid gas dew points and greater than the water dew point of the exhaust gas stream. Frequently, the exhaust gases from hazardous waste incinerators become extremely corrosive when these dew points are reached. Figure 3.6 shows a series of waste heat recovery systems utilizing waste heat boilers on incinerators.

Air Preheaters

Just as in the section above, the waste gases from the incinerator can be used to preheat combustion air. In virtually all cases, the combustion air pre-heater utilizes finned tubes with the incinerator exhaust gas passing across the finned exterior surface of the tubes and the combustion air passing through the interior surface of the tubes. Because of the very large quantity of waste heat available, the combustion air can be elevated to very high temperatures. If liquid wastes are being fired, it is necessary to limit the amount of waste heat which is transferred to combustion air to prevent carbonization of the waste material at the burner.

Condensing Heat Exchangers

Condensing heat exchangers are relatively new systems in industry. With the development of high temperature polymeric materials, including Teflon (registered trademark of E.I. duPont deNemours & Company, Inc.), elevated temperature heat exchangers with corrosion resistant surfaces became economically feasible. One such system is called SUBDEW (registered trademark of Andersen 2000 Inc., a division of Crown Andersen Inc.), and utilizes a thin Teflon extruded film over the surface of a tube. The Teflon is limited to applications on waste heat streams at temperatures of 260°C or less. Thus, it is commonly employed after a waste heat boiler, as shown in Figure 3.6. Because the exhaust gas stream from the incinerator comes in contact with Teflon covered surfaces only, the gas can be taken down to below its acid and water dew points. This allows capture of both sensible and latent heat from the system.

Gaseous, Particulate and Ash Discharges

Incinerators produce gaseous emissions which must be controlled in accordance with air pollution control regulations. They also produce particulate emissions which must be controlled, and ash which must be disposed of. This section deals with these secondary discharges.

Ash Handling and Disposal

In the United States, the ash material remains classified as a hazardous material unless the incinerator operator can prove, via extensive sampling, that the ash contains no hazardous components. Often, it is not practical to analyse these waste materials to prove that they do not contain hazardous components. Thus, the ash is disposed of in a hazardous waste landfill which is isolated from aquifers and from the remainder of the environment. It is always preferred to keep the ash material in a dry form rather than introduce it into a wet ash collection system. Once the ash has been introduced to the wet ash collection system, the potential for contaminated water develops and any soluble materials re-dissolve in the water. This can create a water pollution problem in addition to the secondary ash disposal problem.

Flue Gas Particulate Removal

In most hazardous waste incinerator applications, wet scrubbers are used for particulate removal simultaneously with acid gas removal. To a lesser extent, fabric filter baghouses and electrostatic precipitators have also been employed. The scrubber is often the least expensive but highest energy consuming device available for both particulate and acid gas removal. Figure 3.7 shows a schematic flowsheet for a combination particulate and acid gas removal system utilizing a venturi scrubber followed by a packed tower absorption system. Figure 3.8 shows a similar system which does not utilize an alkaline absorbing material. Figure 3.9 shows a more complex emission control system which incorporates not only the venturi scrubber for particulate removal followed by an acid gas absorber, but a spray dryer to convert the waste liquid from the scrubber to a dry solid ash material which can be disposed of with the ash from the kiln. This system also incorporates a condensing heat exchanger after the spray dryer, since the spray dryer reduces the gas temperature to less than 260°C.

Gaseous Emission Absorption

Table 3.1 shows the products of combustion of various organic compounds. The halogenated waste materials produce acid gases which must be absorbed from the exhaust gas stream to prevent secondary environmental contamination. Nitrogen-containing compounds can burn to nitrogen oxides, which are regulated in many countries. In the case of the halogenated compounds, the acid gas absorption systems discussed in the section on fine gas particulate removal above can be employed to achieve greater than 99% removal. In the case of the nitrogen-containing compounds, the most effective means for preventing nitrogen oxide formation is to control the peak gas temperature and maintain it at the minimum possible level to accomplish complete destruction of the waste materials. This can be done by staging the combustion air introduction to prevent "hot spots" in the flame.

Table 3.1 Incineration of Organic Compounds

Components	*Products of Combustion*
C–H	CO_2 and H_2O
C–C	CO_2
C–O	CO_2
H–C–Cl	HCl and CO_2
H–C–I	HI and CO_2
H–C–F	HF and CO_2
H–C–Br	Br_2, H_2O, CO_2
C–N	NO, NO_2, CO_2, N_2

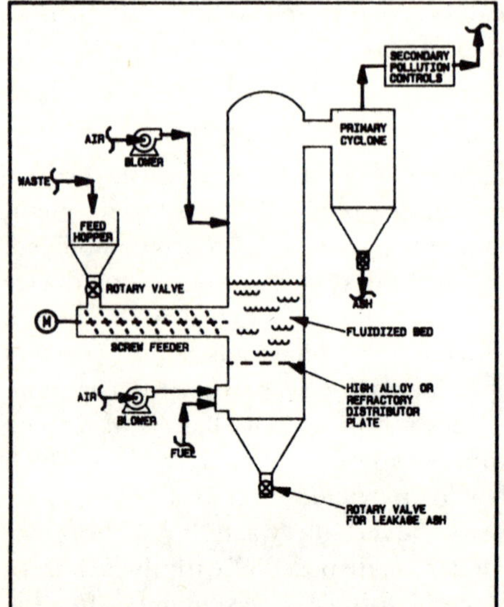

Figure 3.5 Fluid Bed Combustor

Figure 3.6 Typical Waste Heat Recovery Flowsheets for Incinerator and Waste Heat Boiler Exhaust Gases

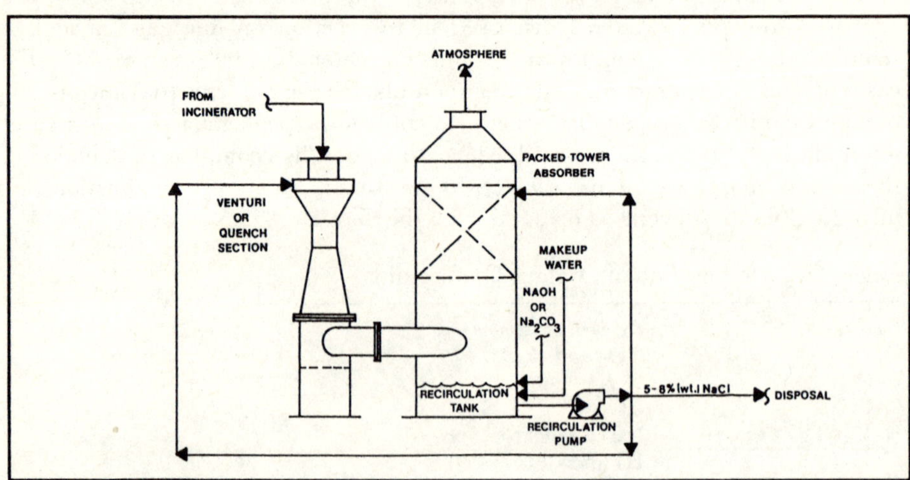

Figure 3.7 Incinerator Scrubbing System with NaOH or Na_2CO_3 Acid Neutralization

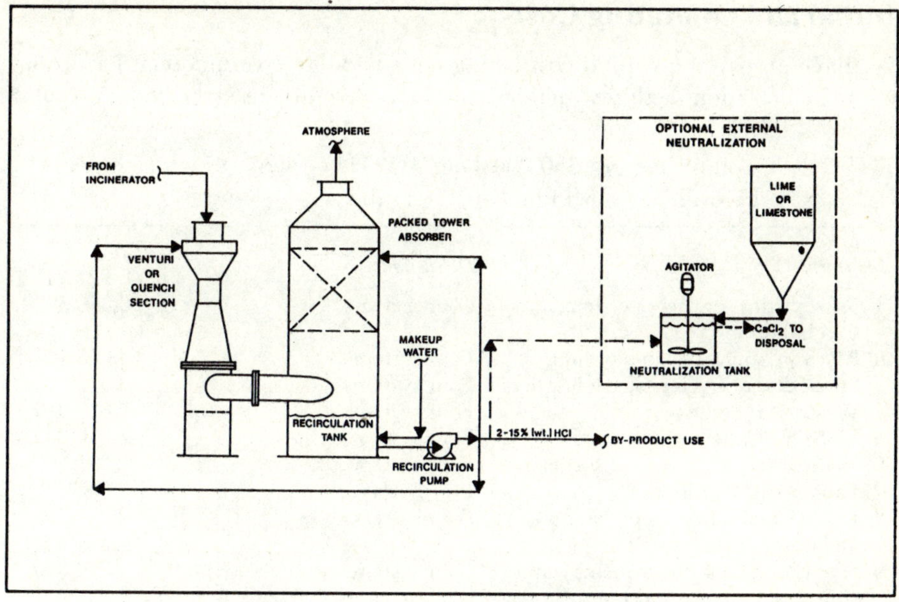

Figure 3.8 Incinerator Scrubbing System to Produce Merchant Grade HCl as a By-Product

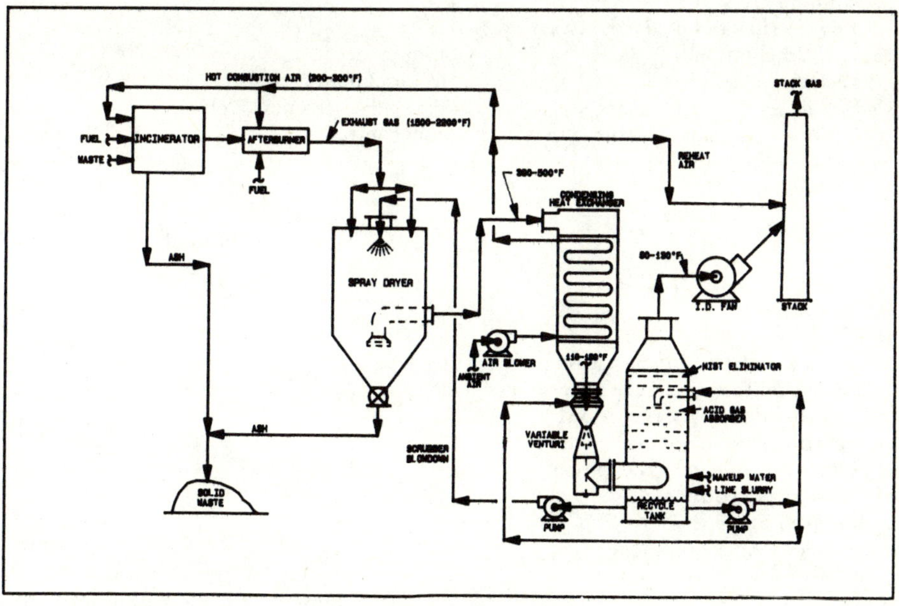

Figure 3.9 Spray Dryer–Condensing Heat Exchanger–Scrubber System for Heat Recovery and Emission Control on Hazardous and Toxic Waste Incinerators

Initial and Operating Costs

We have prepared a capital cost listing for a very large commercial hazardous waste incineration facility which our company recently designed to be installed

Table 3.2 Capital Costs for 350,000 Ton/Year Hazardous Waste Incinerator (4 Parallel 67.8 Million Btu/Hr Trains)

Component	*Discussion*	*Price (US Dollars)*
1. Rotary kilns, burners, feeders, and feed handling	Four systems	6,336,590
2. Afterburners with ash handling	Four systems	5,805,400
3. Spray dryers, cyclones, ash hoppers	Four systems	2,572,250
4. Waste heat boilers	Four systems	1,652,960
5. SUBDEW® heat exchangers	Four systems	1,210,800
6. Venturi scrubbers and absorbers	Four systems	1,893,200
7. Fan and air handling	Process gas and air	882,070
8. Ducts, stacks, reheaters, baghouses and dust collection systems	Four systems	2,400,000
9. Liquid recirculation system, water supply, air supply and pumps	Four systems	760,000
10. Lime storage and feed system		235,000
11. Waste liquid storage and mixing systems, sludge handling equipment, cranes	Tanks, mixers, pumps, delivery, portable systems	780,000
12. Engineering and field expenses		386,600
13. Instrumentation	Computerized	1,447,160
14. Electrical	Power and motor controls	1,955,300
15. Safety, security and emergency	Fires, spills, personnel, vehicles	942,400
16. Sump pumps and controls with piping	Runoff control	64,000
17. Installation		5,864,000
	Subtotal	35,187,730
Optional Items		
18. Office, lab, concrete foundations, water supply, maintenance building, roads, lighting, auxiliary generators, control building, scales, fence, guardhouses, maintenance shop		23,114,000
19. Office furniture and equipment		254,200
20. Laboratory equipment		986,400
21. Shop equipment		298,800
22. Other mobile equipment		207,000
	Subtotal Options	24,860,400
	TOTAL	60,048,130

in the United States. The costs of this plant include support facilities for analysis of the waste materials and incorporate all of the necessary safety systems required by regulatory agencies in the United States. Operating costs for this plant are also presented. Based on present disposal charges of about $800 per ton in facilities of this type, this commercial incinerator will generate annual revenues for the owner of about $280 million.

Plant Installation Costs

Table 3.2 lists all of the capital costs for this 350,000 tons (318,200 metric tons) per year hazardous waste incineration facility. This facility includes four rotary kiln incinerators followed by four vertical afterburners, and including waste heat boilers, spray driers, scrubbing systems, and all necessary dry ash handling. In addition, these costs include the loading and unloading facilities, a

Table 3.3 Operating Costs for 350,000 Ton/Year Hazardous Waste Incinerator (Plant Operates 7500 Hrs/Yr)

Expenses	*Annual Cost (US Dollars)*
1. Payroll, including payroll added costs	4,048,791
2. Travel and travel related costs	550,000
3. Incineration Plant Operations	
A. Supplemental fuel at $4.50/MM Btu and 130.7/MM Btu/hr	4,411,125
B. Electrical power at 5.5¢/kwh and 2840 kwh/h for 7500 hrs/yr	1,171,500
C. Water at $0.01 per gallon	622,710
D. Slaked lime at $140/ton delivered	3,227,474
E. Ash disposal at $100/drum	112,699,800
F. Maintenance parts and consumables (refractory, drums, etc.)	4,100,000
4. Office And Lab Operations	
A. Communications	40,000
B. Office supplies	, 23,000
C. Lab supplies	300,000
D. Liability and casualty insurance	2,000,000
E. Utilities (Non-process)	58,000
F. Accounting	200,000
G. Legal	500,000
H. Professional services	200,000
I. Permits and fees	1,000,000
5. Miscellaneous	3,000,000
Total Expenses	138,152,400
6. Depreciation at 7 year straight line on approx. $35 million of equipment	5,000,000
Total Expenses Including Depreciation	143,152,400

Total Employment = 125 People

complete containment system for the entire operation to prevent any spills from ever entering the environment, and an administration and laboratory facility adjacent to the plant to provide the necessary support services for operation of the plant in a safe fashion.

Operating Costs

For the plant described above, the operating costs are as listed in Table 3.3. Because of the regulatory requirements in the United States for such a facility, both safety and security forces are required in the plant at all times. Thus, staffing costs for the facility are quite high. In addition, we have assumed disposing of waste materials from the plant in a hazardous waste landfill. Thus, the waste disposal costs are also quite high and constitute the largest single operating cost in the system. If these waste materials could be certified as non-hazardous, this waste disposal cost would be eliminated.

4
Incineration Processes for the Disposal of Difficult Hazardous Wastes

Kevin J. Whiting

With national governments and many local regulatory authorities in the European Economic Community (EEC) tightening hazardous waste regulations, many companies are turning to in-house incineration as an alternative to other waste treatment or disposal techniques. As a result of advances in the processing technology of the chemical and petrochemical industries, by-product wastes are being generated. Because of their chemical make-up they are difficult to handle within the incineration system and pose technical and engineering challenges to the incinerator designer.

Obviously incineration systems must be designed to dispose of the hazardous waste in a way that imposes the minimum financial burden on the operating company. Consequently, the optimum system will incorporate energy and by-product recovery equipment to minimize operating costs and, in addition, capital costs for associated auxiliary equipment (such as scrubbers) will be lower because, in general, the volume of flue gases will be reduced. HCl recovery will also avoid a further effluent stream from the absorption system which would require additional treatment.

Waste Characteristics

Understanding the characteristics of different waste streams is necessary to ensure proper selection and design of the incineration system.[1] Based on the thermal ratings, there are two basic waste classifications: high heating value waste, which is usually self-combustible and does not require supplementary fuel, and low heating value waste, which is not normally combustible without the addition of support fuel. Another method of classification is to define the waste according to its chemical make-up. The categories used are: organic, halogenated, metallic, aqueous, nitrogen-containing. They are described below.

Organic waste includes hydrocarbons that contain only carbon, hydrogen, oxygen (sometimes sulphur in small quantities), and are self-combustible. They can be used as fuel and when burned with the correct amount of air will yield carbon dioxide, oxygen, nitrogen, and water vapour (sulphur dioxide) and can be discharged into the atmosphere. Heat in the product gas can be recovered through a boiler or some other waste heat recovery device. When sulphur dioxide is produced, caustic scrubbing or other desulphurization devices are required to ensure clean, acceptable stack emissions.

Halogenated waste includes such chemicals as carbon tetrachloride, vinyl chloride, methyl bromide, and other halogenated materials, such as fluorine and iodine. Heating values of these wastes depend upon the halogen content and may or may not require auxiliary fuel. When oxidized, the product of combustion will contain hydrogen halides and free halogens; the proportion of the latter is governed by the incineration temperature and the hydrogen to halogen ratio in the waste/fuel mixture. The hydrogen halides can be either recovered or neutralized, depending on specific needs.[2]

Metallic waste includes inorganic and organic salts, such as sodium or potassium compounds. When oxidized, the combustion products will contain these salts in the molten state and refractory selection, oxidation temperature, and residence time become prominent design considerations. Since this group of wastes usually bypass the burner, auxiliary fuel is required for complete combustion. Sub-micron particulates and mists in the product gas will require secondary gas treatment equipment and heat in the system can be recovered by preconcentrating the waste. This class of waste may also contain heavy metals such as, lead, arsenic, mercury and vanadium which leads to other design considerations such as the use of bag filters and electrostatic precipitators.

Aqueous waste includes any, or a combination of, the above wastes in a solution of higher than 60 per cent water. Due to low heat of combustion, this group of wastes will not support combustion in the burner and must be injected through atomized sprays downstream of the flame zone.

Nitrogen-containing waste includes organic compounds which have nitrogen atoms bonded directly to carbon, hydrogen or oxygen atoms within the chemical structure. The chemical bonds between the nitrogen atom and the remainder of the molecule are considerably weaker than the bond dissociation energy of nitrogen. Consequently, these organic molecules can yield larger quantities of NO_x during combustion than is derived by the thermal fixation of N_2 alone. To minimize NO_x formation, a two-stage combustion system can be used. The first stage is designed to be operated with a fuel rich condition. The unburned hydrocarbons are then oxidized in a secondary combustion chamber. Alternatively to combustion modification, catalytic NO_x abatement systems can be applied to reduce stack NO_x emissions to acceptable levels.

Incinerator Design: General Considerations

Incineration Temperature

The rate of the combustion reaction is highly dependent on the incineration temperature which must be such that

1. It exceeds the auto-ignition temperature of the waste
2. In conjunction with the specified residence time, it ensures complete oxidation of the waste
3. It is a suitable operating temperature for the specified materials. In addition the incineration temperature influences the occurrence of secondary reactions that occur within the combustion chamber and may thus be used as an important parameter for the control of noxious emissions

Generally, operating temperatures in the oxidation chamber are controlled between 800°C and 1450°C. An upper limit of 1650°C exists to prevent damage to the refractory lining of the incinerator and above 1500°C carbon dioxide dissociates to form carbon monoxide, which will lead to emission problems.

Residence Time

The selection of the residence time will be influenced by the incineration temperature and must be such that the waste is maintained at the desired temperature for sufficient time to ensure complete oxidation of the waste. Increased residence times improve the destruction efficiency of the waste but the following economic penalties result: increased overall mass and size of the unit, and increased quantity of refractory material.

Excess Air

Regulations such as RCRA in the United States and T.A. Luft in the Federal Republic of Germany recommend incinerator operating conditions which include 3 to 6% O_2 as measured in the stack, however, this tends to work against the incinerator designer. The higher oxygen content leads to higher free halogen formation. This is due to the shift in the equilibrium reaction caused by the additional oxygen. Also, it means that the free halogen, usually chlorine that is formed is much more difficult to neutralize and forms the hypochlorite which causes severe corrosion problems. The higher oxygen content also means increased sizes for blowers and scrubbers because of the increased gas flow rate and added kilowatts because of the additional mass required to be pumped. A further problem created by operation with high stack per cent O_2 is that more oxides of nitrogen are formed leading to additional environmental impact.

Vortex Burner

Figure 4.1 shows a cross-sectional view of the vortex burner. Waste is introduced through a nozzle at the centre line of the burner. Combustion air

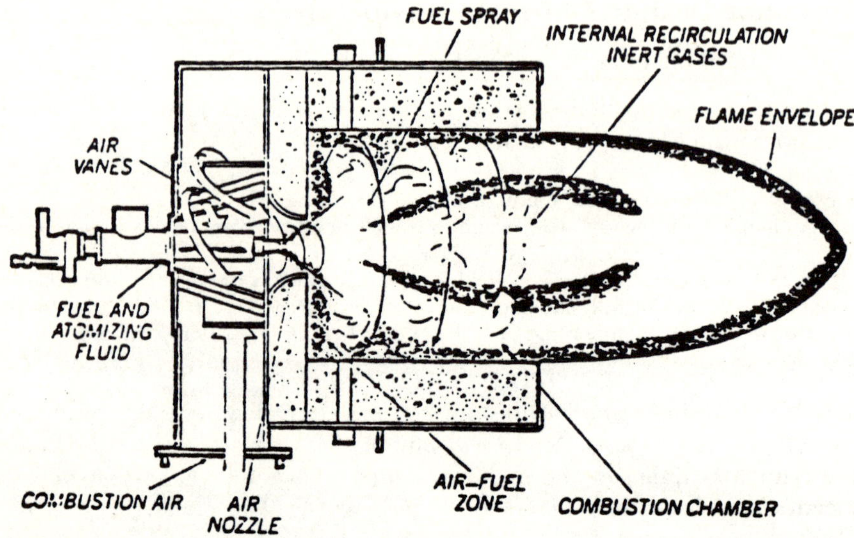

Figure 4.1 The Vortex Burner

enters tangentially into the burner body and passes through swirl vanes which impart rotational energy. A twisting high velocity vortex action results in complete mixing with the fuel spray at its point of injection, and at the same time creates a low pressure zone immediately downstream from this point. As the highly turbulent air-fuel mixture enters the flame zone, the low pressure area causes a recyling of the hot gases of combustion back into the mixture. Thus, the mixture is pre-heated, vaporized and raised to ignition temperature almost instantaneously. The high intensity combustion allows the combustion chamber to be considered as a reaction chamber. The vortex burner will maintain stable combustion without auxiliary fuel with a waste having a heating value between 4500 and 5400 Btu/lb (2500 to 3000 kcal/kg), although it varies depending upon waste composition.

Liquid Atomization

An important design aspect which contributes to the efficient turbulence/mixing characteristics necessary in the incineration chamber to maximize the efficiency of the oxidation reactions is liquid atomization. In order to dispose of liquid wastes, atomizers are used to inject the waste into the combustion zone and/or the oxidation chamber. The main purposes of injecting the liquid as a spray are

1. to break up the liquid into fine droplets
2. to place the liquid in the chamber in a specific location with a specific pattern and with sufficient penetration and kinetic energy
3. to control the rate of flow of the liquid discharged to the nozzle

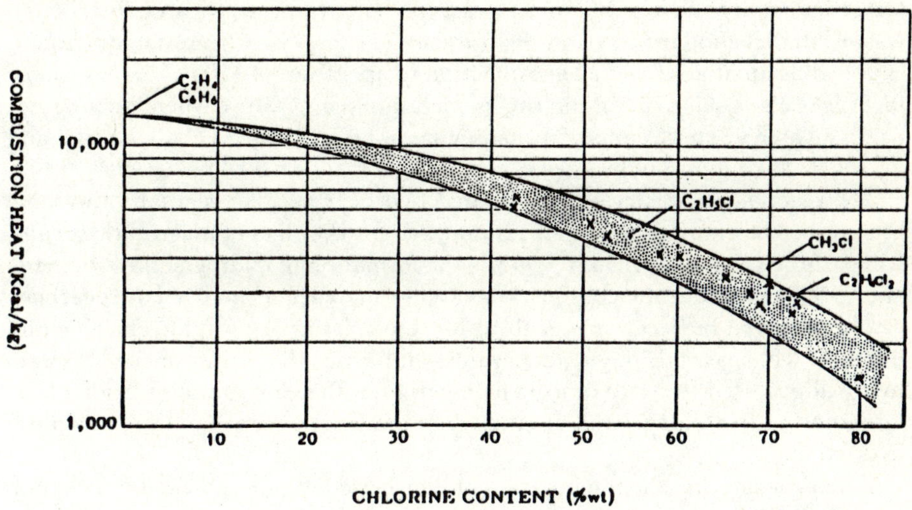

Figure 4.2 Heat of Combustion of Chlorinated Hydrocarbon

In thermal oxidation applications, the primary purpose of the spray atomizer is to permit the minimum droplet size possible to provide maximum surface area for heat and mass transfer. This leads to rapid evaporation, superheating, ignition, and oxidation. This is sometimes not possible with waste liquids containing solids.

It is important that the following information be provided to the incinerator designer so that proper nozzle selection is made: flow rate, specific gravity, viscosity (at two temperature levels), chemical analysis, solids content, and maximum particle size.

Halogenated Hydrocarbons

Figure 4.2 shows the relationship between the percentage of chlorine in various chlorinated hydrocarbon species and their heating value. In order to burn waste containing more than 70% chlorine, it is necessary to use auxiliary fuel with the waste to bring the total heating value to the level mentioned above.

In a practical incinerator the equilibrium ratio of hydrogen chloride to free chlorine is a function of temperature and to generate the highest yield of hydrogen chloride, the incinerator should run at high temperature, with low excess air and with high excess hydrogen levels.[3] In fact, for each waste mixture, there is an optimum point for economical design which can also be affected by the materials of construction of the incinerator.

The conditions within the burner and incineration chamber must be controlled to optimize the thermal oxidation reaction and minimize the formation of free halogens. In the case of both chlorinated and fluorinated wastes, the reaction proceeds in the presence of excess hydrogen (i.e. methane), supplied as auxiliary fuel gas at a reaction temperature of 1200°C. The support fuel gas is also used to maintain the incineration temperature when burning low heating value waste streams. To ensure maximum conversion of Cl_2 to HCl and F_2 to HF a minimum hydrogen to halogen ratio of 3 : 1 is used.

The reaction chemistry for fluorinated hydrocarbons is essentially the same as that for chlorine but the equilibrium constant is orders of magnitude greater which throws the equilibrium towards the formation of hydrogen fluoride. HCl and HF formed during combustion of chlorinated/fluorinated hydrocarbons can be removed by scrubbing with water, but relatively insoluble chlorine and fluorine will pass through the water and into the atmosphere. Sodium hydroxide is used to scrub out the halogens and, therefore, minimization of the formation of free halogens is essential to minimize the consumption of sodium hydroxide.

When considering the incineration of fluorinated wastes, particular attention must be paid to the possible corrosion conditions that may occur within the incineration system. HF is extremely corrosive to metals; material selection is extremely important, and to engineer and operate a system reliably for fluorinated wastes, would require expensive materials, duplication of critical instrumentation and extremely rigorous operating regimes.

Metallic and Aqueous Wastes

If an aqueous waste stream contains from a trace to 100% organics it can be incinerated. The aqueous waste may contain alkali metals, chlorides, carbonates, bicarbonates, sulphides, sulphates, silicates and phosphates, which during combustion will combine to form salts which have low melting points. The waste may also contain heavy metals such as lead, arsenic, titanium and vanadium.

In an incinerator burning this type of waste, inorganic compounds formed during the combustion reaction are molten and would accumulate on horizontal surfaces within the incineration system. On shutdown, the molten salt would solidify and require mechanical removal. Particulates, including molten salts, are also carried out of the incinerator and would require special downstream collection devices.

If an aqueous salt bearing waste was injected directly into the body of the flame, this would quench the flame, resulting in soot formation and other combustion problems. Therefore, aqueous waste is injected through secondary injectors downstream of the flame zone where the waste starts to heat immediately while the products of combustion from the burner are being cooled to the design or oxidation temperature.

Incineration System Configuration

For organic gaseous and liquid wastes (including chlorinated hydrocarbons) the incineration system could be as shown in Figure 4.3.

Incineration Chamber

The cylindrical incineration chamber is mounted horizontally with a vortex burner positioned at one end. A multi-component refractory lining is installed in the incinerator to protect the shell from the high temperatures (1200°C), generated during the combustion reaction yet provide enough heat transfer to keep the shell at a temperature between 135°C and 200°C. The maintenance of the shell within this temperature range is critical to prevent condensation of hydrochloric and/or sulphuric acids which will cause severe damage. The shell temperature would likewise not be allowed to rise past 315°C as dry chlorine attack will occur at this temperature.

Waste Heat Recovery

When designing a system to incinerate chlorinated hydrocarbon wastes, the need for support fuel may form a significant portion of the operating costs for the unit. The economics of recovering a portion of these operating expenses via waste heat recovery may justify the inclusion of a waste heat boiler. Energy recovery from hazardous waste incineration yields other benefits:

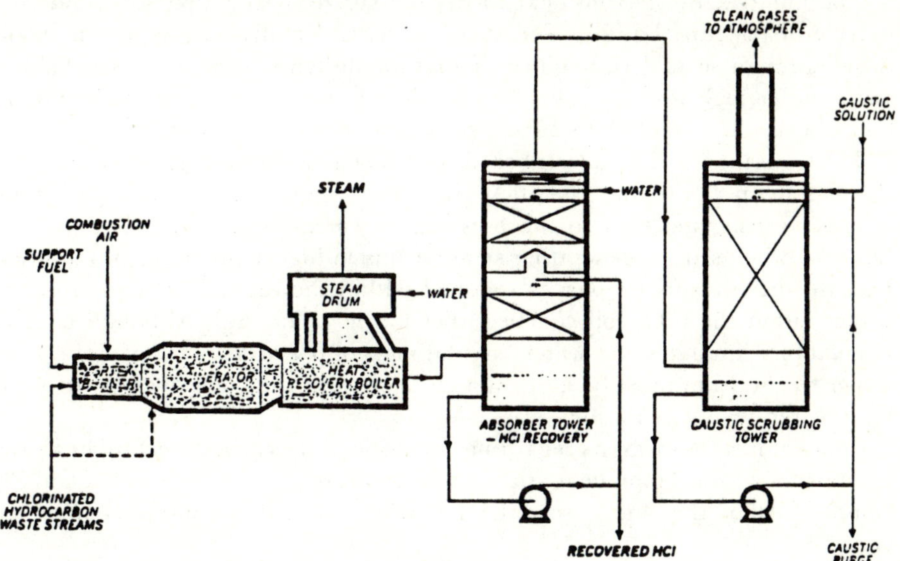

Figure 4.3 Chlorinated Waste Incinerator with Steam Generation and Hydrochloric Acid Recovery

1. Lower flue gas volumes due to the reduced water vapour concentration and flue gas temperature, resulting in smaller downstream scrubbing equipment with reduced operating costs
2. Reduced steam plume from the stack and enhanced water conservation as a result of evaporating less water in the quench column due to the lower flue gas temperatures leaving the boiler

Table 4.1 indicates the levels of net energy available for recovery from a typical flue gas (considering the major components only).[4]

Table 4.1 Net Energy Recovered from a Typical Flue Gas (Source: Novak, R.G., Troxler, W.L., and Dehnke, T.H. 1984. Recovering Energy from Hazardous Waste Incineration. *Chemical Engineering*, March 19: 146–54)

Component	*Energy content (kcal/kg)*		*Energy recovery (kcal/kg)*	%
	At 980°C	*At 260°C*		
CO_2	260.6	55.1	205.5	78.9
N_2	258.2	61.2	197.0	76.3
Excess air	251.6	59.3	192.3	76.4
Water vapour	1083.3	700.6*	382.7	35.3

*includes 588.9 kcal/kg latent heat of vaporization.

In a firetube boiler the high temperature flue gases flow through the inside of a tube bundle, transferring heat to the pressurized water that surrounds the outside of the tube bundle. The steam generated is discharged into a steam drum through several riser pipes. A firetube design is generally used for hot gases under high pressure, or when a gas-tight seal is required. This type of boiler is usually limited to steam pressures of 65 bar gauge.[5]

The watertube boiler has a tubecoil constructed of externally finned tube and the high temperature flue gases flow transversely across the outside of the tube bundle, transferring heat to the pressurized water inside them. This type of boiler can be used for generating steam at much higher pressures, if required, because the pressurized part is retained within the tubes. Circulation of the water within the tube coil can be either forced or natural. Although natural circulation eliminates the water circulating pump and associated piping, the upper steam drum must be high enough to create a sufficient hydrostatic head for the required circulation ratio.

Concern justifiably exists regarding corrosion of the carbon steel tubes by the combustion gases. To prevent attack of the metal by the aggressive HCl and Cl_2 components of the flue gases, the following operating conditions must be ensured

1. The tube inside wall temperature must be maintained above the dew point of the acid gases in the combustion products

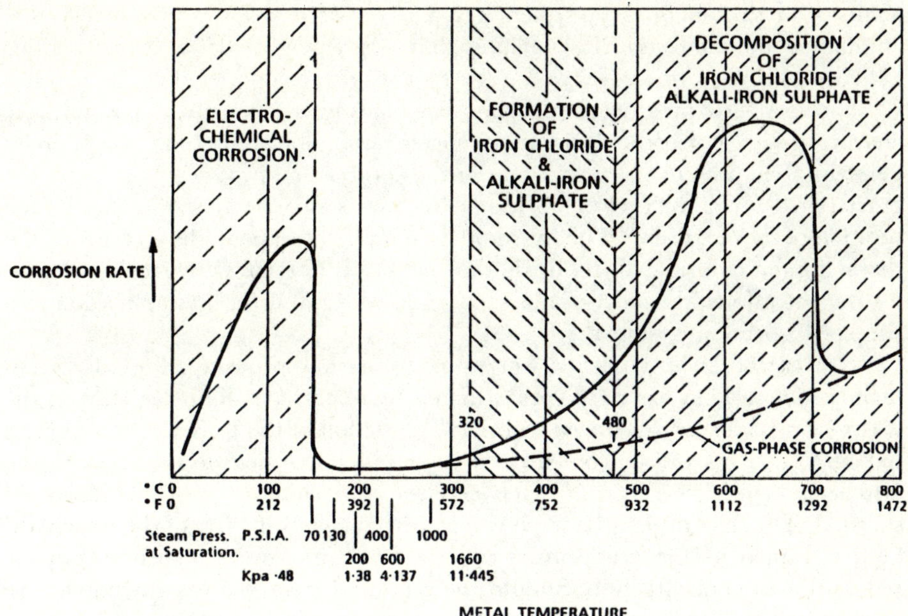

Figure 4.4 Corrosion Rate vs. Metal Temperature

2. The tube inside wall temperature must be maintained below 300°C; above this temperature the tube materials will exprience so-called "dry chlorine attack"

This range of temperatures is referred to as the operating window and is shown as the unshaded area in Figure 4.4.[6]

HCl Recovery Systems

As has been previously discussed, the incineration of chlorinated hydrocarbon waste requires that the operating parameters are appropriately controlled to ensure the complete destruction of the hydrocarbons and to minimize the presence of residual chlorine. The exhaust products leaving the incinerator will be at temperatures in excess of 1200°C and will contain hydrochloric acid which must either be neutralized or recovered prior to the exhaust gas being discharged to atmosphere.

Both the recovery and neutralization process require cooling of the combustion products to temperatures between 50 and 100°C. For processes in which waste heat boilers have been included, the remainder of the cooling being accomplished via direct contact cooling and quenching either in a submerged type quench tank or in a tower. Alternatively, for processes in which no heat recovery equipment has been included, the entire required heat removal must be accomplished by direct contact cooling and quenching. This is generally achieved in a submerged type quench tank as opposed to a tower, as this system eliminates the mechanical problems that are encountered at the interface between the very hot combustion gases and the cold quench system.

At atmospheric pressure, hydrochloric acid/water systems display azeotropic characteristics at hydrochloric acid concentrations of approximately 21 mass per cent. The presence of inert gases in the combustion products will lower the partial pressure of the hydrochloric acid/water system and thus shift slightly the position of the azeotrope. This phenomenon makes it impossible, irrespective of the hydrogen chloride concentration in the gases, to recover hydrochloric acid in concentrations in excess of the azeotropic concentration by application of a simple absorption system.[7]

Recovery of 20% HCl by weight is the usual process adopted for an incineration system destroying chlorinated hydrocarbons. The gas leaving the waste heat boiler is introduced in the lower section of the tower – "the quench section". The gas will be cooled by releasing its latent heat to the evaporating liquid and by direct contact between the recirculating concentrated acid solution. The flow rate of the fresh water introduced at the top of the tower will be determined by the evaporation experienced in the quench section and the required acid concentration. Should the quantity of fresh water introduced be insufficient to satisfy the direct contact cooling requirements, it will be necessary to provide a cooler in the quench recycle line.

The gas discharge from the top of the tower is composed primarily of nitrogen, carbon dioxide, water vapour and oxygen. However, there are also traces of hydrogen chloride vapour and chlorine which pass through the absorption tower and require further treatment. Thus the cases leaving the absorption tower will be fed to a scrubber where the HCl and Cl_2 will be scrubbed out via an alkaline solution to meet emission requirements.

In vinyl chloride monomer production plants there is a requirement for the recovery of two grades of hydrochloric acid. One is to recover 35% acid solution which may be fed to the anhydrous hydrogen chloride production plant and the other is to recover anhydrous hydrogen chloride directly for use in the production of vinyl chloride monomer. Figure 4.5 is a flowsheet of a typical process for the recovery of 35% hydrochloric acid. A quench tank has been used instead of the quench section in the absorption tower. The basic principle for the recovery of the 18% hydrochloric acid solution is as previously discussed. The overflows from the absorber and quench tank will be fed to the stripper feed tank. An extractive solvent, for example calcium chloride, will be used to degrade the azeotrope which disappears when concentration of calcium chloride exceeds 30%. Calcium chloride is recycled to the quench tank where it is concentrated by evaporation of water in the quench. Sufficient calcium chloride must be introduced to the quench tank to ensure that the calcium chloride concentration in the stripper feed tank exceeds 30%. The mixture from the stripper feed tank is fed to the top of the stripper whose reboiler provides sufficient heat to recover hydrochloric acid vapour at concentrations above the

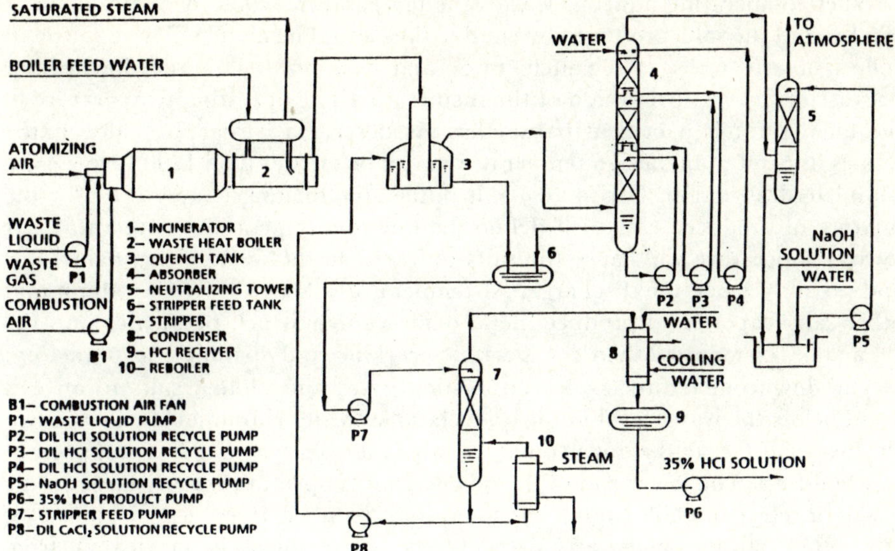

Figure 4.5 Process for Recovery of 35% Hydrochloric Acid

azeotropic concentration, approximately 60%. The concentrated hydrochloric acid vapour from the tower overheads is condensed and diluted to the required concentration, in this case 35%. The tower bottoms contain diluted calcium chloride solution, which is recycled to the quench tank where it is concentrated as previously discussed. The stripper will be a rubber lined steel tower, lined with acid resistant brick and the internals will be ceramic "Raschig" rings as the packing material. Impregnated graphite will be used for the reboiler and condenser.

The process for the recovery of anhydrous hydrochloric acid is the same as for the 35% acid recovery with the additional processing to reduce the water content of the distillate by multistage cooling with water and then brine. The resulting condensate is refluxed to the tower from the condensate receiver.

Submerged Combustion Incinerator

As has been previously discussed, the incineration of difficult waste such as fluorinated hydrocarbons and aqueous wastes containing inorganic salts and/or heavy metals poses severe engineering and operational challenges to the incinerator designer and operator respectively. For these arduous duties, we would select the submerged combustion (Sub–X) incineration system. A cross-sectional view of the incinerator is illustrated in Figure 4.6. The incinerator chamber is vertically mounted with a downward fired vortex burner mounted at the top.

When incinerating aqueous wastes the temperature would be maintained at 1000°C and the salts produced by combustion would be molten. These salts will collect on the walls of the incinerator and flow down into the downcomer because of the vertical design of the incinerator. The operating temperature of the incinerator is a critical parameter. As described earlier for halogenated wastes, the optimum design temperature must be greater than 1200°C to ensure complete destruction. For molten salt duties, the melting temperature for the mixture of produced salts will define the minimum operation temperature. A lower temperature will cause the salts to freeze on the walls of the incinerator and possibly block the discharge. At temperatures above 1200°C, sodium and other salts vaporize to produce metal oxide vapour which permeates into the refractory, condenses, then revaporizes, cracking and spalling the refractory.

The downcomer directs the combustion gases and molten salts to an exit point below the water level in the quench tank. Water continuously flows down the inside surface of the downcomer to protect it from the hot gases and prevent salt build-up. The lower end of the downcomer is open in case any large pieces of salt or refractory falls from the incineration chamber. If halides, phosporus or any other strong acids are present, the downcomer is fabricated from Hastelloy–C, however, if the salts present are of weak acids, often stainless steel can be used.

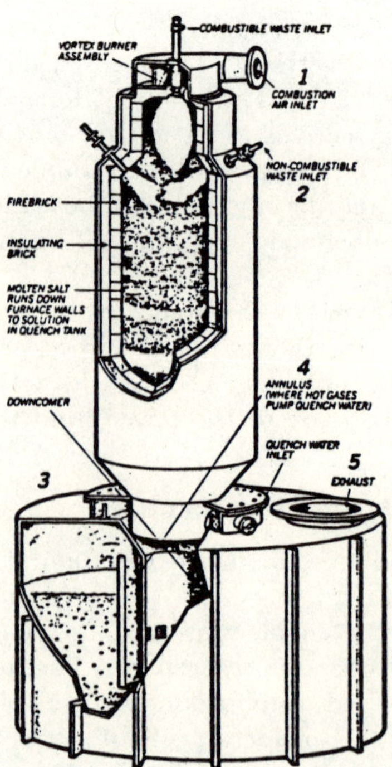

Figure 4.6 Submerged Combustion Incinerator

The hot gases and salts pass from the downcomer into the quench water. Molten salts from the incinerator flow directly into the bottom of the tank. If the salts are water soluble, they will be dissolved into water through the agitation of the gas lift action of the exhaust gas. The incinerator exhaust gases exit through openings in the downcomer tube into the bath. The gas will break into tiny bubbles which have a surface area permitting complete heat and mass transfer in a very short time.

It has been estimated that 0.1 m^3 of gas when broken into bubbles of 25μm diameter, represents a surface area of 23,619 m^2 available for heat transfer. As a result of the direct contact between the high temperature gas and the quench water, heat transfer efficiency is almost 100%. For halogenated hydrocarbon waste incineration duties, the downcomer/quench tank assembly is designed for shallow submergence. The combustion products are forced through a 4–6 inch depth of water which is sufficient to scrub about 65% of the hydrogen halides from the gas stream and significantly reduce the temperature of the gases.

The quench system for an aqueous salt system is designed to give deep submergence of 24 inches of water in order to maintain a longer residence time within the quench liquid. Most of the molten salts and large-size particles will be scrubbed out in the quench tank. Only particles in the size of 2μm and smaller will exit from the quench tank with the saturated exhaust gases. A weir is provided in the tank, which together with the gas lift action caused by the gas sparging effect, creates turbulence and mixing between the gases and the quench liquid. The exhaust gas from the quench chamber is in equilibrium with the solution in the tank and, thus, is saturated with water vapour.

Installed in the top of the quench tank and surrounding the downcomer are a series of spray nozzles through which water is sprayed. This provides secondary cooling to the downcomer and final quench of the products which pass upward through the sprays to exit the quench tank near the top.

The Sub–X quench tank is a well proven high temperature safety device for this system. When the supply quench water is lost, the tank will act as a heat sink until all the water above the openings in the downcomer tube is evaporated. Usually, a minimum of 30 minutes is available for the operator to correct the system malfunction without shutdown being necessary or damaging of downstream equipment. The quench tank is constructed of corrosion resistant materials such as epoxy-lined carbon steel or fibre reinforced plastic (FRP), to withstand the acids generated when the hot combustion products are quenched. Since the temperature of the saturated gases leaving the quench tank is lower than 95°C, the downstream pollution control equipment can be fabricated from low temperature construction materials.

An incineration unit destroying fluorinated hydrocarbon waste would require at least a caustic scrubber, but possibly a water wash primary absorber as well, depending upon the guaranteed emission levels required. The blowdown from the quench tank and absorption columns will contain hydrofluoric and hydrochloric acids which must be further treated before disposal to drain

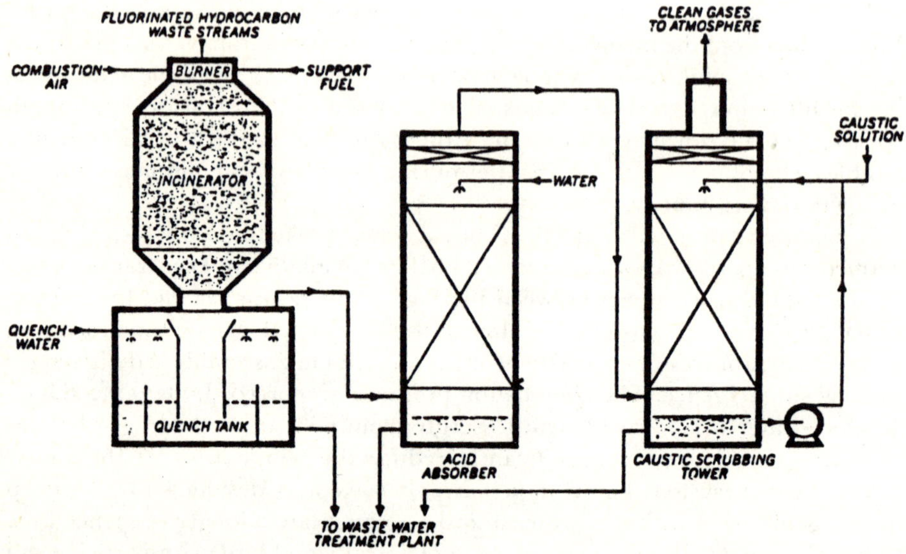

Figure 4.7 Fluorinated Waste Submerged Combustion Incinerator with Gas Treatment Facility

(Figure 4.7). Aqueous salt incineration systems also employ a venturi scrubber which is designed to remove the salt particulates carried over from the quench tank.

Conclusions

At the present time, incineration is not an exact science. Temperature, residence time, excess air and mixing are so interrelated that theoretical calculation models are not yet available. It is possible to specify residence time, incineration temperature and excess air requirements as design parameters because these variables can be reduced to finite numbers. But the mixing aspects of an incinerator and waste material are empirically different for each waste and each incinerator. Thus, specification of time, temperature and excess air is necessary but not sufficient to ensure satisfactory waste disposal. A reliable pilot plant test burn is necessary for each new waste material to determine the incineration characteristics. Previous experience on the incineration of similar wastes is also a source of valuable information. A combination of pilot plant data and previous experience is necessary to ensure the design of an incinerator which will ultimately dispose of hazardous organic wastes satisfactorily and within the framework of environmental regulations. Finally, hazardous waste disposal systems can be profitable if designed and installed with the total economics included in the facility design (see Appendix).

Acknowledgements

I would like to express my sincere thanks to T-Thermal Inc. of Conshohocken, Pennsylvania, United States and Nittetu Chemical Engineering Ltd of Tokyo, Japan for their permission to publish certain technological aspects.

Appendix: Typical Process Economics for a Chlorinated Waste Incinerator with Waste Heat and Anhydrous HCl Recovery

The economics of any hazardous waste disposal system is an important factor that has to be taken into account at the design stage. In this section an assessment is made of the costs involved and overall operating gain for a typical chlorinated waste incinerator with waste heat and anhydrous HCl recovery, given the following plant specifications

Waste:	Mass flow rate=4000 kg/hr of liquid chlorinated hydrocarbons Chlorine content=70%
Thermal Oxidizer:	Thermal duty=5.63 MW (19.2×10^6 Btu/hr) (Combustion of waste self-supporting at 1200°C)
Waste Heat Boiler:	Saturated steam produced=7000 kg/hr at 18 bar(g)
HCl Recovery Unit:	Mass of anhydrous HCl produced=2744 kg/hr (98% recovery)

Operating Costs

Total operating costs for the process consist of utility costs, depreciating capital cost, interest and maintenance costs, and labour costs. The figures for these costs are expressed in £/hr.

Utility Costs

Item	*Unit Cost**	*Consumption*	*£/hr**
Demineralized water	1.25×10^{-3} £/kg	18,500 kg/hr	23.13
Process water	0.15×10^{-3} £/kg	8000 k/g/hr	1.20
Cooling water	0.03×10^{-3} £/hr	90,000 k/g/hr	2.70
Steam	7.60×10^{-3} £/kg	6300 k/g/hr	47.88
Electrical power	0.035 £/kW	300 kW	10.50
Neutralizing agent	2.08×10^{-3} £/kg	120 kg/hr	0.25
			85.65 £/hr

*typical European costs

Cost of Depreciating Capital

Total capital investment (including foundations, site preparation, structural, piping, electrical, etc):	£4,000,000.00

Capital shall be depreciated by a fixed amount (10%) annually over a seven-year period:

Cost = 4,000,000×0.9×1/7×1/8000 = £64.29/hr

Interest and Maintenance Costs

Interest rate	=	11% per annum of total installation cost =	£55.00/hr
Maintenance	=	3% per annum of total installation cost =	£15.00/hr
Other expenses	=	3% per annum of total installation cost =	£15.00/hr
			£85.00/hr

Labour Costs

No. of operators = 2.5 persons = 0.5 person per shift × 3 shifts + 1 person

Average annual wage	=	£18,000/yr
Cost per hour	=	£5.63/hr

therefore:

Total Operating Costs

Utility costs	=	£85.65
Depreciation	=	£64.29
Interest	=	£85.00
Labour	=	£5.63
		£240.57/hr

Recovered Costs

Total recovered costs consist of steam generation, HCl and alternate disposal costs as follows:

Steam Generation

Mass of steam generated	=	7000 kg/hr
Unit cost	=	7.6×10^{-3} £/kg
Recovered cost	=	£53.20/hr

HCl Recovery

Mass of HCl recovered	=	2744 kg/hr
Unit cost	=	160×10^{-3} £/kg
Recovered cost	=	£439.04/hr

Alternate Disposal Costs

Mass of Waste for disposal	=	4000 kg/hr
Estimated disposal cost	=	11.20×10^{-3} £/kg
Recovered cost	=	£44.78/hr

Total Recovered Costs

Steam generation	=	£53.20
HCl recovery	=	£439.04
Alternate disposal	=	£44.78
		£537.02/hr

The operating gain for a typical process based on the specifications above is therefore £296.45/hr (i.e. total recovery costs of £537.02/hr minus total operating costs of £240.57/hr). The capital payback period is 1.7 years.

References

1. Kiang Y-H. 1980. Incineration of Hazardous Organic Wastes. ASME Proc. Nat. Waste Proc. Conference.
2. Santoleri, J.J. 1973. Chlorinated Hydrocarbon Waste Disposal and Recovery Systems. *Chemical Engineering Prog.*, 69, 1: 68–74.
3. Mitachi, K. *et al.* Treatment of Liquid Waste Derived from Chlorinated Hydrocarbon Plants. Technical Report. Nittetu Chemical Engineering Ltd.
4. Novak, R.G., Troxler, W.L., and Dehnke, T.H. 1984. Recovering Energy from Hazardous Waste Incineration. *Chemical Engineering,* March 19: 146–54.
5. Csathy, D. 1967. Evaluating Boiler Designs for Process Heat Recovery. *Chemical Engineering,* June 5.
6. Fassler, K., Leib, H. and Spahn, H. 1968. Corrosion in Refuse Incineration Plants. *Mitteilungen der VGB,* 40, 2: 126–39.
7. Oldershaw, C.F., Simenson L., Brown, T., and Radcliffe, F. 1947. Absorption and Purification of Hydrogen Chloride from Chlorination of Hydrocarbons. *Chemical Engineering Prog.*, 43, 7: 371–8.

5

Hazardous Solid Waste Disposal in the Geological Environment

A.J. Egger

Unprecedented economic growth since World War II has resulted in record production of industrial and consumer goods, including huge quantities of chemicals for various end uses. This has led to an increased output of waste materials which cause air, soil and water pollution. The amounts of waste have become so large that new industries and specialist engineering expertise have had to be developed to minimize waste generation and to establish acceptable disposal systems.

Air and surface water pollution are the most conspicuous types of pollution and therefore receive prime attention by the public. In contrast, soil and groundwater pollution are concealed from direct observation and become public issues only after massive intoxication has already taken place and/or hazardous conditions for human health have been confirmed by analytical evidence.

There are many sources of pollution and they vary according to the country and the location. Single polluting compounds are generated in gaseous, liquid or solid form. Many of them are rather inert to physico-chemical or biological degradation and thus pose no particular threat to the natural environment. Others, however, are highly reactive and must be treated and disposed of safely to minimize possible environmental impact.

Waste products which cannot be further reduced and treated end up in the form of solid waste on landfill sites. Often these sites are a great nuisance because of their ugly appearance and molestation by smoke, odour and windblown debris. More important is the potential danger posed by the unknown amounts of hazardous components frequently contained in household and industrial wastes. In the form of leachates they may find their way into the groundwater, thus violating one of the principal rules of safe waste disposal management, namely the prevention of water resource contamination.

Such contamination can only be avoided if waste products and potentially toxic leachates are safely contained within the boundaries of the landfill site and no further contact occurs with the enclosing geologic environment. This condition should be fulfilled as far as possible for inert wastes, but it constitutes a mandatory provision if hazardous waste is deposited.

Definition and Identification of Topics

Hazardous Solid Waste

Definition

What is hazardous solid waste? As yet there is no common definition because considerable disagreement exists about what constitutes "hazardous" waste. To step barefoot on a piece of broken glass may be just as hazardous as drinking contaminated water or eating poisoned fish. But while the glass sherd constitutes a single and calculable risk, water and food chain poisoning exert a far more serious, multiple, often unpredictable and perhaps irreversible public health risk.

Depending on the status of the actual environmental situation and the alertness of the public to accept or to oppose pollution of the natural environment, more or less stringent measures for its protection have been adopted by different countries. What is considered "hazardous" in one country may not be hazardous in another. Thus, a variety of hazardous or special waste regulations exist, making it very difficult to compare waste types and waste volumes on an international level. For the purpose of this paper hazardous solid waste is defined as comprising all solid refuse which by its bio-relevance poses a potential risk to human health. Nuclear or radioactive waste, which requires special treatment and disposal technologies is excluded from this arbitrary definition.

The term bio-relevance deserves particular attention. In brief, it refers to the ability, quickness and ease by which a given chemical substance is absorbed in plant or animal tissue in such high concentrations as to endanger human health through contamination of the natural food chain. Disregarding radiation and conventional air pollution, it is obvious that contaminated surface water and groundwater as well as contaminated soil constitute the main sources for potentially hazardous trace element intake by plants and animals. Consequently, these media need to be protected efficiently against uncontrolled sludge and solid waste disposal.

Generation and Disposal

By far the largest portion of hazardous waste is generated by industrial production. Recent information indicates that in 1983 almost 50% of the 266 million tons of hazardous waste generated in the United States pertained to chemical and allied products (Table 5.1).[1] Although it has been possible in the meantime to halve this figure, the per capita generation is still about 0.6 tons. It can be assumed that other industrialized countries feature similar source and type patterns. For instance, the results of a systematic waste disposal survey carried out in Austria in 1984 indicate the generation of some 33.5 million tons of total waste and an estimated 2.7 million tons of hazardous waste, corresponding to per capita quantities of 4.4 tons and 0.4 tons respectively.[2]

Table 5.1 Industrial Hazardous Waste Generation in the United States, 1983

Rank	*Major Industry*	*Percent of Total Waste Generated*
1	Chemical and allied products	47.9
2	Primary metals	18.0
3	Petroleum and coal products	11.8
4	Fabricated metal products	9.6
5	Rubber and plastic products	5.5
6	Miscellaneous manufacturing	2.1
7	Nonelectrical machinery	1.8
8	Transportation equipment	1.1
9	Motor freight transportation	0.8
10	Electric and electronic machinery	0.7
11	Wood preserving	0.7
12	Drum reconditioning	0.1
	Total	100.0

Table 5.2 Generation of Solid Wastes in Selected Industries in India

Industry Branch	*No. of Units*	*Tons*		
		Total	*Hazardous*	
Chloralkali	9	9681	9022	
Drugs and Pharmaceuticals	20	20,171	16,250	(?)
Dyes and pigments	19	48,654	48,654	
Engineering	4	57,871	49,873	
Nitrogenous fertilizer	4	5735	433	
Phosphatic fertilizer	5	413,320	—	
Inorganic chemical industries	27	412,188	62,364	
Organic chemical industries	30	34,445	19,974	
Pesticides	15	5088	5088	
Total		1,007,153	211,558	(21.01%)

Little is known about waste management in developing countries. Several case studies, however, suggest that hazardous waste generation in selected industrialized areas has already reached significant levels and will cause serious problems if timely counteraction is not taken (Table 5.2).[3] It is estimated that about 68% of hazardous waste generated in Canada and the United States and almost 50% handled in Europe are still landfilled. Since open dumps are the least expensive form of solid waste disposal, it can reasonably be expected that largely uncontrolled dumping is the predominant form of hazardous waste disposal in developing countries.

In the past, numerous adverse environmental impacts originated from such "wild" landfill sites. Often they contain not only ordinary domestic refuse of rather low bio-relevance but also a wide spectrum of no longer needed consumer items such as chemicals, paints, batteries, spent lubricants, pharmaceuticals, putrid matter, which are hazardous to the natural environment if not properly disposed of. Highly toxic industrial and commercial wastes deposited simultaneously and unlawfully together with domestic refuse led in many cases to severe groundwater contamination and closure of water supply wells.

Abandoned open dump sites, therefore, constitute a permanent potential source for water resource contamination, thus requiring careful investigation. The magnitude of the problem is apparent from a comparison of managed, unmanaged and abandoned waste disposal sites in Austria and points to the necessity of controlled solid waste disposal on well investigated and properly engineered disposal sites (Figure 5.1).

Geological Environment

Definition and General Considerations

In a broad sense, the geological environment is defined by the three-dimensional setup of rock formations and their weathering products at a certain time in a given area or location.

The three-dimensional setup is reflected by the presence of a variously composed and variously structured rock sequence belonging to the crustal part of the lithosphere. In the uppermost portion, this sequence is in direct contact with the pedosphere, hydrosphere, biosphere and atmosphere respectively, including the activities of man. The lower portion constitutes a largely self-contained unit, influenced only by endogenous geological processes, except deep reaching descending waters and occasional perturbation by deep drilling and deep mining.

The rock formations occurring in a specific area or site location generally comprise several rock types which may be of sedimentary, igneous or metamorphic origin. Mineralogical composition as well as the general physical–chemical properties of principal rock types are well known and need no further comment. What needs to be emphasized, however, is the fact that rocks are the result of a dynamic formation process. Their composition and technical characteristics are subject to often abrupt changes within short

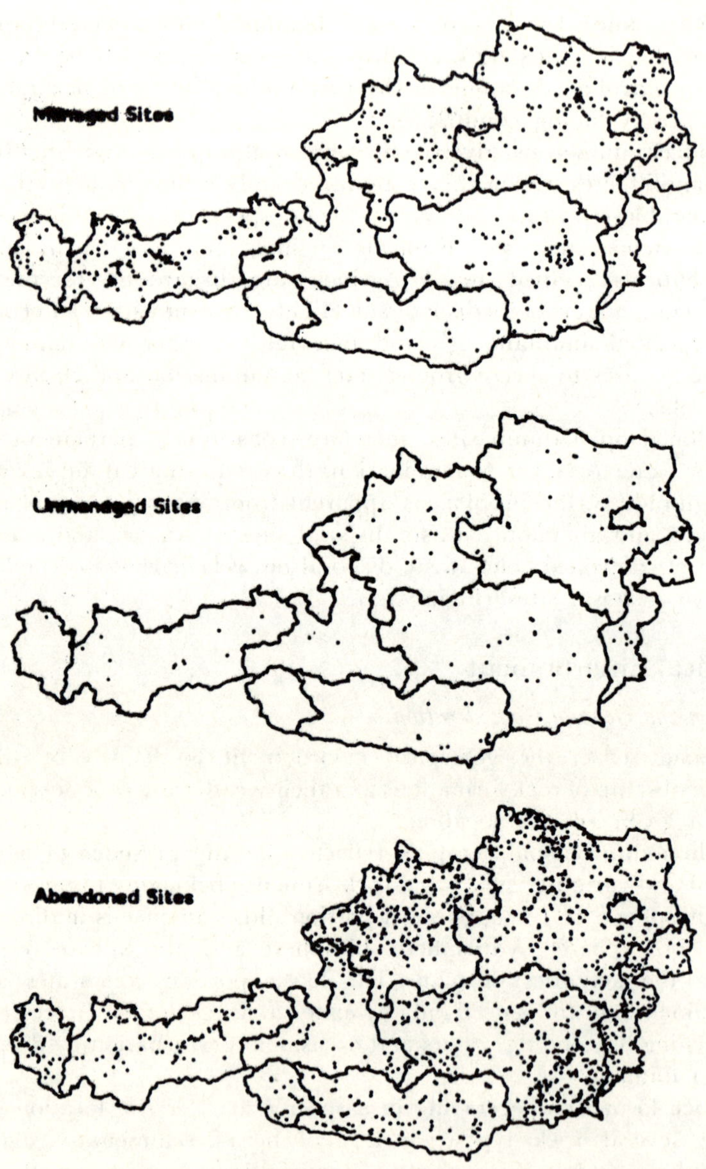

Figure 5.1 Waste Disposal Sites Reported in Austria for 1983 (Source: Österreiches Bundesinstitut für Gesundheitswesen, 1984)

distance. It is therefore difficult to appraise the technological suitability for a specific purpose with sufficient accuracy, especially since a number of additional important geological features must be observed and evaluated, for instance structural disturbance, complex folding, differential bedding and facies changes.

Weathering products are the accumulated debris resulting from the physical, chemical and biological disintegration of rocks. Depending on source rock, landform, climatological condition, geographical location and degree of disintegration they occur in many forms, their ultimate product being mature soil. Since weathering products lie on top of bedrock, there generally exists an intimate relationship with the hydrosphere (through the presence of circulating waters) and biosphere (by means of plant and animal life). Soils and other surficial weathering products, therefore, are very sensitive against environmental burdening.

The time/space relationship refers to the dynamic evolution of the geological environments. Time-wise it is measured in millions rather than thousands of years but in the context of this paper it is of relevance only with respect to recent impacts like earthquakes, volcanic events as well as catastrophic floods, inundations and rock/landslides which cannot be controlled by man. Concerning space, the specific uniqueness of each geological site cannot be overestimated. Even within similar and comparable environments, each individual site features distinctive properties which need to be carefully investigated and evaluated before laying claim to a particular land use.

Geological Appraisal

The main objective of the geological work is to improve site knowledge until safe conclusions on project viability are possible. Depending on the quantity and quality of information available, one or two surveys will have to be carried out (i.e. a reconnaissance survey on map scales 1:50,000 or 1:25,000 and/or detailed surveys at 1:10,000 and 1:5000 and even smaller scales).

In the course of the reconnaissance survey the geological environment will be appraised on the basis of lithological, stratigraphical and structural evidence from existing geological maps, satellite imagery, aerial photographs, geophysical survey data and field inspection of selected areas. If no geological mapping has been carried out, the necessary basic data must be established by frequent checks to support remote sensing information.

The findings of this initial project stage are expected to identify the geological environment in at least general terms with regard to the following:

1. Identification of principal rock types
2. Areal distribution of principal rock types
3. Identification of rock sequence
4. Determination of structural pattern
5. Strike and dip

6. Folding pattern (type, direction, plunge)
7. Fault pattern (type, direction, displacement)
8. Major lineaments (earthquake potential)
9. Identification of weathering products (type of thickness of overburden, degree of consolidation)
10. Observation of possible land and rock slides
11. Observation of springs and water seepages

Geophysical work and drilling are not carried out during the reconnaissance stage on a routine basis but orientation sampling is a valuable tool to support preliminary surveying results and to facilitate the selection of promising target areas for detailed investigations.

Detailed surveys are limited to defined target areas. Their main objective is to provide sufficient additional information to appraise the given geological environment. Besides the detailed routine work, a number of special methods of investigation are employed:

1. Air photo interpretation
2. Geophysical ground surveying
3. Drilling and borehole logging
4. Detailed sampling
5. Mineralogical–petrographical determinations
6. Rock and soil testing
7. Chemical analyses

Air photographs assist in the identification of fault structures which may be traced into the proper project area where they may not be recognizable because of extensive overburden. Small-scale, high resolution photographs also yield valuable information on the detail geology and special imagery often indicates features directly related to subsurface hydrological patterns. Geophysical ground surveys are a well established hydrogeological investigation tool, electrical resistivity measurements and seismic sounding being the preferred methods. More recently, specially adapted magnetometric and radiometric surveys on very narrow grid layouts are also applied.

Drilling and borehole logging are the principal methods to obtain reliable subsurface information. In combination with the surface mapping a three-dimensional picture of the underground thus can be constructed. Drill cores, sludge samples and waters retrieved from various borehole depths can be compared with surface samples and geophysical logging results to give a valid model of the underground.

Detailed and specific sampling serves the purpose of thoroughly investigating the different rock types and rock formations with regard to their mineralogical–petrographical composition as well as their physical and chemical properties. Rock and soil testing are necessary to determine certain geotechnical properties, which must be known if engineering work is to be carried out safely on a particular geological site.

Hydrogeological Appraisal

The main objective of waste disposal in the geological environment must be the protection of the hydrosphere against hazardous pollution. Such pollution, in fact, touches the central nerve of our biological cycle and must be avoided under all circumstances (Figure 5.2). Hydrogeological investigations, therefore, assume priority in the entire site evaluation scheme.

During the preliminary project stage (reconnaissance) only some general assumptions will be possible on the basis of limited geological knowledge and by application of standard hydrological parameters to identified rock types under existing geological conditions. To obtain the required detailed hydrological information a well conceived programme of soil-, rock-, and groundwater drilling must be carried out in follow-up of detailed geological mapping and geophysical surveying.

Upon conclusion of this programme, valid information on several essential subsurface geological and hydrological conditions must be presented, namely

1. Type and thickness of overburden
2. Type and sequence of bedrock
3. Facies relationship of sediments
4. Presence of aquifers and aquitards
5. Type and fluctuation of water table
6. Hydraulic conductivity and permeability
7. Groundwater flow
8. Groundwater composition

Overburden in general comprises unconsolidated fluvial, colluvial, aeolian and glacial deposits but also different weathering products, including mature soil. Owing to its position on the top of the geological environment, overburden of very different hydrological characteristics is a frequent location for hazardous waste disposal.

Bedrock consists of well consolidated, layered, foliated or dense mineral assemblages of sedimentary, metamorphic and igneous origin. If unfractured, these rocks generally exhibit good uniformity and predictability with regard to their hydrological behaviour. However, whereas unfractured metamorphic and igneous rocks as well as dense shales are a favourable disposal site, limestone and dolomite – particularly karst limestone – permeable basalt and certain sandstones are quite unfavourable because of differentiate, inhomogeneous bedding, solution cavities, irregular grain size and varying degrees of induration.

Facies relationship applies primarily to unconsolidated and poorly consolidated sediments of detrital composition in fluvial and deltaic environments of deposition. Individual beds exhibit a pronounced lenticular form and may vary considerably from each other with respect to grain size and amount of clayey matter. In consequence, their hydrological properties are highly inhomogeneous and subject to sudden changes within short distance.

Such formations are difficult to predict and should be avoided if possible. Similar conditions may develop when different types of limestone and/or dolomite are deposited in alternate layers and preferential solution takes place.

Aquifers and aquitards are permeable rocks or rock sequences holding exploitable water resources or being deficient of water supply potential respectively. Quite naturally, they are situated below the groundwater table. The presence of a good aquifer generally is indicated with a hydraulic conductivity of 10^{-5} m/s or higher, regardless whether the water-bearing formation is sedimentologically or structurally controlled. In both cases hazardous waste disposal must be avoided unless a thick aquitard or aquiclude (hydraulic conductivity lower than 10^{-9} m/s) separates the waste volume from the aquifer.

The water table in simple terms is the boundary between the water saturated zone below and the unsaturated zone above. It may fluctuate for a wide variety

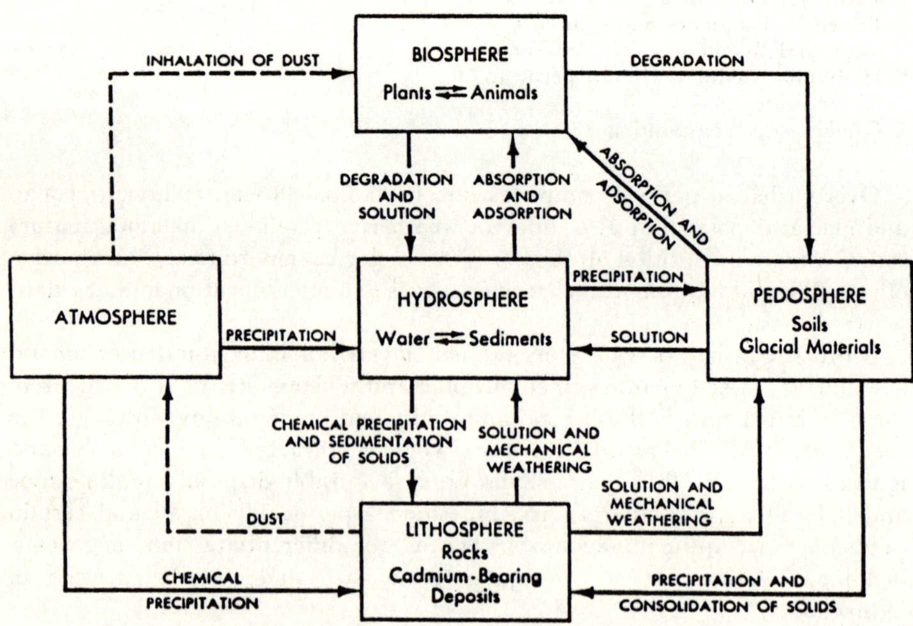

Figure 5.2 Central Status of Hydrosphere within the Biological Cycle (Source: Thornton, I.: Geochemistry of Cadmium, in: Cadmium in the Environment, edit. by H. Mislin & O. Ravera. Birkhäuser Verlag, 1986)

of reasons and also, depending on whether it is confined or unconfined, in diurnal, seasonal or long-term intervals. Hazardous waste disposal should occur above the highest possible water level.

Hydraulic conductivity and permeability probably are the most often used hydrological parameters. Both terms are closely related and indicative of the speed by which a certain liquid – say water – will pass through a natural or artificial medium. Figure 5.3 indicates the range of values of hydraulic conductivity (also "K-value") and permeability in five different systems of units for a wide range of geological materials.[4]

The first impression is that hydraulic conductivity varies over a very wide range. More important is the fact that within a well-defined rock type the K-value may differ to such extent as to make one sandstone conductive (10^{-5} m/s) and the other sandstone practically nonconductive (10^{-10} m/s). On the other hand, unconsolidated gravel has the same high hydraulic conductivity as karst limestone or permeable basalt. Metamorphic and igneous rocks occupy almost the entire conductivity scale, depending whether they are unfractured (low K-values) or fractured (high K-values).

Thus, the influence of rock facies and structural deformation on hydraulic conductivity is clearly shown and the difficulty to attach a certain K-value to a certain rock type clearly indicated. Rocks seemingly appropriate for hazardous waste disposal should have a K-value of 10^{-9} m/s or lower.

Groundwater flow and groundwater composition are other properties which must be appraised through detailed hydrogeological investigation. Whereas flow direction indicates the plume of possible groundwater pollution, the chemical composition is the standard against which any contamination originating from hazardous waste disposal can be measured in the monitoring wells.

Geotechnical Appraisal

Even under the rare premise that all geological and hydrogeological conditions are being met by a specific hazardous waste disposal site, a substantial amount of earthwork and/or quarrying may be required to transform the natural surface into the planned design model. In this process hereto stable soil and rock formations are likely to be disturbed, resulting in unstable artificial slopes and undesirable flows of groundwater.

It is mandatory that a well engineered disposal site must be safe from unexpected soil and rock slides, foreseeable subsidence, ground upheaval and the like for its entire lifetime. To meet this requirement, the various site investigation results need to be appraised in geotechnical terms and by application of special *in situ* and laboratory testing methods. This appraisal, though, has no direct influence on site selection but is part of the engineering effort to implement a given waste disposal project.

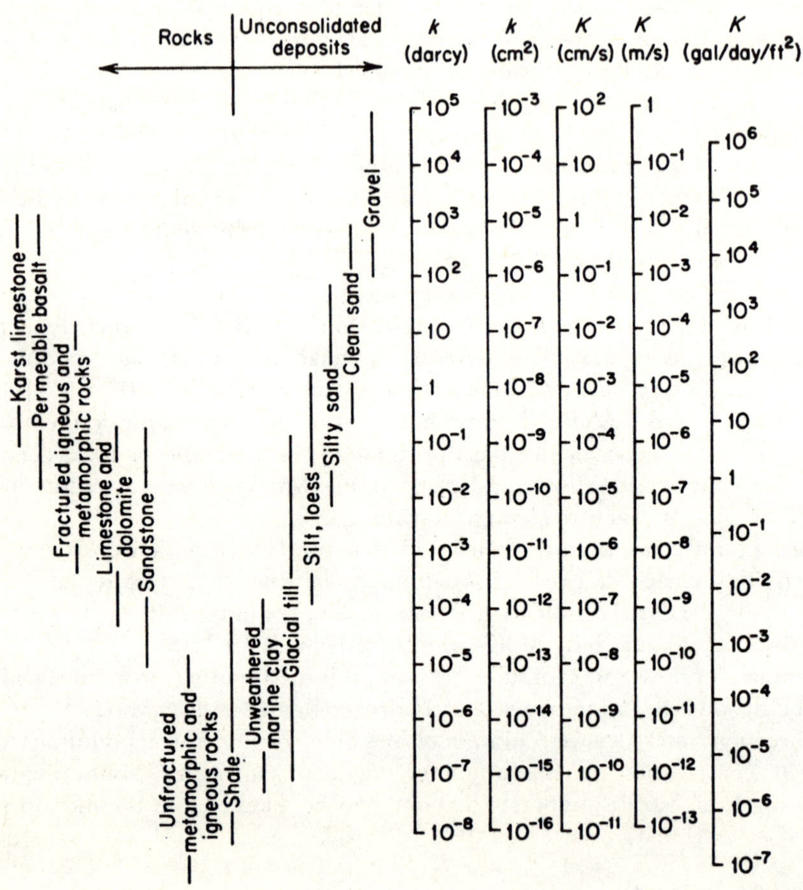

Figure 5.3 Range of Value of Hydraulic Conductivity and Permeability (Source: Freeze, R.A. and Cherry, J.A. *Groundwater*, Prentice Hall, Inc., 1979)

Site Selection

The selection of a site for hazardous waste disposal must be considered a complex undertaking involving different personnel from the public administration sector and from the technical community. In general, it is carried out in several stages and may last some years until one or several potential site locations are sufficiently investigated to satisfy the technical demands and to be accepted by the public.

Experience has shown that public acceptance is a significant factor whether a site choice can be realized as a waste disposal site or not. It therefore, is highly advisable to explain to the public the necessity of such disposal sites, to offer understandable and convincing evidence regarding their safety, if properly selected and engineered, and to seek public support as early as possible. Ignorance is the biggest enemy of progress and well-being, and emotionalism causes the end of any fruitful discussion.

Although a hazardous landfill may be active for 10 to 20 years only, its entire lifetime (i.e. the time span during which it emits toxic leachates into the natural environment) may be a hundred years or more. This implies that the site selection must be based on a long-term municipal and industrial development concept, including population growth, expansion of housing areas, provision of infrastructure, protection of exploitable water resources, disposition of recreation areas.

For reasons of safety and economy, waste disposal in the form of landfilling should occur close to the generating sources. A thorough and reliable source study, taking into account waste projections for the active lifetime of the envisaged disposal site to determine provision of waste treatment facilities and land requirement, should be the starting point for all further considerations.

Once a wider region is defined as a potential siting area, promising locations are singled out on the basis of geological–hydrogeological setting, geomorphology and landscape, actual or future land use, including important restrictions like water reserve areas, outstanding historical and natural monuments, land ownership, various anticipated development schemes and accessibility. With a stepwise approach the number of locations is reduced by a more detailed appraisal of the natural and (what may be called) social site characteristics until one or several sites are identified which merit special technical investigations.

Such final project sites, among which one location will be implemented as a hazardous waste disposal site, must offer to the largest possible degree the geological–hydrological precondition for safe hazardous waste disposal, namely an adequate natural setting to preclude the contamination of surface and groundwater resources. In the ideal case the disposal site rests on a thick layer of clay, shale or unfractured massive igneous/metamorphic rock, well above the groundwater level and at a reasonable distance on the leeward side of the prevailing wind direction from the next settlement. It is clear that this condition cannot be met in many instances. A listing of suitable and unsuitable geomorphological and geology-related site locations is given in Table 5.3.

Depending on the actual optimum geological environment for hazardous waste disposal, a smaller or larger amount of site engineering will have to be carried out to improve the natural premises. The type of work will be governed by local site circumstances but always must aim to fulfill its principal objective: to keep the hazardous waste and the leachate products safely within its natural and artificial containment. One precondition which should always be met is the presence of at least one 1–2 metre thick clayey layer with low hydraulic conductivity at the bottom of the disposal site. If it is absent, an equal thickness of similar material must be spread out and be reinforced with artificial liners to provide the required safety standards.

Table 5.3 Geology Related Suitability of Waste Diposal Sites

Suitable	*Unsuitable*
Geomorphological Features	
Plains	Valley bottoms
Gentle slopes	Steep slopes
Desert bedrock	Hilltops and ridges
Above drainage level	Broken terrain
	Natural depressions
	River terraces and alluvial fans
	Beaches
	Tidal flats and flood plains
	Karst
	Colluvial deposits
Geological/Lithological Features	
Unfractured rocks	Fractured rocks
Unfaulted terrain	Faulted terrain
Unfolded or gentle folding	Intense folding
Massive and dense formations	Differential bedding and foliation
Surficial salt domes	Structurally disturbed salt formation
Crystalline basement rocks	Unconsolidated fluvial deposits
Predominantly shale/clay formation	Unconsolidated aeolian deposits
Homogeneous lithology	Irregular sedimentary deposition
Thick, homogeneous soil cover	Strong facies changes
	Irregular weathering products
	Karst formation
Igneous rocks	Gravel, sand, silt
Metamorphic rocks	Poorly consolidated sandstone
Shale	Karstic limestone and dolomite
Clay	Permeable basalt and similar rocks
	Young lava flows
	Volcanic agglomerates and tuffs
Hydrological features	
Low groundwater table	High groundwater table
Absence of aquifer near surface	Presence of aquifer near surface
Low hydraulic conductivity	High hydraulic conductivity
Poor recharge potential	Strong recharge potential
Unfractured ground	Fractured ground
Dry climate	Wet climate

Appendix: Geological and Hydrogeological Appraisal of Waste Disposal Sites

A. Geological Appraisal

Reconnaissance Information

- Identification of principal rock types
- Areal distribution of principal rock types
- Identification of rock sequence
- Determination of structural pattern
  - Major lineaments
  - Strike and dip
  - Folding pattern
  - Faulting pattern
- Identification of weathering products (type, thickness, consolidation)
- Observation of possible land and rock slides
- Observation of springs and water seepages

Detail Investigation Methods

- Air photo interpretation
- Geophysical ground surveying
- Drilling and borehole logging
- Detailed sampling
- Mineralogical–petrographical determinations
- Rock and soil testing
- Chemical analyses

B. Hydrogeological Appraisal

Reconnaissance Information

- General hydrogeological evaluation

Detail Investigation

- Type and thickness of overburden
- Type and thickness of shallow bedrock
- Facies relationship of sediments
- Presence of aquifers and aquitards
- Type and fluctuation of water table
- Hydraulic conductivity and permeability
- Groundwater flow
- Groundwater composition

References

1. Freeman, H.M. and Eby, E. 1987. A Review of Hazardous Waste Minimization in the United States. US Environmental Protection Agency, March (unpublished).
2. Laub, B., *et. al* 1984. Abfallerhebung. Österreichisches Bundesinstitut für Gesundheitswesen, Wien 1985.
3. WHO/UNEP/WB. 1987. Technical Manual for the Safe Disposal of Hazardous Wastes with Special Emphasis on the Problems and Needs of Developing Countries. PEP/87.2 Draft Report, Vol. I.
4. Freeze, R.A. and Cherry, J.A. 1979. *Groundwater*. Prentice Hall.

6 The Monitoring of Hazardous Waste Repositories and the Prediction of Contaminant Distribution

J.H. Meyer

The monitoring of hazardous waste repositories forms an inherent part of the safety requirements governing their operational and post-operational periods, the latter covering the period commencing with the closure and sealing of their access-ways.

Recent years have seen considerable activity world-wide to find a solution to the problems of storage and/or disposal of a category of hazardous wastes, namely the radioactive wastes arising mainly from the operation of nuclear reactors and to a lesser extent from the application of radioisotopes in industry and medicine. While a number of alternatives for the management and disposal of this type of hazardous wastes have been investigated, the disposal option appearing to be the most viable is their underground isolation in suitable geological environments.

The nature and duration of monitoring programmes to ensure their continuing, safe isolation depends on a number of parameters, including their physical condition, composition and the nature of the geological host formation. Much has been learnt from the necessary investigations that is applicable to the problem of the safe disposal of other, non-radioactive industrial hazardous wastes, which are significantly greater in quantities and which have a toxicity which, contrary to radioactive wastes do not decay with time.

For the storage or disposal of radioactive wastes, the regulatory authorities impose conditions and limitations on waste classification, type, quantity and level of activity for each particular repository. The repository operators are required to conduct monitoring programmes to detect any failure in waste containment systems and transport of nuclides to the biosphere, so that remedial action can be taken.

The time spans over which repositories require surveillance over and beyond their active use could range from a few years to many thousands of years, depending on the rate of radioactive decay of the particular nuclide and the sorption/precipitation and diffusion conditions of the waste and the host rock environment.

In contrast, some toxic, non-radioactive chemical wastes may require perpetual isolation, but in some cases their toxicity is modified by sorption and biodegradation through bacterial activity, the latter being particularly prevalent in organic wastes.

Monitoring Activities during Repository Operation

Although radioactive wastes may require somewhat different preconditioning and containment than other hazardous wastes, the type of surveillance activities during the operational stage is quite similar. The following activities may be included in monitoring activities during the operational stage:

1. Determination and observance of the limits of exposure of operating personnel
2. Monitoring of effluents from the repository site
3. Monitoring of the environment
4. Measurement of hydrogeological parameters: groundwater flow, acidity, analysis of solutes
5. Meteorological observations
6. Morphology: extent and nature of erosional processes
7. Ground stability: changes of a seismic nature, erosional processes, changes due to operation of the repository (e.g. surface slump caused by excavations). Verification by inspection of geophysical data, including remote sensing and satellite imagery
8. Rock mechanics parameters: loads on linings, integrity of grouting, loss of integrity of repository cover caused by soil pressure, rock failure, decomposition of (organic) waste components and matrix/packing materials. Fissure formation and fractures possibly caused by waste settling; access of surface waters to the waste and resulting increased leaching of pollutants

With some exceptions, such as arid regions, the principal pathway by which hazardous substances can migrate from the repository to the biosphere is through streaming groundwater. To a lesser extent other factors, such as vegetation, biological species and evapo-transpiration can sometimes be significant.[1]

Because of the complexity of the reactions possible between solutes and host environment, diverse migration paths and transfers between pathways, the use of computer codes is usually advisable. This could include the carrying out of sensitivity analyses to take into account the uncertainties and variabilities in the parameters describing the geological and environmental features at the particular site. The use of computer codes in the prognostication of pollution distribution will be described in a later section of this paper.

The preliminary hydrogeological investigations, carried out partially by drilling before the repository site is readied, will have determined the probable rate and direction of the groundwater flow. The layout of groundwater monitoring stations is determined with the use of this data (at the same time ensuring that the groundwater flow is not adversely affected by any sub-surface investigations, such as drilling). The type and spacing of such stations is site-specific. If the repository is brought into operation, air-sampling in the area may have to be carried out to test for airborne contamination arising during waste emplacement operations.

The results of the monitoring programme are recorded on maps, sections and graphically. Any evident trends in the values recorded should become apparent; anomalies in these values should be investigated immediately and, where appropriate, remedial action taken. As monitoring is an iterative process, the results are periodically reviewed and the siting of sampling points and periodicity of sampling is adjusted accordingly. As an initial reference, data should always be compared to baseline values (i.e. those obtained prior to the commencement of operation). At the time of closing the repository, the impact of closing operations, such as the use of backfilling materials, the sealing of access openings and of engineering barriers should be monitored.

Monitoring during the Post-sealing Period

The type of post-operational monitoring required is determined by site conditions, the interpretations of mathematical models, such as groundwater 3-D models, models on solute transport and migration and models of accident scenarios. In addition, monitoring may be required to allay public concern about the safety of the closed-down repository.

By the end of the operating period sufficient knowledge of the particular site may have been gained to enable predictions on repository performance during the period in which institutional control is foreseeable to be made with a high degree of confidence.

Obviously, although repositories may require isolation from the biosphere for many thousands of years, monitoring programmes can only be planned for the foreseeable future. This dictates the necessity to select a site with optimum geological, tectonic and seismic conditions. The effect of the variation in parameters such as the bulk-K coefficient, fissure spacing and porosity on the time taken for solute travel is shown in Table 6.1 below, showing data and interpreted results from the variation of hydrogeological parameters of fissured rocks, in this case sandstones and mudstones.[2]

The use of mathematical models and the interpretation of geological, meteorological and hydrological events, using the principal of uniformitarianism aids in the prognostication of repository behaviour over long time spans. Geologists, working according to this principal, under which existing geological processes are clues to the past and by extrapolation to the future, can have reasonable confidence that formations that have been stable for millions of years can be expected to continue so in the future for a further

Table 6.1 Hydrogeological Properties of Fissured Rocks and the Resulting Solute Travel Times (Source: Black, J.H. 1981. The Geology and Hydrogeology of West Cumbria. British Geological Survey. Report No. ILWGDS 81, October : 28)

Bulk K_{-1} *(m sec)*	*Fissure spacing* *m*	*Fissure aperture* *m*	*Fissure porosity* $\emptyset_f$	*Water velocity (V)* *m sec*	*Travel time of indicated concentrations (years)* *10%*	*50%*	*90%*
1×10^{-5}†	1	2.5×10^{-4}	2.5×10^{-4}	4.0×10^{-5}	105	141	188
	10	5.4×10^{-4}	5.4×10^{-5}	1.9×10^{-4}	4.5	28.1	446
	100	1.2×10^{-3}	1.2×10^{-3}	8.6×10^{-4}	0.05	0.3	7.9
	porous medium flow: porosity=0.15		6.7×10^{-8}	143	= same	= same	
1×10^{-6}†	1	1.2×10^{-4}	1.2×10^{-4}	8.6×10^{-6}	1430	1430	1430
	10	2.5×10^{-4}	2.5×10^{-5}	4×10^{-5}	446	1257	2508
	100	5.4×10^{-4}	5.4×10^{-6}	1.9×10^{-4}	4.5	28	793
	porous medium flow: porosity=0.15		6.7×10^{-9}	1430	= same	= same	
1×10^{-7}‡	1	5.4×10^{-5}	5.4×10^{-5}	1.9×10^{-6}	14,300	14,300	14,300
	10	1.2×10^{-4}	1.2×10^{-5}	8.6×10^{-6}	10,500	14,100	18,800
	100	2.5×10^{-4}	2.5×10^{-6}	4.0×10^{-5}	450	2810	44,600
	porous medium flow: porosity=0.15		6.7×10^{-10}	14,300	= same	= same	
1×10^{-8}‡	1	2.5×10^{-5}	2.5×10^{-5}	40×10^{-7}	143,000	143,000	143,000
	10	5.4×10^{-5}	5.4×10^{-6}	1.9×10^{-6}	143,000	143,000	143,000
	100	1.2×10^{-4}	1.2×10^{-6}	8.6×10^{-6}	44,600	125,700	250,800
	porous medium flow: porosity=0.15		6.7×10^{-11}	143,000	= same	= same	

* This is the time taken for a given concentration of leached nuclide to travel to the sea floor from a repository 300 m below.
† Value of hydraulic conductivity likely to be typical of a sandstone (in this case St. Bees Sandstone).
‡ Value of hydraulic conductivity likely to be typical of a mudstone (in this case Mercia Mudstone).

100,000 years or so, barring completely unforeseeable events such as man-made impacts on the environment.

The principal uncertainties are likely to be the effects of climatic changes, with their effects on erosional processes and changes in sea levels. Other possible negative parameters are seismic events and soil erosion. The latter could in the long run have deleterious effects on the protective overburden of repositories, and mathematical models may be useful there to estimate the rates at which such processes could take place.

In general, some type of monitoring programme is advisable for the post-closure period and is sometimes even mandatory for the early stages at least. The most salient reason for post-closure monitoring is to ensure that exposure limits to man are not exceeded. Ideally, the locations and distances between monitoring boreholes/wells and the frequency of sampling should be directly related to the velocity of groundwater and surface flows and other soil interactions. They should be concentrated on the predicted significant pathways of exposure arising from the release of pollutants. Arrangements and layouts of monitoring sites should be reviewed in the light of data collected. The siting of sample points is an iterative process and should be adapted to the particular site.

In one example, a land-fill repository of solid radioactive wastes (low and intermediate level) in the United States, an extensive network of groundwater monitoring wells was drilled in the solid waste burial area.[3] Five wells were put down along the perimeter of the 31 ha. site, a further six being added after it was determined that groundwater flow was to the Southwest. A total of eleven dry borehole wells were drilled to a depth of 15 m through the actual waste emplacement. The latter are not open to soil water, being intended to ensure that wastes remain where they were originally installed. A number of trench wells were put down to sample any perched water collected through percolating rainwaters in the backfilled trenches. In addition, 66 wells were drilled at 60 m centres, penetrating 3 m into the mean water table, the perimeter wells penetrating 6 m. The grid and perimeter wells are sampled quarterly and annually respectively, (for tritium). There is no routine monitoring of soil or air, but samples of vegetation are taken annually. It was found that of a total estimated 1.9×10^{6} GBq (50,000 Ci) of tritium in the groundwater beneath the site, about 2% had migrated in groundwater into a 7 ha. area beyond the fence bordering the site to the Southwest.

An example of an extensive programme of investigation to monitor the nature and extent of groundwater pollution caused by the disposal liquid industrial wastes comes from the United Kingdom.[4] The wastes were discharged into lagoons excavated beneath the water table. This particular site was selected for investigation because it is an example of a most unfavourable case, where wastes were being introduced directly into the saturated zone of an intergranular aquifer and where potentially beneficial effects of an unsaturated zone were absent. The site provides the opportunity to study the behaviour in

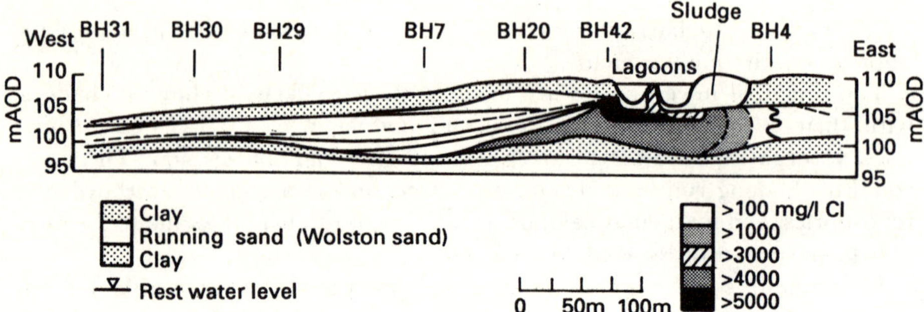

Figure 6.1 Cross-section through Aquifer showing Chloride Concentration (mg/l)

the saturated zone of heavy metals, phenols, solvents, acids and a variety of miscellaneous wastes and to attempt to define the major controls on contaminant mobility. It was found that although heavy metals comprised a major part of the pollutants, their concentration in the aqueous phase of the lagoon is low, suggesting that at or near neutral pH they are present in an insoluble form. It is likely that, under the reducing conditions prevailing in the lagoon and groundwater, heavy metals are precipitated as sulphides and carbonates.

The natural background concentrations of chloride being low (about 20 mg/l), its occurrence and concentrations were used as indicators of groundwater pollution concentrations. Figures 6.1 and 6.2 show the distribution of the chloride in the development of a pollution plume to the West

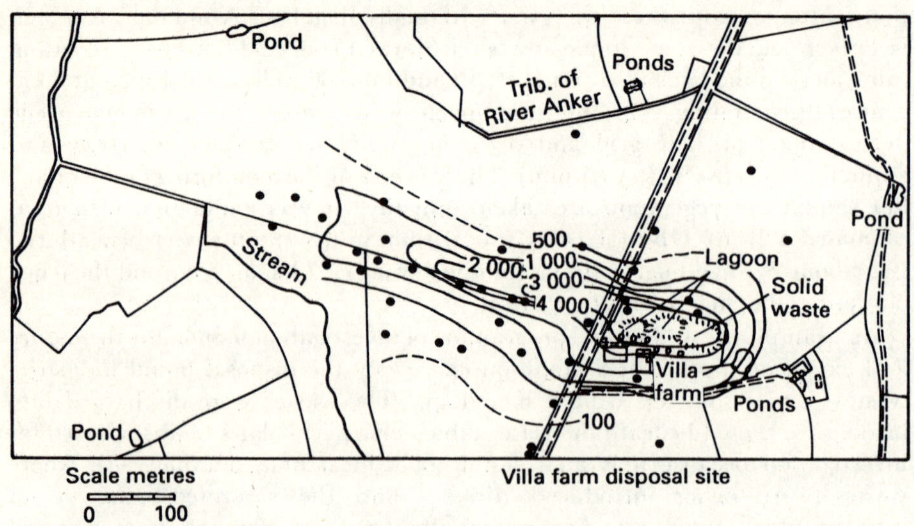

Figure 6.2 Chloride Concentration in Groundwater (mg/l)

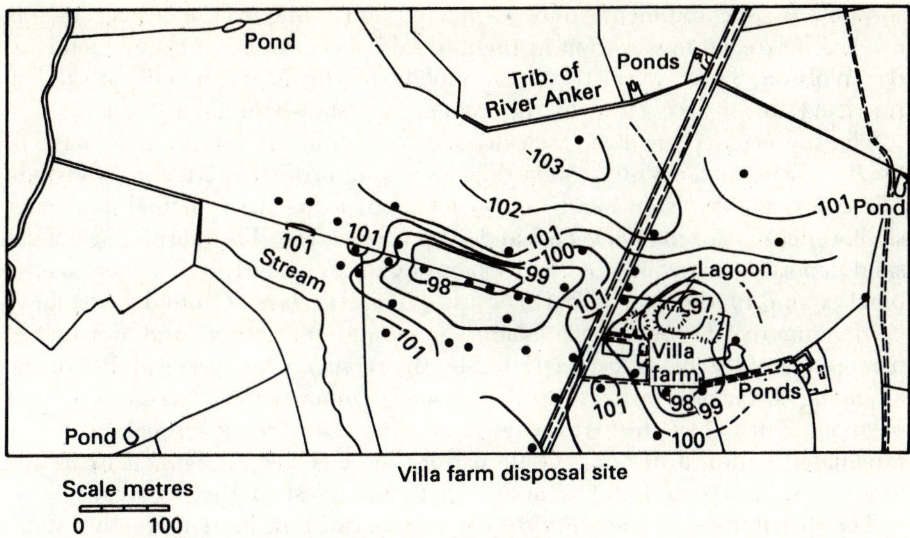

Figure 6.3 Base of Running Sand (m AOD)

of the lagoons. Groundwater flows mainly through the Wolston Sand. Infiltration and downward percolation of leachates are restricted by the low permeabilities of the over- and underlying clays. It is apparent that the plume has developed along the direction of the highest transmissivity (Figures 6.2 and 6.3). Additional attenuation of solutes may occur through the exchange or

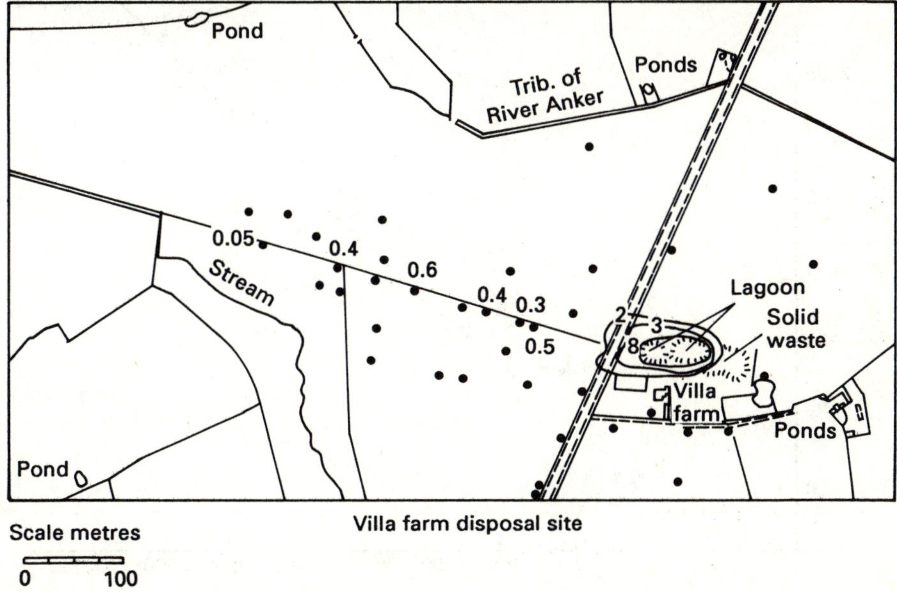

Figure 6.4 Distribution of Nickel in Groundwater (mg/l)

sorption by the sediments and co-precipitation with iron, however this is masked by the large variation in the natural concentration of heavy metals in the Wolston Sand. Even the most mobile of the heavy metals, nickel, is restricted to an area close to the lagoons, as shown in Figure 6.4.

The site geology has been well defined by drilling. It comprises a range of Quaternary sediments up to 50 m thick overlying bedrock of Mercia Mudstone (Triassic). The Wolston Sand (unconsolidated) forms the principal aquifer at shallow depth, varying between 3 and 12 m in thickness. The morphology of the sand deposit is controlled by the topography of the underlying clay surface. It pinches out to the North between two boulder clays. Groundwater flows predominantly through the Wolston Sand and infiltration and downward percolation of leachates are restricted by the relatively low permeability of the adjoining clays. Infiltration from the lagoons dominates the recharge, a mound of about 3 m above the natural groundwater levels being formed, which is attenuated within a distance of 10 m from the lagoons. A regional hydraulic gradient of 1:300 to 1:500 is indicated to the West of the lagoons.

The distribution of the chloride ion can, because of its non-reactive state, provide a useful index of dilution from other attenuation processes. With simple dilution, the concentration ratios of contaminant and chloride will remain constant, lying close to the mixing line between leachate and natural groundwater (Figure 6.5). However, relative concentration ratios between total organic carbon (TOC) and chloride for the same sample which show little

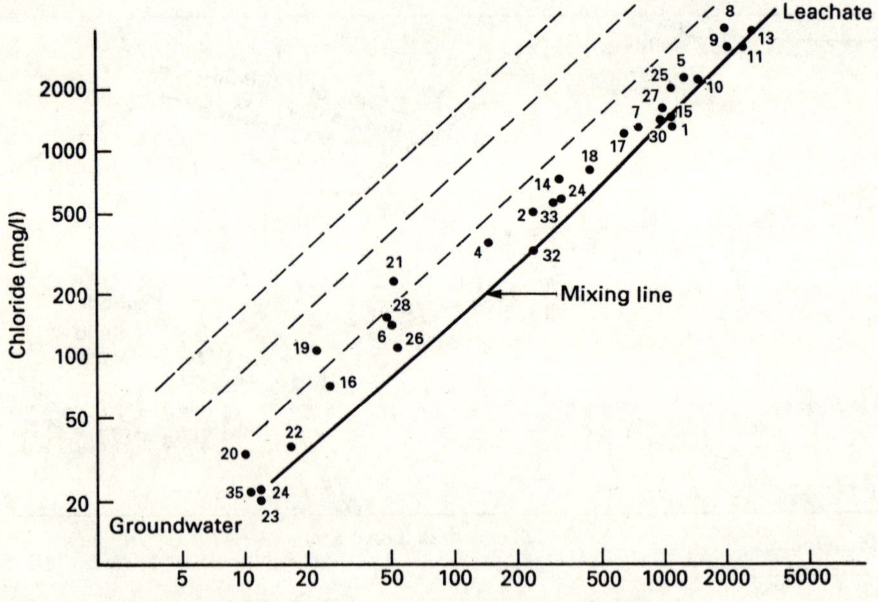

Figure 6.5 Relationship between Sodium and Chloride

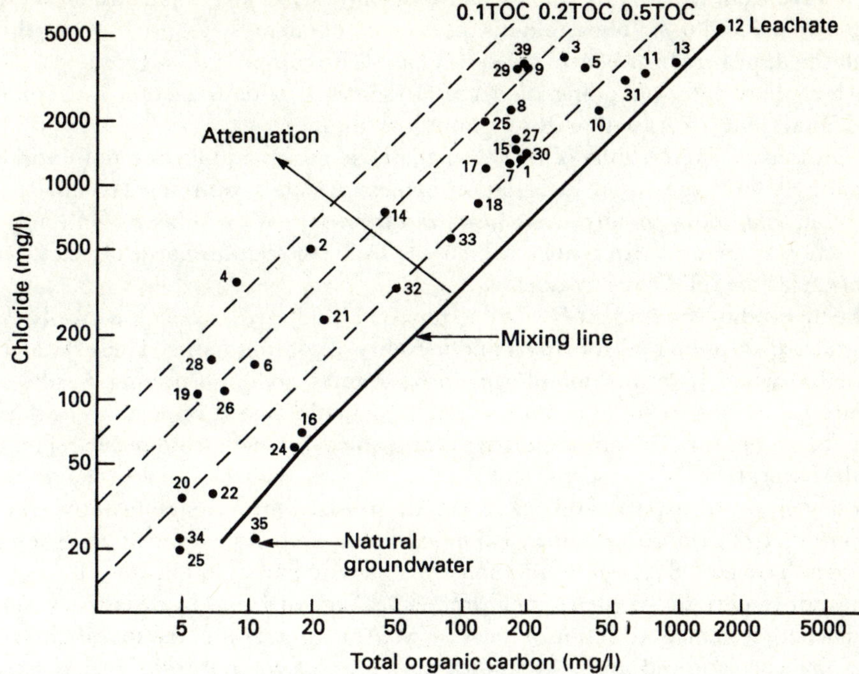

Figure 6.6 Relationship between TOC and Chloride

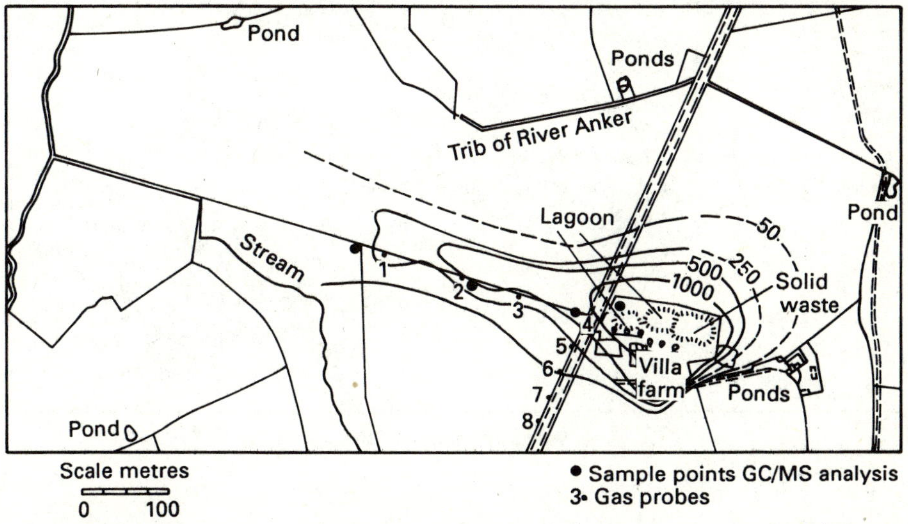

Figure 6.7 Maximum Concentration of TOC in Groundwater (mg/l)

departure from the mixing line for sodium, show a marked departure for TOC (Figures 6.6 and 6.7). The rapid disappearance of phenols (Figure 6.8) together with the appearance of benzoic acids and alicyclic compounds suggests that the leachate may be undergoing biodegradation, with potential both for aerobic and anaerobic degradation throughout the aquifer.

Elevated concentrations of nickel at the base of the aquifer are not entirely consistent with the rapid attenuation of heavy metals near the lagoon. It is possible that heavy metals are remobilized from the sediment as a result of the reduction of Fe and Mn oxides, which are likely to contain naturally elevated concentrations of heavy metals.

Such studies can identify major physical and geochemical controls on migration, sorption and transport and mobility of contaminants. These data are valuable when planning monitoring programmes and interpreting results at future hazardous waste repositories. The use of field data is important, as these may be an order of magnitude higher than values obtained from measurements in the laboratory. Macroscopic features (e.g. cracks, fissures and root holes) can affect solute transport and contamination distribution significantly. The importance of calibrating numerical models with known field behaviour should be emphasized. Such investigations are particularly applicable to older repositories of liquid wastes to determine if leakage to groundwater is occurring and whether remedial action is necessary. Among preliminary investigations that may be required are field studies to determine the nature of unsaturated

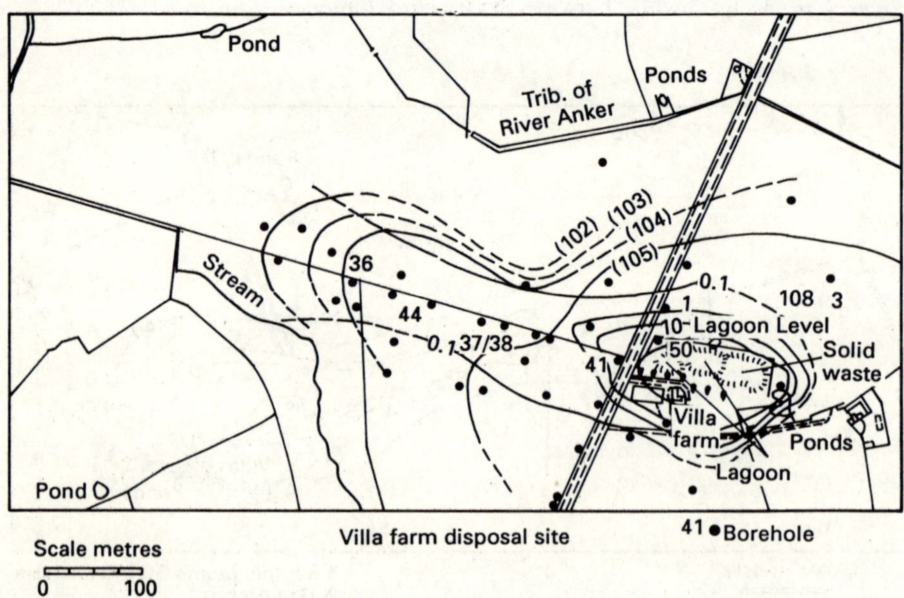

Figure 6.8 Concentration of Phenol in Groundwater (mg/l) under 1982 Average Rest Water Levels (m AOD)

zones, boring shallow and deeper holes and subsequent geophysical logs. The latter investigations could, for example, include all or part of the techniques outlined in Table 6.2.

Table 6.2 Techniques for Borehole Testing and Logging

Required Information	*Widely Available Logging Techniques*
Lithology and Stratigraphic correlation of aquifers	Electric, sonic or caliper logs made in open holes; nuclear logs made in open or cased holes
Total porosity or bulk density	Calibrated sonic logs in open holes; calibrated neutron or gamma-gamma logs in open or cased holes
Effective porosity or true resistivity	Calibrated log-normal resistivity logs
Secondary permeability, fractures, solution openings	Caliper, sonic or borehole televiewer or television logs
Moisture content	Calibrated neutron logs
Direction, velocity and path of groundwater flow	Single-well tracer techniques, multiwell tracer techniques

In order to apply the results of such investigations to predict the movements of contaminants at other sites to be verified by field monitoring investigations, reactions that attenuate various components of the waste in different geochemical zones must be quantifiable. This is particularly important for organic species, which break down in a number of ways into a series of compounds, each degrading at a different rate, depending on intrinsic properties, the nature of the microbial population and the physio-chemical environment. Predicting the movement, nature and distribution of contaminants based on such studies as described above is a useful exercise for laying out monitoring programmes at other repositories. In order to do so, three dimensional models of solute transport calibrated by using known data of groundwater velocity and mass should be developed.

Only by detailed study of the behaviour of contaminants and the ability to model their movement at existing sites will it be possible to improve confidence in predicting the impact of future disposal sites on the environment in general and groundwater quality in particular.

References

1. International Atomic Energy Agency. 1982. Consultants Meeting on Operational Experience in Shallow Ground Disposal of Radioactive Wastes, December.
2. Black, J.H., *et al.* 1981. *The Geology and Hydrogeology of West Cumbria.* British Geological Survey. Report No. ILWGDS 81, October : 28.
3. Ibid. 1982. *Site Investigations for Repositories for Solid Radioactive Wastes in Shallow Ground.* Technical Report Series No. 216. Vienna.
4. Williams, G.M., *et al.* Controls on Contaminant Migration at the Villa Farm Lagoons. *Quarterly Journal of Engineering Geology,* 17: 39–55.

7

Dechlorination of PCBs, Dioxins and Difurans in Organic Liquids

P.F. van den Oosterkamp, L.J. Blomen, H.J. ten Doesschate, A.S. Laghate and R. Schaaf

Chlorinate waste is a name for a wide variety of waste materials. An important distinction exists in the way in which chlorine is chemically present. In chloroform and monochlorobenzene it is bound to an organic molecule, whereas in sodium chloride or copper chloride it is present as an inorganic salt.

Another distinction that can be made is the phase of the waste (i.e. if it is a solid, a liquid or a gas). Liquid wastes can be further divided into aqueous- and non-aqueous. If the chlorine in an aqueous liquid waste is bound to hydrocarbons it can be recovered as a non-aqueous stream by means of phase separation processes such as extraction and absorption.

In this paper we focus on organic liquid streams containing chlorinated hydrocarbon compounds. At present, most of this type of waste is disposed of by incineration in specially designed furnaces to minimize the formation of toxic compounds such as dioxines.

At Kinetics Technology International (KTI), another process for the dechlorination of organic liquid streams was developed as one of the environmental product lines. One of these product lines is concerned with the refining of waste lube oil streams. This process, the KTI Relube Process, is presently being operated at a number of sites worldwide. One of the features of the Relube process is the removal of sulphur – and chlorinated compounds. Further research on this specific aspect has resulted in a second process. This process, the KTI Chloroff Process, is quite similar to the Relube Process and incorporates a catalytic hydrodechlorination step in which chlorinated hydrocarbons can be effectively removed. In Table 7.1, the development of the two processes is shown.

The Chloroff Process is a recycling process particularly suitable for moderately contaminated liquid waste with 0.1–10 wt% of chlorine. The

Table 7.1 Development of KTI Relube and Chloroff Processes

KTI Relube Process	
for waste lube oil treatment by a catalytic hydrogenation process:	
Pilot Plant FRG	(1972)
Plant Greece	(1982)
Plant Tunisia	(1983)
Plant FRG	(1983)
Plant USA	(1985)
KTI Chloroff Process	
Catalytic hydrodechlorination:	
Semi pilot plant in The Netherlands	(1984)
Plant FRG	(1985)
Several designs	

products are hydrogen chloride, pure or as a concentrated aqueous solution, and the dechlorinated hydrocarbons. Extensive testing of the Chloroff Process has been carried out on a laboratory scale and some of the results will be presented below. First, however, a brief outline of chlorinated waste streams will be given, focusing mainly on PCBs, TCDDs and TCDFs. When waste streams containing chlorinated hydrocarbons are concerned, we mostly refer to waste streams which are contaminated with polychlorobiphenyls (PCBs), polychlorodibenzodioxins (PCDDs) and/or polychlorodibenzofurans (PCDFs).

Chlorinated Waste Streams

PCBs

PCBs are synthesized by the reaction of biphenyl (C_6H_5–C_6H_5) with chlorine. Replacement of one or more hydrogen atoms by chlorine takes place and, theoretically, 209 different isomers can be synthesized in this way. Mostly, compounds with 4, 5 or 6 chloratoms predominate. Figure 7.1 gives the general basic structure for PCBs.

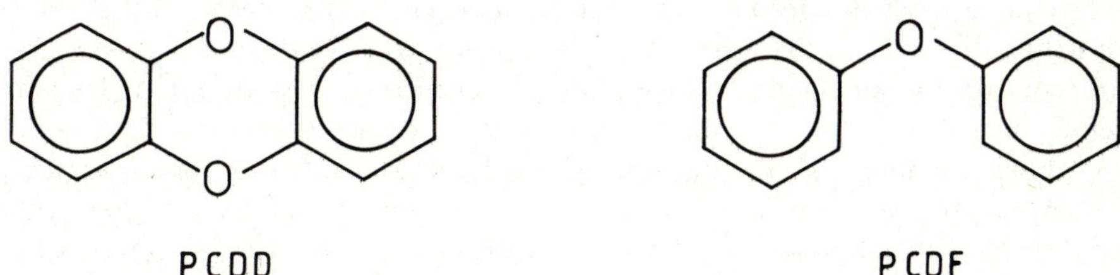

Figure 7.1 General Chemical Structure of PCBs

They are characterized by exceptionally high chemical and thermal stability. They are of moderate to low volatility (decreasing with increasing chlorine content), are relatively nonflammable, stable to oxidation at elevated temperatures, and have excellent electrical insulating characteristics. PCBs are insoluble in water, are soluble in most common organic solvents, and are relatively non-hygroscopic. PCBs are among others being used as insulating liquid in transformers and capacitors. Internationally, about 350,000 tons of PCBs are still in use for this purpose. Cumulative worldwide production of PCBs is around 1 to 2 million tons. It is estimated that since 1929, when chemical production began, some 250,000 tons of PCBs contaminated the environment. The largest portion is present in water, where adsorption on sludge takes place.

They easily accumulate in organisms, especially since they are biologically difficult to degrade and soluble in fats. Accumulation mainly occurs in the aquatic environment, because of the intensive contact between organisms and their (polluted) environment.

Toxicological research has revealed that it is not their acute toxicity that endangers organisms, but their sub-acute and chronic toxicity. This is expressed mainly in the inhibition of growth and reproduction, liver damage, hormonal disturbances and behaviour changes. Table 7.2 reviews the effects of PCB concentrations in water on aquatic organisms.

Table 7.2 Effects of PCBs on Aquatic Organisms (Source: Roberts, Rogers, Baily and Rorke (1978) and EPA report no. 560/6–75-004)

PCB Concentration in Water (μg/l)	*Species*	*Effect*
10–100	Shellfish	Death
	Fish	Death, liver lesions
1.0–10	Fish (mainly young animals)	Death
	Oysters	Inhibited shell growth
0.1–1.0	Shrimps	Death, increased virus infection

PCBs are most dangerous compounds, especially when orally fed. This in contrast to inhalation, because the PCB concentration in the air is so low that the effect on human beings is negligible. Exposure via the air might be only significant under certain labour conditions. The higher the chlorine content of the biphenyl compound, the more toxic it is liable to be. Oxides of chlorinated biphenyls are more toxic than unoxidized materials. PCBs are generally considered to be carcinogenic. The human body reacts in two distinct ways to the chlorinated biphenyls: the skin is affected, the skin lesion which is the result is called chloracne, and there is a toxic reaction on the liver.

Knowledge of the toxic properties of PCBs is based on experiments with guinea pigs as well as the exposure of larger numbers of people in accident situations. In persons who have suffered systematic intoxication, the usual signs and symptoms are nausea, vomiting, loss of weight, jaundice, oedema and abdominal pain. Where the liver damage has been severe, the patient may pass into a coma and die.

Based on this information, the value of 0.0001 mg/kg body weight/day in various countries is mentioned as the possible limit of an acceptable daily intake by humans. Such an intake limit could, however, not be determined accurately because of the lack of certain toxicity data. Even for technical mixtures, these data are not available. The presence of very toxic dibenzofuran impurities in PCB liquids hampers the interpretation of animal toxicity experiments. The National Institute of Occupational Health and Safety (NIOSH) proposes a limit of 1 $\mu g/m^3$ for occupational exposure (1979).

PCDDs and PCDFs

Polychlorodibenzodioxins (PCDDs) and polychlorodibenzofurans (PCDFs) consist of two aromatic rings, bridged by 1 or 2 oxygen atoms. Totally, 75 PCDD isomers and 135 PCDF isomers can be found. Toxicity of the various isomers depends largely on the amount and position of the chlorine atoms. Most toxic compounds are reviewed in Figure 7.2.

It should be noted that a prerequisite for toxicity of dibenzodioxins and dibenzofurans is the presence of chlorine on positions 2, 3, 7 and 8 of the molecule. The chlorine atoms are then positioned on the angular points of a rectangle of 3 × 8 Angström, a structure which is apparently determining for

Table 7.3 Relation between the Chemical Structure of PCDDs and PCBs and LD_{50} for Guinea Pigs

No.		LD_{50}*
1.	2,3,7,8-tetra-CDD	2
2.	2,3,7,8-tetra-CDF	7
3.	1,2,3,7,8-penta-CDD	3
4.	1,2,3,4,7,8-hexa-CDD	73
5.	3,3′,4,4′,5,5′-hexachlorbiphenyl	223
6.	2,3,3′,4,4′,5,5′-heptachlorbiphenyl	3000
7.	2,3,4,7,8-penta-CDF	10
8.	1,2,3,4,6,7,8-hepta-CDD	600
9.	2,3,4,6,7,8-hexa-CDF	120
10.	2,3,7-tri-CDD	29,444
11.	2,8-di-CDD	300,000

*LD_{50} in µg per kg of body weight.

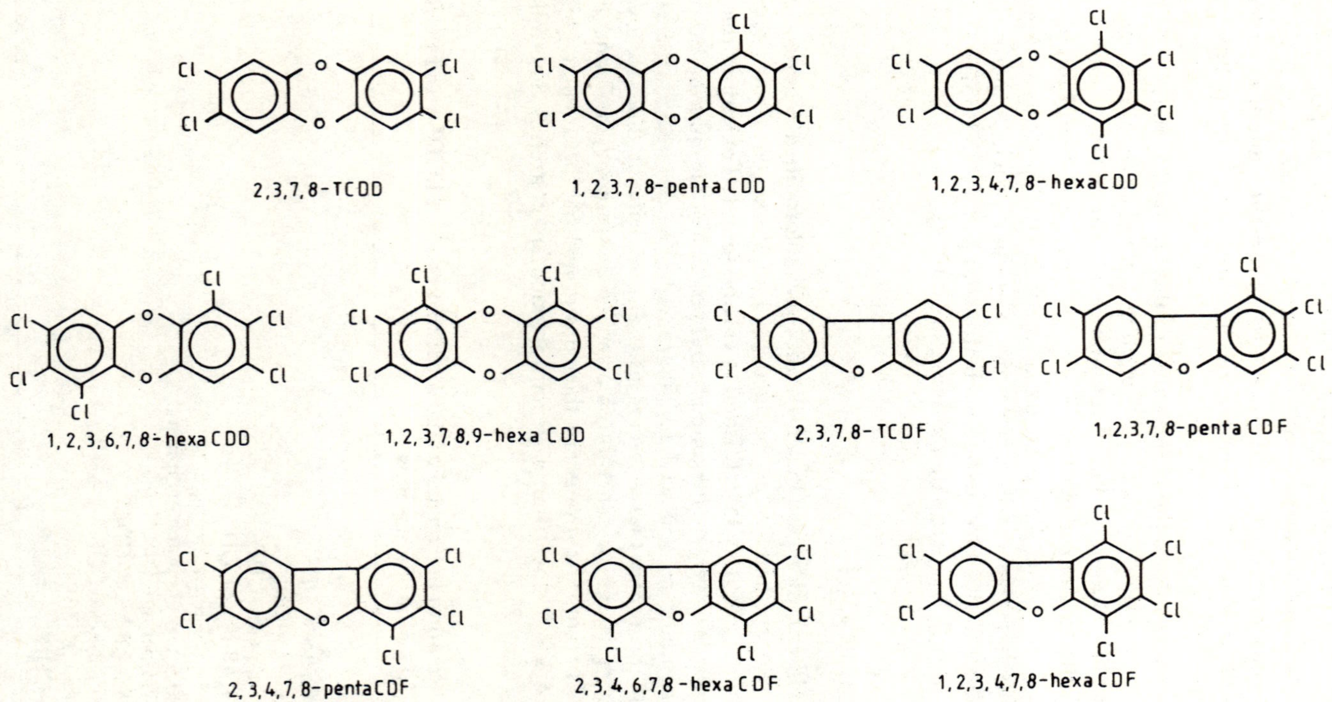

2,3,7,8-TCDD

1,2,3,7,8-pentaCDD

1,2,3,4,7,8-hexaCDD

1,2,3,6,7,8-hexaCDD

1,2,3,7,8,9-hexaCDD

2,3,7,8-TCDF

1,2,3,7,8-pentaCDF

2,3,4,7,8-pentaCDF

2,3,4,6,7,8-hexaCDF

1,2,3,4,7,8-hexaCDF

Figure 7.2 The Most Toxic PCDDs and PCDFs

toxicity. From these compounds, the 2, 3, 7, 8–TCDD or "Seveso dioxine" can be considered as the most toxic compound ever produced by man. Toxicological levels vary among different countries. Annual Daily Intake (ADI) level in The Netherlands is 240 pg/day (TCDD). In Table 7.3, the toxicity levels of various PCDDs and PCBs are given.

It should be stressed that toxicity reports vary greatly and toxicity data between animals species also vary.

Method and Materials

The principle of the Chloroff Process is the catalytic hydrogenation whereby chlorine is removed from the chlorinated hydrocarbon compound. The heart of the process is formed by a trickle phase, high pressure reactor in which the pretreated feed is dechlorinated and partially hydrogenated according to:

$$RCl_n + xH_2 \text{ ---- } RH_xCl_{n-x} + xHCl$$

H for this exothermic substitution reaction lies between -40 and -90 kJ/kmol. Reaction takes place at elevated pressure and a temperature between 250 and 400°C over a heterogeneous catalyst.

Not only chlorinated compounds are treated in this way, other heterocompounds behave in a similar manner: sulphur is converted to hydrogen sulphide, nitrogen to ammonia and oxygen to water. For example, with conversion of dioxins, not only dechlorination but also deoxygenation presumably takes place.

Besides hydrodechlorination, some cracking reactions take place, whereby light compounds are produced. Reaction conditions determine the extent to which these cracking reactions occur. After reaction, a separation step takes place, where excess hydrogen is recycled and the HCl removal via a washing procedure with a basic solution is established. The resultant chlorine-free hydrocarbons can be fractionated or be burnt in order to supply the energy demand of the system.

Conversion of contaminated hydrocarbons has been tested at the Technical University of Eindhoven. Figure 7.3 gives a description of the microflow reactor system used for the experiments. The contaminated stream is introduced into the system with a syringe high-pressure pump, mixed with hydrogen, preheated and fed into the microflow reactor. Reaction products are cooled and separated in a liquid- and gasphase, respectively.

Analyses of reaction products were performed offline and were carried out by several laboratories in both the Netherlands and the Federal Republic of Germany. PCB analyses were performed by the Technical University of Eindhoven using High-Pressure-Liquid-Chromatography and GC–MS (Gas Chromatograph–Mass Spectrometer). Most of the PCDD and PCDF analyses were performed by the Toxicological Department of Amsterdam University, using the method described in Table 7.4.

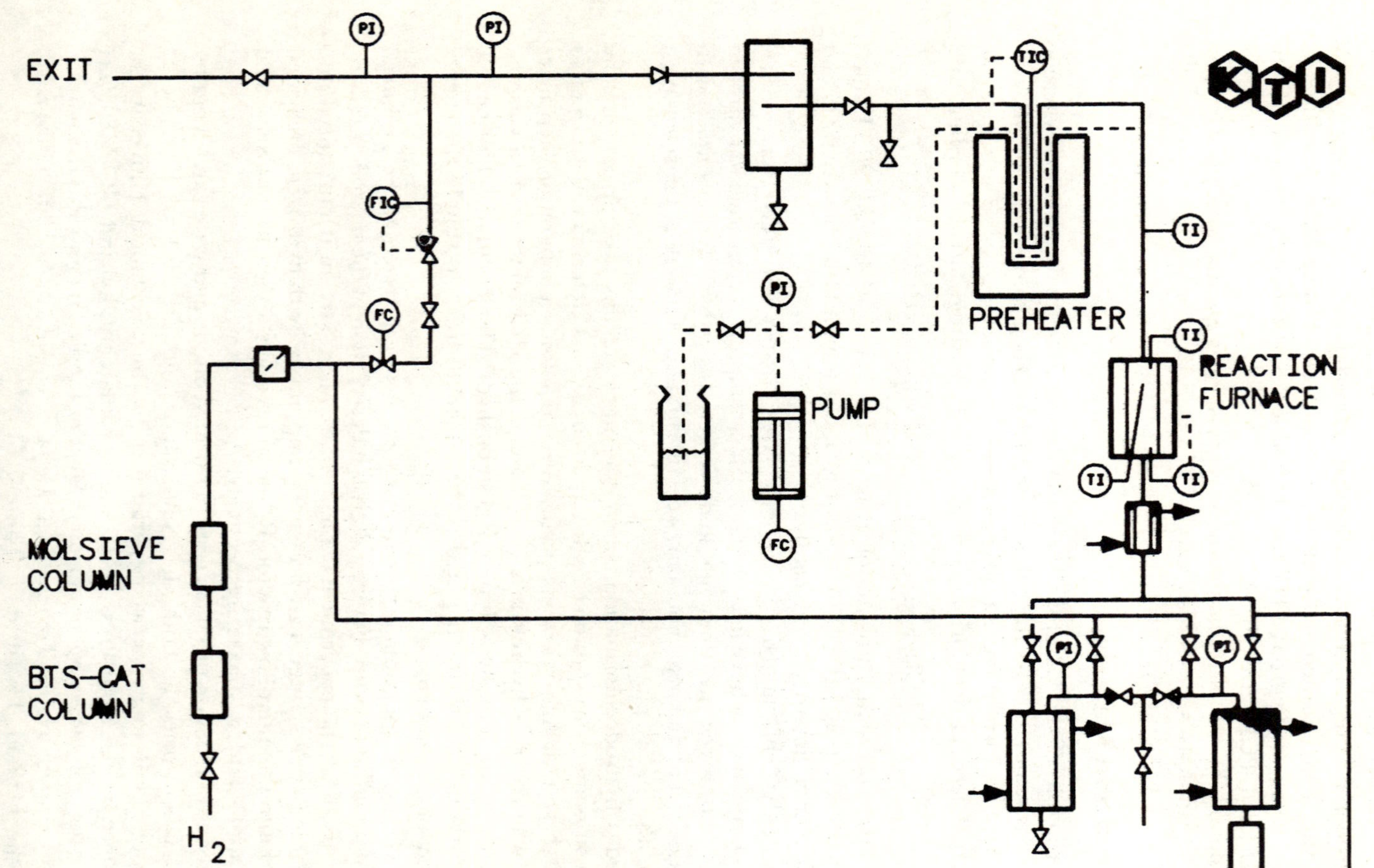

Figure 7.3 The High Pressure Microflow reactor System for Testing of the Chloroff Process

Table 7.4 Principle of Chlorinated Hydrocarbon Analysis (University of Amsterdam)

1. Addition of Standard mixture with
   2,3,7,8-^{13}C TCDD,
   1,2,3,7,8-^{13}C PCDD and
   1,2,3,6,7,8-^{13}C HCDF
2. Dilution with hexane
3. Clean-up procedure via open column chromatography:
   silica H_2SO_4/NaOH macro column
   Al_2O_3 – "high aspect" column
   silica – $AgNO_3$ – Al_2O_3 – "high aspect" column
4. Fractionation by reversed phase HPLC
5. Analysis with GC–MS (HP–5970–MDS) with a SP 2331 fused silica column (50 m)

The following contaminated streams have been processed:

1. Gasoil with different levels of chlorobenzene and PCB
2. Percolation oil from land fill deposit site
3. Waste stream from pesticide production site
4. Lube oil, contaminated with PCDDs and PCDFs
5. Gasoil, spiked with PCDDs and PCDFs

A description of the hydrodechlorination experiments is given below.

Results and Discussion

Conversion of Gasoil Contaminated with Chlorobenzene and PCB

In Table 7.5, the results of the conversion experiments with a contaminated gasoil are presented. As can be concluded from the figures in Table 7.2, concentrations of PCB around 500 ppm are reduced to less than 0.1 ppm, a reduction of more than 99.9%. The same conversion of monochlorobenzene is observed.

Table 7.5 Conversion Experiments with Contaminated Gasoil

Component	*Untreated Sample mg/kg*	*Treated Sample mg/kg*	*Detection Limit mg/kg*
PCB	530	nd	0.1
Chlorobenzene	0	–	–
PCB	530	nd	0.1
Chlorobenzene	4600	nd	10

As can be concluded from the figures in Table 7.2, concentrations of PCB around 500 ppm are reduced to less than 0.1 ppm, a reduction of more than 9.99%. The same conversion of monochlorobenzene is observed.

Conversion of Percolation Oil from a Landfill Deposit Site

In Table 7.6, conversion of percolation oil from a landfill deposit site is presented. As can be concluded from the figures in Table 7.6, conversion factors for both dibenzofurans and dibenzodioxins are mostly more than 99%.

Table 7.6 Conversion of a Contaminated Percolation Oil

Component	*Untreated Sample* µg/kg	*Treated Sample* µg/kg	*Conversion* %
2,3,7,8-tetra-CDD	26	0.1	99.6
Sum tetra-CDD (other)	58	1.5	97.4
Sum penta-CDD	196	7.5	96.2
Sum hexa-CDD	1820	4.1	99.8
Sum hepta-CDD	3840	3.6	99.9
Octa-CDD	7610	4.1	99.9
2,3,7,8-tetra-CDF	nd	nd	–
Sum tetra-CDF	298	1.5	99.5
Sum penta-CDF	1080	1.6	99.9
Sum hexa-CDF	1650	1.3	99.9
Sum hepta-CDF	3100	1.9	99.9
Octa-CDF	4980	1.6	9.99

Conversion of Waste Stream from Pesticide Production Site

In Table 7.7 the conversion of a contaminated waste stream from a pesticide production site is given. As can be concluded from the data in Table 7.7a, PCDD and PCDF concentrations were reduced to below detection limits.

Table 7.7 Dechlorination of Residual Mixture (pesticide production) in the Chloroff Reactor

Component	*Untreated Sample* µg/kg	*Treated Sample* µg/kg	*Detection Limit* µg/kg
a. Dibenzodioxins and furans			
2,3,7,8-tetra-CDD	19	nd	0.02
Sum tetra-CDD (other)	2	nd	0.05
Sum penta-CDD	7	nd	0.05
Sum hexa-CDD	25	nd	0.05
1,2,3,4,6,7,9-hepta-CDD	20	nd	0.1
1,2,3,4,6,7,7-hepta-CDD	23	nd	0.1
Octa-CDD	64	nd	0.2
2,3,7,8-tetra-CDF	0.1	nd	0.02
Sum tetra-CDF	9	nd	0.05

Sum penta-CDF	8	nd	0.005
Sum hexa-CDF	7	nd	0.05
Sum hepta-CDF	7	nd	0.1
Octa-CDF	12	nd	0.2
b. Chloroaromatics and – alicyclics			
2,3,4-trichlorophenol	0.1	nd	0.1
2,3,5-trichlorophenol	0.1	nd	0.1
2,3,6-trichlorophenol	0.8	nd	0.1
2,4,5-trichlorophenol	10.9	nd	0.1
2,4,6-trichlorophenol	0.1	nd	0.1
3,4,5-trichlorophenol	0.1	nd	0.1
2,3,4,5-tetrachlorophenol	0.2	nd	0.1
2,3,4,6-tetrachlorophenol	0.1	nd	0.1
2,3,5,6-tetrachlorophenol	0.3	nd	0.1
Pentachlorophenol (PCP)	0.1	nd	0.1
1,2-dichlorobenzene	7.7	0.6	0.1
1,3-dichlorobenzene	10.3	nd	0.1
1,4-dichlorobenzene			
1,2,3-trichlorobenzene	72.0	nd	0.1
1,2,4-trichlorobenzene	297	nd	0.1
1,3,5-trichlorobenzene	14.2	nd	0.1
1,2,3,4-tetrachlorobenzene	76.9	0.1	0.1
1,2,3,5-tetrachlorobenzene	53.6	0.1	0.1
1,2,4,5-tetrachlorobenzene			
Pentachlorobenzene	2.0	nd	0.1
Hexachlorobenzene (HCB)	0.5	nd	0.1
alpha-HCH	0.1	nd	0.1
beta-HCH	0.1	nd	0.1
gamma-HCH	0.1	nd	0.1

Table 7.7b shows that chlorinated aromatic compounds like chlorophenols and chlorobenzenes can be removed effectively. Aliphatic ring compounds like hexachlorohexane behave in a similar manner. Exact conversion factors could not be established due to the relatively high detection limits of the method of analysis. However, conversion factors are in general more than 98%. Only conversion of 1,2-dichlorobenzene (Table 7.7b) shows a relatively low value of 92.2%.

Conversion of Lube Oil, Contaminated with PCDDs and PCDFs

Conversion of lube oil, spiked with some PCDD and PCDF compounds in the ppb concentration range and PCB in the ppm concentration range is presented in Table 7.8. As can be seen, conversion factors for the PCDD and PCDF compounds are at least larger than 95.5%. For the elimination of PCBs a conversion factor of 93.3% can be calculated. Because of the fact that the detection limit of the analysis for PCDDs and PCDFs was around 0.2 μ/kg (corresponding with approximately 200 ppt), we were not able to calculate exact conversion figures. For this reason, we performed another experimental series and used the facilities and specialized knowledge of the Toxicological Chemical Laboratory of Amsterdam University for the analysis of chlorinated hydrocarbons in general and PCDD/PCDF analysis in particular. In this experiment, a contaminated lube oil fraction was processed and very sensitive analytical techniques were employed. Typical detection limits were now around 20 ppt and more exact conversion data could be calculated. Elimination of PCDD and PCDF compounds was achieved with an average conversion factor of about 99.2% (Table 7.9).

Table 7.8 Conversion of Contaminated Lube Oil

Component	*Untreated Sample μg/kg*	*Treated Sample μg/kg*	*Detection Limit μg/kg*
1,2,4–Tri–CDD	4.5	nd	0.2
1,2,3,4–tetra–CDD	19.5	nd	0.2
Sum penta–CDD	nd	nd	0.2
Sum hexa–CDD	5.2	nd	0.2
Sum hepta–CDD	17.3	nd	0.2
Octa–CDD	64.1	nd	0.2
Octa–CDF	40.4	nd	0.2
PCB	15,000	nd	100

Table 7.9 Elimination of Chlorinated Hydrocarbons from Lube Oil Samples

Component	*Untreated Lube Oil (ppt)*	*Hydrotreated Lube Oil (ppt)*	*Elimination (%)*
Sum tetra-CDD	39,622	9	99.98
Sum hexa-CDD	17,937	488	97.28
Sum hexa-CDF	3605	467	87.05
Sum hepta-CDD	61,112	394	99.36
Sum hepta-CDF	5880	162	97.24
Octa-CDD	103,418	168	99.84
Octa-CDF	3597	33	99.08
Total	235,171	1721	99.26

Conversion of Contaminated Gasoil

For kinetic reasons, experiments have been performed with gasoil, spiked with dibenzodioxin compounds and samples have been processed in the microflow reactor. In Table 7.10, the results of a conversion experiment with gasoil contaminated with OCDD are shown.

Table 7.10 Conversion of Gasoil Contaminated with OCDD

Component	*Untreated Gasoil (ppt)*	*Hydrotreated T=225°C*	*Gasoil (ppt) T=250°C*
Sum tetra-CDD	–	7227	80
Sum penta-CCD	–	5468	80
Sum hexa-CDD	–	3351	70
Sum hepta-CDD	–	1129	70
Octa-CDD	550,000	298	199

In addition to the elimination of OCDD, which increases from 68% to 99% (based on the sum of concentration of dioxines) with the small rise in reaction temperature, we observe a redistribution of dioxins with a lower amount of chlorine per molecule. In fact, at the higher temperature, levels of these smaller molecule compounds strongly decrease.

In a second experiment, a gasoil spiked with 2,3,7,8-TCDD, the most notorious dioxine compound, at a concentration of 80,000 ppt was processed at different reaction temperatures (Table 7.11). Also, we observe here some redistribution of chlorine and an increase in conversion with temperature. Note that at or above 250°C dioxine levels are very low indeed. As an illustration, Figure 7.4 shows the conversion of TCDD as a function of the reaction temperature, we observe a redistribution of dioxins with a lower amount of a temperature above 250°C.

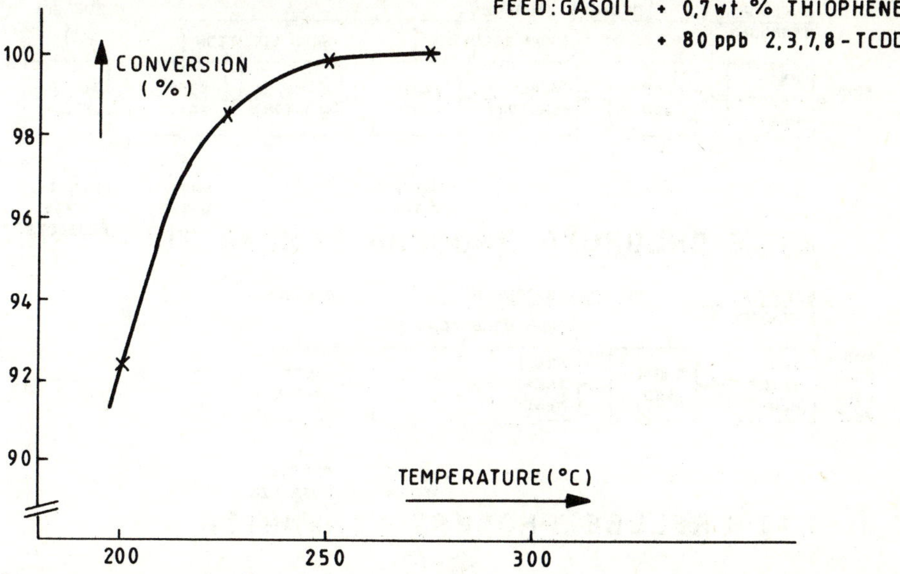

Figure 7.4 Conversion of 2, 3, 7, 8-TCDD as a Function of Reaction Temperature

Table 7.11 Conversion of Gasoil Contaminated with 2,3,7,8-TCDD

Component	*Untreated Gasoil (ppt)*	*Hydrotreated Gasoil (ppt)* 200°C	225°C	250°C	275°C
2,3,7,8-TCDD	80,000	6107	1152	56	40
OCDD	–	75	58	26	20

Description of the Chloroff Process

Based upon the results of the pilot plant study and the experiences of the similar Relube Process, we developed the Chloroff Process (Figure 7.5). Both processes feature a pre-treatment part, a process part and an after-treatment part.

The pre-treatment for the Relube Process consists principally of dewatering and removal of light ends, gasoil removal and high vacuum distillation.

For the Chloroff Process, the pre-treatment consists mostly only of a filter step where particulates are removed. Both processes in the process part apply a guard bed to remove impurities which might poison the catalyst, a conversion step consisting of a trickle-phase-reactor and a separation step where water is removed. For complete conversion of chlorinated hydrocarbons, two catalytic conversion reactors are foreseen. After-treatment for both processes consists of a distillative step.

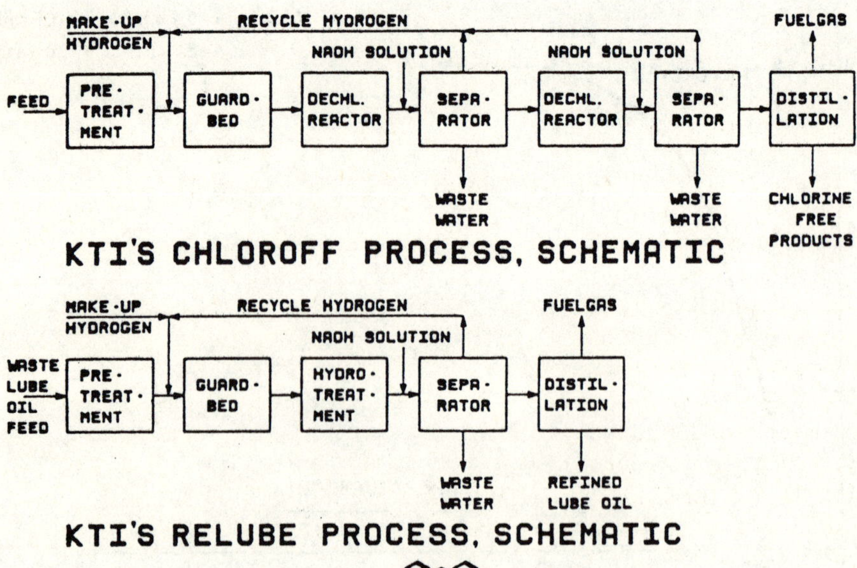

Figure 7.5 The KTI Chloroff and Relube Processes

For the time being, smooth treatment of chlorinated hydrocarbons up to 5% w/w chlorine can be guaranteed. When a waste stream with a higher content of chlorine has to be treated, dilution with a suitable solvent will be performed. Waste streams with more than 5% w/w chlorine have been treated satisfactorily in our labscale pilot plant, but more experience has to be gained on a larger scale.

An indicative economic evaluation based upon the features of the process is given in Table 7.12.

Table 7.12 Indicative Economic Evaluation of the Chloroff Process

Basics:	10,000 t/yr of waste stream, chlorine content less than 5% w/w Operating capacity 7800 h/yr
Main variable costs:	Hydrogen: 5–7 kg H_2/ton feed about Dfl 20/ton feed Catalyst: about Dfl 40/ton
Fixed costs:	Personnel (manager + 2 operators): 400 kDfl/yr Capital cost (depreciation + interest): 1350 kDfl/yr Maintenance: 3% or 270 kDfl/yr
Total fixed costs are 2020 kDfl/yr or Dfl 202/ton.	

As can be seen in Table 7.12, total annualized costs for the Chloroff Process amount to approximately 262 Dfl/ton. It should be noted that operating costs can be considerably higher for lower capacity plants, because they have relatively high capital costs. Credit of the processed product can be as high as Dfl 100–1000/ton.

Conclusions

The results of our experiments indicate that the catalytic process described above is able to eliminate chlorinated hydrocarbons from a wide range of origins with a remarkably high efficiency. The preliminary features of the process are: reaction temperature 250–300°C, reaction pressure 50–60 bar.

The Chloroff Process has many characteristics in common with the KTI Relube Process, which has already been applied worldwide. The Chloroff Process is a sophisticated application of the Relube Process and is therefore not an essentially new process design. Chlorinated hydrocarbons will be converted in a closed system – unlike the incineration of chlorinated hydrocarbons. Flue gases, which form a central problem in the incineration process, will not be present. We are therefore of the opinion that the Chloroff Process is a significant contribution to the treatment of chlorinated waste stream in an environmentally safe and acceptable way.

8

Treatment and Recycling of Waste Oils and Solvents

K.J. Youtsey, T.N. Kalnes, R.W. Johnson and D.R. Hedden

The vast majority of organic waste liquids have an origin in petroleum or coal-based hydrocarbon matrices. Some typical waste liquids derived from petroleum or coal base stocks include dielectric fluids (transformers and capacitors), heat transfer fluids, used lubricating and used hydraulic oils, used cutting oils, used solvents, still bottoms from solvent recycle operations, coal tars and PCB-contaminated oils. Lubricating oils, as an example, are produced by fractionating crude oil into narrow boiling ranges of specific viscosity. Upgrading is achieved by such unit operations as solvent extraction, dewaxing, and hydroprocessing to give satisfactory lubricant properties.

The waste liquids typically do not lose the original qualities of their virgin stock; many are high quality distillates or specialty chemicals with significant value-added characteristics. In particular the waste liquids generally retain their original Btu value. However, once utilized, these materials become hazardous wastes for the purpose of disposal due either to their basic constituents, as with degreasing solvents (e.g. trichloroethylene, perchloroethylene), or to cross-contamination by service or disposal practices as with used lubricating oils (lead, zinc, copper, PCBs). Some examples of the types of contaminants that make organic waste liquids hazardous are polychlorinated biphenyls and askerels, chlorinated solvents, fluorinated solvents, oxygenates (cresols, ethers, ketones), sulphur and heavy metals (lead, zinc, copper).

UOP has developed decontamination technology that allows the valuable hydrocarbon and specialty chemicals of the waste oils to be recovered. The technology can either convert or segregate the waste liquid's contaminated or toxic components producing, as a minimum, a clean fuel product. The heart of UOP's decontamination technology is catalytic hydrogenation, known generally as hydroprocessing.

Catalytic treatment of waste oils and spent solvents provides an environmentally sound, commercially viable alternative to high temperature incineration or chemical treatment. Through the UOP decontamination process, selective hydrogenation can be utilized to transform regulated hazardous materials contaminated with substances, such as PCBs or heavy metals, into usable hydrocarbon products.

This process is an outgrowth of petroleum refining and petrochemical technology which UOP has commercialized throughout the world during the last 70 years.[1] Typical applications for UOP's hydrogenation technology range from the desulphurization/demetallization of residual oils to the production of specialty chemicals. Table 8.1 summarizes some of UOP's existing hydrogenation processes.

Table 8.1 UOP Hydrogenation Technology

Petroleum Refining	*UOP* Process*
Production of low aromatic solvents	AH Unibon*
Desulphurization of residual oils	RCD Unibon*
Production of LPG	LPG Unibon*
Hydrogenation of transportation fuels	HD Unibon*
Chemical Industry	
Conversion of benzene to cyclohexane	HB Unibon*
Conversion of butadiene to butene	Huels SHP

* UOP, AH Unibon, LPG Unibon, HB Unibon and HD Unibon are registered trademarks and service marks of UOP Inc.

UOP's wide range of catalytic hydrogenation technology is now extended into hazardous waste management through the UOP decontamination process. Some examples of the potential applications of this process are provided in Table 8.2.

Table 8.2 Potential Applications of the UOP Decontamination Process

Pollutant Chemical	*Transformed Hydrocarbon*
Trichloroethylene	Ethane
Tetraethyl Lead	Ethane
Cresol	Toluene
PCB	Biphenyl
Polluted Stream	*Treated Product(s)*
Still Bottoms from Solvent Recycle	Solvent/Fuel
Used Motor Oil	Lube Base Stock
Spent Solvent	LPG/Fuel
Coal Tar	Fuel Oil
Transformer Oil	PCB-free Mineral Oil

This paper describes the UOP decontamination process and presents examples of typical waste stream treatment options.

Process Description

Catalytic hydrogenation of contaminated organic waste streams is carried out at moderate temperatures and pressures. Proprietary UOP catalysts are utilized to effect rapid reaction of hydrogen with the pollutant molecule(s). Reactor design configuration is based on waste oil composition, desired end products, regulatory specifications and overall processing flexibility requirements. Within the scope of the UOP decontamination process, additional unit operations such as phase separation, extraction, adsorption, evaporation and distillation are integrated with catalytic hydrogenation to meet necessary pre- and post-treatment requirements.

The hydrogenation section of this process is shown in Figure 8.1. The waste oil is heated and combined with a gas stream that is rich in hydrogen. This mixture is charged to a reactor containing proprietary catalyst(s) where pollutants such as PCBs are converted. Fresh hydrogen gas is charged to the inlet of the reactor to balance consumption.

The effluent from the catalytic reactor is cooled and combined with an aqueous solution of a basic compound. The stream is then transferred to a product separation system. The gas phase is recycled to the reactor inlet to provide a large excess of hydrogen gas in the catalytic reaction zone. The

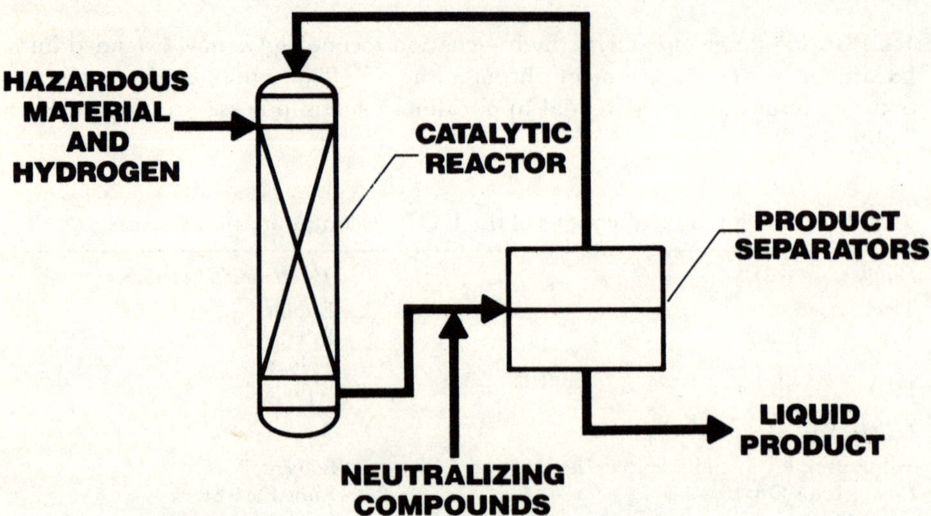

Figure 8.1 Hydrogenation Section of the UOP Decontamination Process (Source: OUP 1451–16)

aqueous fraction is recovered as a solution of non-hazardous salts. The treated organic phase is generally suitable for re-use as a fuel oil. The organic product may also be converted into a slate of higher value products depending upon the nature of the original waste oil. Products are collected in receiver tanks and certified prior to removal from the treatment site. Post-treatment required to produce specific products varies with the charge stock and site-specific economic considerations.

Hydrogenation technology can be applied to almost any organic hazardous waste stream. Catalytic reaction kinetics are favourable over a wide range of pressures, temperatures and flow rates. Operating range for the technology is primarily constrained by economic, not technical, criteria. Treatment plant cost is proportional to the amount of flexibility incorporated into its design basis. A reasonable amount of flexibility can be achieved within the following physical constraints: operating pressure – 14–70 bar; operating temperatures – 65–454°C. A typical treatment plant capacity would be 40 m^3/d but could be easily scaled upward or downward. Viscosity of the waste stream is not critical unless it prevents the material from being pumped up to the required operating pressure. If required, a diluent oil can be utilized as a carrier solvent for the pollutant(s) or to enhance unit operations on highly viscous materials.

Process Advantages

Environmental and Safety

The UOP decontamination process inherently produces no acutely toxic byproducts. For example, chlorinated dioxins and furans are always potential products of an oxidative process such as the incineration of PCBs[2,3]. The formation of these byproducts is fundamentally inconsistent with operation under the reducing conditions present in the UOP process. In addition, the opportunity to certify the conversion products before release from the treatment facility, and to reprocess the materials in the event of a plant upset, offers extra protection to the environment not possible with incineration.

Hydrogenation catalysts offered by UOP minimize undesirable side reactions involving the hydrocarbon matrix, thus eliminating many of the safety concerns associated with other technologies. For example, water is known to react violently with metallic sodium, another reducing medium utilized commercially. There is no reaction between water and hydrogen. The UOP process handles waste streams containing entrained water without pre-treatment. The acid gas formed by the hydroconversion of halogenated molecules, such as hydrogen chloride from PCB conversion, is neutralized in a uniquely engineered product separation system, which minimizes the corrosivity and volatility of the product streams while emphasizing safety of operation.

Flexibility/Product Re-use

The UOP decontamination process allows treatment of many hazardous materials in a cost-effective environmentally acceptable manner. Typical organic waste streams, such as transformer oils, still bottoms, waste solvents, cutting oils and used lubricating oils all fall within the scope of this technology. The reactivity of heteroatoms, such as nitrogen, chlorine, sulphur, oxygen and metal complexes, under catalytic reducing conditions, provides a powerful tool for conversion of chemical waste streams into hydrocarbon products that have value as petrochemical charge stocks, solvents, lubricants or fuels.

Case Study: Used Lubricating Oils

Used lubricating oils represent a major portion of the volume of organic waste liquids generated worldwide. The majority of used lube oils are petroleum derived and originally produced through multiple refining steps including vacuum distillation, solvent extraction, dewaxing, acid and clay treatment, hydrofinishing, and compounding. As a result of these many refining steps a premium grade product is produced having certain key properties, such as low pour point, high saturate content, high flashpoint, high hydrogen content, proper viscosity. Table 8.3 summarizes some typical waste lubricating oil properties.

Table 8.3 Typical Waste Oil Composition: Used Motor Oil

Bulk Properties	
Specific Gravity at 15.5°C	0.9
Flash Point, °C	>100
Pour Point, °C	35
Viscosity at 38°C	150–350
Viscosity Index	>100
Hydrocarbon Breakdown	
Paraffins and Naphthenes, wt-%	77
Aromatics, wt-%	23
Hydrogen Content, wt-%	13.5

Used lube oils retain the overall petroleum characteristics of newly refined lube oils but are contaminated from such undesirable sources as leaded gasoline, diesel fuel, metallic wear particles and byproducts of oxidation. Table 8.4 lists some typical ranges of contaminant levels found in used oils. These contaminants degrade the lubricating ability of the original material, complicate any re-refining or recovery operations, and potentially endanger the environment when disposed of haphazardly. Typically the used oil is released into the environment by dumping it into landfills, pouring it into sewers or domestic drainage systems, or applying it to roads to reduce dust problems. The oil in turn leaches into the soil and water supplies despoiling the water's taste, jeopardizing the biota, and releasing hazardous metals into the environment.

Table 8.4 Typical Waste Oil Contaminants: Used Motor Oil (Source: Cotton, F.O., Whisman, M.L., Goetzinger, J.W. and Reynolds, J.W. 1977. Analysis of 30 Used Motor Oils. *Hydrocarbon Processing*. September)

Contaminant	*Concentration Range*
Water	0–34%
Fuel Dilution	0–10%
Sulphur	0.35–0.5%
Nitrogen	0.07–0.1%
Ash	0.2–1.5%
Toxic Species	
PCB	0–200 ppm
Lead	20–10,000 ppm
Zinc	200–1500 ppm

Pilot Plant Operation

A waste lubrication oil obtained from a contaminated oil recycling site was continuously processed in a UOP pilot plant to demonstrate the process feasibility. The oil was PCB contaminated with 270 wt-ppm of Aroclor 1260 along with 750 wt-ppm of lead and an ash content of 0.32 wt-%. Additional properties of the untreated waste oil can be found in Table 8.5.

Table 8.5 Pilot Plant Charge Stock and Product Analyses

Analyses		*Charge Stock*	*Product*
Specific Gravity	0.887	0.875	
Distillation, °C			
	IBP	163	122
	50% over	416	405
	FBP	485	545
	% residue	20	6
Viscosity Index	128	126	
PCB Concentration, wt-ppm		270	No PCBs
Aroclor Type		1260	Detected
Sulphur, wt-%		0.49	0.13
Nitrogen, wt-ppm		600	250
Lead, wt-ppm		750	5.5
Zinc, wt-ppm		285	1.7
Vanadium, wt-ppm		2.9	<0.1
Nickel, wt-ppm		3.8	2.6

The oil was processed without dilution in a pilot plant designed to evaluate catalytic activity, stability and selectivity. To eliminate scale-up assumptions demonstration data were obtained using a recycle gas flow configuration. Both the product liquid and gas streams were subjected to a variety of detailed analyses to quantify pollutant conversion, hydrogen consumption, and hydrocarbon product quality. Reduction in metals, ash, sulphur and other undesirable compounds was monitored using a combination of ASTM, UOP and USEPA derived test methods.[4] Monitoring the level of waste oil PCB contamination was by a gas chromatography/electron capture detection method (GC/ECD). By this method the PCB contamination was identified as Aroclor 1260.

Chemical reduction by hydrogen removed chlorine atoms from the PCB molecule and eliminated the GC/ECD Aroclor pattern in the product oil. Confirmation of the product oil's PCB concentration was performed using a gas chromatography/mass spectrometry method (GC/MS) to provide the necessary resolution for detecting any residual partially dechlorinated molecular species. The limit of detection was established using state-of-the-art techniques and found to be less than 2 wt-ppm per PCB congener.

Conclusions

The basic characteristics of the waste oil processed in the pilot plant are typical of those often found in spent automotive motor oil as shown in Table 8.6. As illustrated in Table 8.5, the treated product contained no PCBs at a 2 wt-ppm level of detectability. In addition, the lead content of the raw oil was reduced to

Table 8.6 Comparison of UOP Pilot Plant Charge Stock (after Several Years of Storage in an Open Tank) and Typical Used Motor Oil

Property	*Pilot Plant Charge Stock*	*Typical Used Motor Oil**
Specific Gravity	0.89	0.9
Flash Point, °C	90	>100
Pour Point, °C	−30	−35
Viscosity at 38°C, SUS	167	150–350
Viscosity Index	128	>100
Fuel+Water Dilution, vol-%	15	0–44
Toxic Contaminants, wt-ppm		
PCB	270	0–200
Pb	860	20–10,000
Zn	400	200–1500
Ba	75	60–600
Cu	22	5–40

* From Cotton,F.O., Whisman, M.L., Goetzinger, J.W. and Reynolds, J.W. 1977. Analysis of 30 Used Motor Oils. *Hydrocarbon Processing*. September.

less than 10 wt-ppm. Desulphurization was approximately 73% with an overall ash reduction estimated at greater than 90%. As the viscosity indices of the raw and treated oil indicate, processing retained the excellent viscosity characteristics of the starting material making the product an excellent candidate for re-use. Viscosity Index (VI) is an empirical number inversely related to the amount of cycloparaffins and aromatic hydrocarbons present in an oil, and is used to indicate how the viscosity of the oil changes with temperature. A high VI changes viscosity to a lesser extent with temperature than a low VI oil.

The value of a waste oil after treatment can be as either a high Btu content, clean burning fuel or a lube base stock equal to a highly refined virgin oil. The choice of product slate is dependent on level of cross-contamination with other non-lubricating waste oils and overall site specific economies. Table 8.7 provides a comparison of two typical processing options.

Processing option A produces a high yield of clean (metals-free) fuel oil from a typical used motor oil. Overall processing severity depends on the level of halogenated pollutants present in the waste oil. For low levels of halogen contamination (i.e., <50 wt-ppm PCB), a single stage operation achieves the desired objective at reduced capital/operating costs.

Processing option B represents a more severe two-stage operation capable of handling high levels of halogenated pollutants and producing a mixed yield of clean fuel oil and lube oil base stock. Higher processing costs are off-set by the production of higher value products.

Table 8.7 Used Motor Oil: Expected Yields and Product Qualities

Processing Option	*A*	*B*
Treated Product Yields, vol-%		
Water	5–10	5–10
Clean Fuel Oil	80–90	10–15
Lube Base Stock	—	70–80
High Ash Residue	5–10	5–10
Clean Fuel Quality		
Heating Value, Btu/lb	19,000	19,000
Ash Content, wt-%	<0.1	<0.1
Heavy Metals, wt-ppm	<10	<1
Lube Base Stock		
Viscosity, SUS at 100°C	—	150–160
Viscosity Index	—	125–130
Pour Point, °C	—	−30

In both options, 5–10% yields of by-product water and high ash residue are typical. Disposal/re-use of these streams is dependent on both the level and type of contaminants present in the starting material. Where required, additional unit operations can be integrated with the hydroprocessing step(s) to further stabilize heavy metals present in the residue product or produce a water effluent meeting local discharge standards. The overall objective of waste minimization is applied to each specific design case.

While the processing options illustrated in Table 8.7 specifically refer to the used lubricating oil case study, design parameters can be selected that allow a wide range of waste stream processing flexibility. UOP's current pilot plant experience includes trichloroethylene-contaminated still bottoms from a solvent recycling operation, PCB-contaminated transformer oils, PCB-contaminated heavy fuel oil, and both chlorinated and non-chlorinated waste solvents.

Process Economics

In general, the added value derived from converting a hazardous waste oil to a usable hydrocarbon product provides the economic driving force when considering alternatives to incineration. Hydrogen, as the reducing agent for the UOP decontamination process, is less expensive than sodium or sodium-based compounds and acts to upgrade, not degrade, the treated product. Overall processing costs are highly competitive with other existing, environmentally acceptable technologies.

Pro forma economics for the UOP decontamination process are presented in Table 8.8. The calculations are based on a plant size of 40 m^3/d operating 255 days per year and conservative assumptions with respect to capital and operating costs. On stream factors of greater than 90% should be easily achievable. The positive contributions to this payback calculation arise from avoidance of disposal costs and from product credits, and are based on current competitive values in the United States.

Table 8.8 Payback Economics of UOP Decontamination Unit (Assuming Standard Plant Design 40 m^3/d Capacity, 70% Utilization)

Waste Type	*Payback (yrs)**
Fluorocarbon Still Bottoms	0.7
Waste Lube Contaminated with PCBs	1.3
Askarel	0.6
Highly Chlorinated Still Bottoms	1.3
Spent Chlorinated Solvent	<1
Low Chlorine Still Bottom[†]	3.5

* Revenue in payback calculation based on avoidance of current disposal costs and credit for reusable product

[†] Less than 0.5 wt-% chlorine.

Commercially attractive paybacks can be associated with a wide range of organic waste streams.

State of Development

UOP catalytic hydrogenation processes have received wide acceptance from the petroleum and chemical industries over the past three decades. State-of-the-art processing technologies coupled with over two dozen proprietary hydrogenation catalysts provide a firm basis for the commercialization of the UOP decontamination process. Pilot plant confirmation of the treatment of many different waste oils provides the detailed data required to scale-up plant design(s) for application to today's waste management problems.

UOP has licensed more than 570 hydrotreating units representing approximately 280 million gallons per day of feedstock capacity around the world. Of these units, about one-third are designed for hydrotreating kerosene, middle distillates and heavier oils, with the remaining two-thirds dedicated to extremely low level decontamination of gasoline base stock petroleum fractions.

In the case of lube oil production, an aspect of primary importance is catalyst selection. Over the years UOP has developed a wide range of catalysts applicable to lube service. These catalysts are characterized by excellent hydrogenation activity, with the ability to selectively convert low VI feed components.

Finally, a highly qualified engineering staff make possible the custom design of treatment plants to satisfy a wide variety of environmental and industrial needs. Plant start-up, operator training and troubleshooting activities are all fully supported by UOP's technical service staff of trained engineers.

References

1. Achilladelis, B. 1975. History of UOP: From Petroleum Refining to Petrochemicals. *Chemistry and Industry*, 6: 337.
2. Olie, K., Vermeulen, E.E., and Hutzinger, O. 1977. Chlorodibenzo-p-dioxins and Chlorodibenzofurans Are Trace Components of Fly Ash and Flue Gas of Some Municipal Incinerators in the Netherlands. *Chemosphere*, 6, 7: 455.
3. Weerasinghe, N.C.A., Meehan, J.L., Cross, M.L. and Harless, R.L. 1985. The Analysis of Tetrachlorodibenzo-p-dioxins and Tetrachlorodibenzofurans in Chemical Waste and in the Emission from Its Combustion. *Chlorinated Dioxins and Dibenzofurans in the Total Environment II*, Keith, L.H., *et al.*, eds. Boston, Mass.: Butterworth, 425.
4. United States Environmental Protection Agency. 1981. *The Determination of Polychlorinated Biphenyls in Transformer Fluid and Waste Oils*, Method EPA-600/4-81-045.

Part II

Economic, Environmental and Institutional Issues

9

Elements of a Balanced Approach Towards Hazardous Waste Management

Amir A. Metry

Since the start of the industrial revolution, industrial waste management practices have primarily involved the disposal and use of either on-site or off-site landfills. Industrial wastes were often co-disposed using typical so-called sanitary landfill practices. Many of these land disposal sites were not designed or operated to handle these industrial residues. As a result, leachate resulting from these wastes was not properly contained or controlled and has migrated out of the landfill into underlying soils and contaminated groundwater or nearby surface water and drinking water supplies.

Several hazardous waste management alternatives are available other than land disposal, and these include:

1. Recovery and re-use
2. Waste minimization and reduction
3. Low-waste generation technology
4. Destruction via incineration
5. Biological degradation and treatment
6. Chemical detoxification
7. Thermal treatment and destruction
8. Stabilization/solidification
9. Above-ground, long-term storage

Most of these technologies will generate a process residue (i.e., ash, sludge) which may ultimately require land disposal. However, this residue is typically non-hazardous, less leachable or less toxic, and is a smaller quantity than the original waste material. A properly designed, secure landfill may be an acceptable option for handling these types of treatment residues.

To complement the advancement in landfill designs, emphasis must be placed on pre-treatment of waste including stabilization through the use of cements, pozzolanic or bituminous materials, thermal treatment, dewatering and moisture removal, and detoxification. Emphasis must be placed on minimizing the amount of liquids contained in the waste material which is deposited in the landfill along with reducing overall toxicity of materials which are disposed.

Many new and efffective technologies are being developed for addressing the environmental problems attributable to past landfilling practices. In general, they fall into two broad categories of: source control and migration control. This is the key to avoiding many of the environmental problems that have plagued many industrialized countries. Some of the key elements of such an approach include the following:

1. Promulgation of realistic and achievable environmental regulations that are based on cost-effective, environmentally sound expectations
2. Focusing on means of avoidance. This includes selection of low-waste generating processes, in-plant recycling, re-use, closed-loop materials handling, etc.
3. Integrating waste management, treatment and disposal with the economic product life cycle to avoid a negative drain on productivity and the economy
4. Application of less complex and rudimentary technology that requires less capital and is easier to operate with average skilled labour
5. Remediation of highly polluting facilities at earlier stages to avoid environmental degradation and costly remediation in the future

Special emphasis must be placed on the application of clean technologies that produce no waste or hazardous byproducts and on waste minimization and control at the source of its generation.

Hazardous waste management practices for major industries have undergone significant changes within the last five-year period. These changes have been driven by public demand for environmental regulations, increasing cost at off-site disposal facilities, closing of commercial facilities, and increasing awareness of environmental problems caused by current and past disposal practices. It is expected that this trend of major changes in the waste management field will continue for the next ten years or longer.

Traditional methods for managing industrial waste are being phased out or are undergoing significant upgrading/modification in response to the more stringent regulations. Industrial firms are also working to minimize their waste and anticipate potential future liabilities due to environmental and public health concerns related to hazardous waste management practices.

Examples of several of the major industrial sectors confronted with significant changes in the management of their hazardous wastes include: petroleum refining, petrochemicals, pharmaceuticals and electronics.

In the petroleum-refining industry, wastes typically fall into both the inorganic and organic categories. Inorganic residues include spent catalysts from process reactor units. Past disposal practices for these spent catalysts typically utilized uncontrolled landfill or recovery/regeneration of the catalysts. Other inorganic residues include spent caustic, lime sludge, boiler ash, and filtering materials.

Organic residues have typically been treated by discharging to the process sewer which historically has treated the wastewaters through oil separation and/or an effluent settling pond system. Major upgrading of the wastewater treatment systems have been undertaken by the refining industry in response to clean water laws and now include biological treatment, physical/chemical treatment and enhanced oil separation and recovery techniques. Organic sludges such as API separator sludge are generated from the wastewater system and these have typically been land-disposed or land-applied. Other organic sludges include tank bottoms from tank cleaning operations. Current trends in managing these organic residues are by either thermal destruction or land application for biological degradation.

The petrochemical industry is also undergoing major changes with respect to hazardous waste management. The types of waste generated by the petrochemical manufacturer vary widely from plant to plant and are highly dependent upon the type of process and plant production. A portion of the wastes has been handled via the process sewer and a wastewater system consisting of oil/water separation, and settling of solids in a pond/lagoon system. Major upgrading of wastewater treatment facilities is underway to incorporate biological treatment in addition to physical/chemical treatment processes such as filtration and carbon adsorption. Residues are generated from the wastewater treatment system along with the plant process units that are typically produced in the form of residues and sludges. Historically, these have been handled via land disposal; however, the current trend is toward thermal destruction, biological degradation, and stabilization. Thermal destruction processes involve incinerators capable of handling both liquid and solid materials and typically incorporate heat recovery units. Chlorinated and non-chlorinated solvents are major waste streams in the petrochemical industry, and the current trends for managing these materials involve thermal destruction or solvent recovery.

The major waste stream from the pharmaceuticals industry is typically wastewater from production operations. This has historically been handled either through discharge to a municipal sewer or through an on-site facility. Current practices and trends involve installation of pre-treatment facilities prior to sewer discharge and upgrading of the existing treatment facilities. This upgrading involves the construction of physical/chemical treatment modules or biological treatment. Other pharmaceutical waste materials are generated by laboratory chemicals from research, organic sludges from wastewater treatment or fermentation steps, spent solvents, off-spec products, and other biological materials from research. Current management trends for these materials include thermal destruction, solvent recovery, destruction via incineration and waste reduction.

The proper application of technology and institutional measures, including environmental regulations, is the key to protecting our environmental resources for future generations.

10
Is Prevention of Pollution More Economical? The Dutch Situation

F. Langeweg

For many years the Netherlands government has been attempting to promote the development and application of clean(er) technology. One way of achieving this goal is the so-called environmentally sound technology scheme of the Ministry of Economic Affairs, supported by the Ministry of Housing, Physical Planning and Environment. Other schemes are also helpful in the clean technology field, such as the biotechnology innovation programmes within a special subprogramme for environmental biotechnology. Before discussing the Netherlands situation, it is necessary to define what is meant by clean(er) technology. Environmental policy was traditionally directed at the control of waste streams after they were generated. The large-scale construction of purification plants for domestic and industrial wastewater, begun in the 1970s, is an example of this. Such a sanitary approach was also initially followed for reducing air pollution. The soil-cleansing operation in the Netherlands is a consequence of past irresponsible waste disposal.

However, such an approach offers only a partial solution to the problems: they are frequently shifted, following the law of misery conservation. What does emerge is an environmental paradox: it takes resources to counteract pollution, resulting in waste products which require yet more resources for their removal, thereby producing further pollution. This paradox can only be avoided by preventing the generation of waste products whether they be gaseous, liquid or solid. The development and application of clean(er) technology will therefore have to concentrate on prevention. This involves actions within the company aimed at the limitation, prevention or elimination of the creation of waste products. Actions directed at the treatment, processing or disposal of waste products after their generation therefore lie outside the scope of prevention-oriented clean(er) technology. They are part of the overall environment-technological measures which can be taken when the existing pertinent technology does not offer a solution. Prevention is, both from the viewpoint of environmental health and of economy, an attractive option. The generation of waste products is in principle a sign of inefficiency in the production process. Prevention can consequently lead to the conservation of raw and other materials that are scarce, expensive or of strategic importance. The industrial companies themselves should attach great importance to this. Prevention-oriented thinking is not always an integral part of the management strategy. It is not uncommon for companies to implement a subsequent environmental policy: how can (new) environmental requirements be met through the treatment of waste streams?

However, prevention as part of the management strategy calls for more. First, there must be good insight into the materials balance of the company. The starting points for prevention should be clearly identified in such a materials balance. The costs associated with waste streams, both for the long and the short term, must be made visible in a consistent form. Thus, a kind of environment-inclusive thinking develops within a company. How far do we follow this process of prevention-oriented thinking in industry in the Netherlands? Are differences discernible between various types of industry? These are questions which will be discussed below.

Pollution Load Caused by Dutch Industry

The Indicative Multi-year Environmental Protection Programme (IMP-M) gives a list of priority substances, consisting of: acidifying and fermenting substances, metals and metalloids, organic compounds (halogenated and non-halogenated), and others.[1] An attempt can now be made to obtain a rough picture of the contribution of various industrial activities to the environmental pollution load from priority substances.

The Dutch emission registry contains data on the pollution of water and air by such substances. There are no such classified data on soil contamination, waste and products. The recourse in such cases is to use incidentally available data. In the selection of priority substances, account was taken of the volume of emission, on the one hand, and risk to the environment, on the other. It can therefore be assumed that waste-reduction policies should pay equal attention to all the priority substances, regardless of the amount emitted. As a result, the reciprocal of the amount can serve as a weighing factor in the determination of the stress on environmental media from all the priority substances together. Another method was followed for ground contamination, based on a risk assessment of activities posing a threat to the soil.

Table 10.1 shows the results of the analysis performed for a number of industrial activities. These results are rough estimates to be made more accurate by additional research. A plus sign means that an activity contributes about 2% to the total national waste burden. Each additional plus sign represents a doubling of this contribution. The plus signs have a different meaning for the stress on the ground, where each additional plus sign means that the risk of soil pollution is ten times higher.

It is not only the current contribution from the activities mentioned to the environmental pollution load that is important. The opportunities for reducing this load also play an important role. Table 10.2 summarizes the reduction in environmental pollution that has already been achieved.

Table 10.1 Contribution From a Number of Industrial Activities to the National Environmental Pollution Load (where + indicates an addition of 2% to the total national waste burden; for soil risk each additional + indicates a 10% increase)

	Air Emission	*Water Emission*	*Soil Risk*	*Problem Waste*	*Problem Products*
Chemical Sector					
phosphorus fertiliser		+++	+	++	
plastic resins (PVC)	++	++	+		+
salts	+	++	++	+	
phosphoric acid		+	++	++	+
ethene	++	++	++++	+	
propene	+	++	++++	+	
other organic commodity chemicals	++	++	++++	++	+
paint, dye, printing ink		++	+++		+
medicines	+	++	++		
pesticides	+	+++		++	+
synthetic yarns/fibres	+	+			
rubber	+		++		
Metal Sector					
base metals	+	++	+++++	+++	
nonferrous metals	++		+++++	+++	
metalworking	++	+	+++	+++	+
electrotechnology	+		++		++
shipyards			+++	++	
Others					
garages/petrol stations		+	+++++	++	
storage and transfer	++	++	+	+++	
waste processing	+++		+++++	+++	++
textile refinement		++	++++		
wood preservation			++++		
food industry	+	+			+
refineries	+	+	++	+	+
dry cleaning	+		+++		
building industry			++	+++	+
power stations	+		+	+++	

In view of the current relatively high pollution level and the as yet only small reductions achieved, metalworking companies and garages/petrol stations qualify for in-depth investigation. This applies to a lesser extent to the ethene and propene industries, the other commodity chemicals and to storage and transfer because for these cases a reasonable reduction in the environmental pollution load has already been accomplished. As far as base metals, nonferrous metals and waste processing is concerned, a substantial reduction of more than 50% in the environmental pollution load has been achieved. Because of their still large contribution to the pollution load an investigation should be

Table 10.2 Reductions Achieved in the Environmental Pollution Load

	Reduction Achieved in the Environmental Pollution Load		
	>50%	*20–50%*	*<20%*
Chemical Sector			
phosphorus fertiliser		+	
plastic resins (PVC)	+		
salts		+	
phosphoric acid			+
ethene		+	
propene		+	
other organic commodity chemicals		+	
paint, dye, printing inks			+
medicines			+
pesticides	+		
synthetic yarns/fibres			+
rubber			+
Metal Sector			
base metals	+		
nonferrous metals	+		
metalworking			+
electrotechnology		+	
shipyards		+	
Others			
garages/petrol stations			+
storage and transfer		+	
waste processing	+		
textile refinement		+	
wood preservation			+
food industry		+	
refineries		+	
dry cleaning		+	
building industry		+	
power stations	+		

conducted into whether further reduction is feasible. The activities focusing on the manufacture of phosphoric acid, paint, dye and printing ink, and of medicines, as well as wood preservation, make a more or less average contribution to the environmental pollution load but few measures have been taken to date, so that there may be attractive opportunities for improvement in this area. For that purpose, a preventive approach will be the preferred choice. The possibilities for prevention are discussed below.

The Main Features of a Preventive Approach

There are several ways of preventing waste by technological means. They are:

1. In-process recycling of (potential) waste products at the site of their generation
2. Improvements in the process-technological sense in the manufacturing process, thereby altering the primary source of waste products
3. Improvements in plant operations such as good housekeeping, preventive maintenance, better monitoring systems
4. Replacing raw materials by less harmful ones or using smaller quantities of such substances
5. Redesign of reformulation of end products

The Office of Technology Assessment (OTA) found that these five approaches had been implemented in the order mentioned by 314 specified companies in the United States where a waste reduction policy is pursued.[2] This is outlined in Figure 10.1 where the distribution over the major industries of the cases described is also given. Apparently, in-process recycling and integrated change in the production process are the first effective options to be recognized and implemented. The chemical and the metalware industries together account for over half the number of case studies of prevention-oriented approaches. The factors that determine the opportunities and willingness for waste prevention in a company according to the OTA are:

1. The nature of the industrial process
2. The size and structure of the firm
3. The technology and information available to the company
4. The economics of prevention
5. The attitude of the government

The most important factor is probably the character of the industrial processes. A summary is given in Table 10.3.

If high product quality demands are made, such as for medicines, then the opportunities for prevention are limited to changes in the production process. Other waste-reducing measures are difficult to introduce because adapting these to the product requirements set by the government is a very lengthy process. In the mature processing industry it is often difficult to change the product and the production method, among other reasons, because of strong

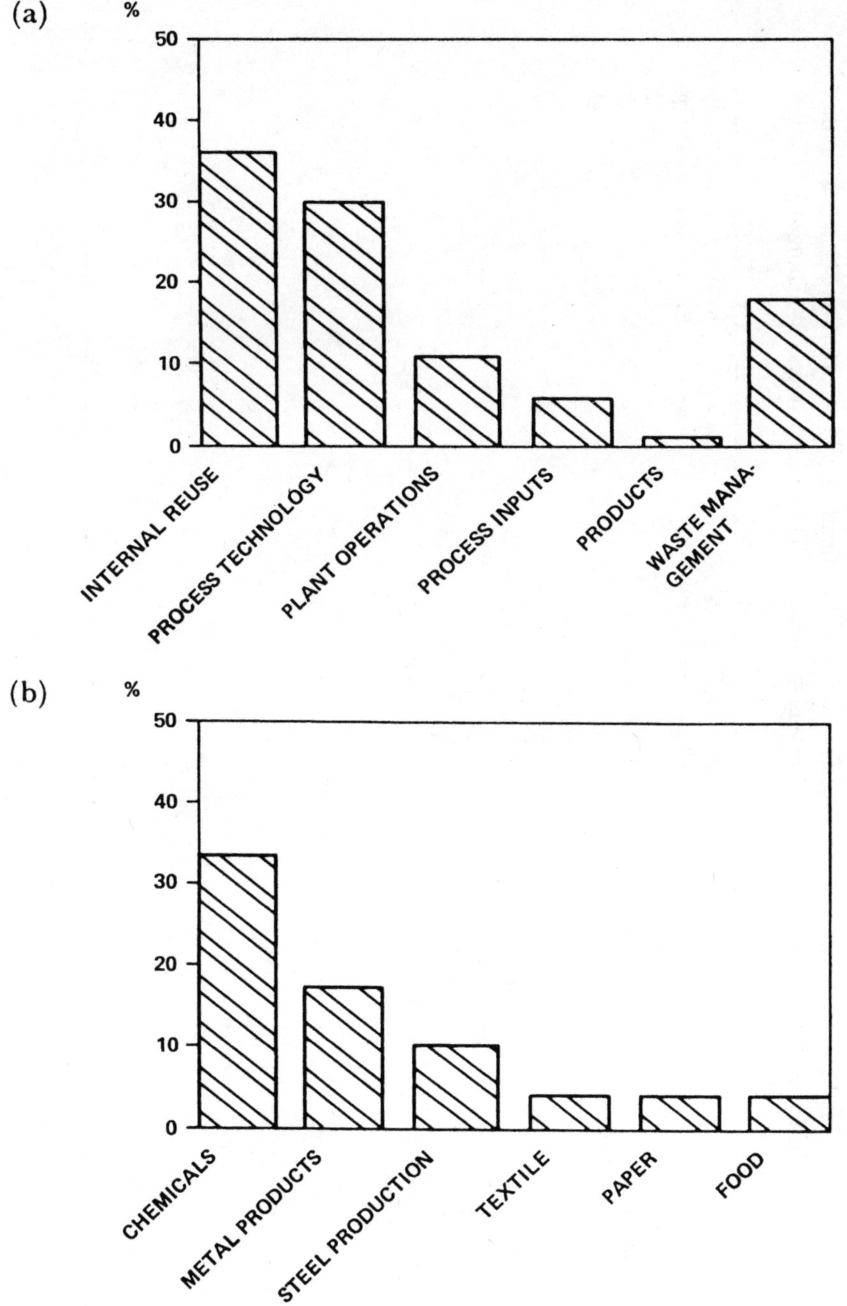

Figure 10.1 Types of Waste Prevention in 314 US Companies, (a) by Approach, (b) by Industrial Branch (Source: OTA)

Table 10.3 Potential for Preventing the Generation of Environmentally Hazardous Waste (Source: Congress of the United States, Office of Technology Assessment. Serious Reduction of Hazardous Waste. Washington, DC: US Government Printing Office)

Characteristics	*Examples*	*Operations Changes*	*In-process Recycling*	*Process Changes*	*Input Substitution*	*End Product Changes*
Stringent product specification/quality products and high cost/ high profit products	Pharmaceuticals weapons, robotics speciality chemicals	+				
Large-scale mature process technology	Rubber, oil commodity chemicals paper products lumber	+	+			
Manufacture of materials for commodity goods	Steel, nonferrous metals, textiles	+	+	+	+?	
Frequently changing production of materials for industrial use (high tech)	Electronic components medical instruments	+	+	+	+	
Processing/manufacture of different industrial products	Electroplating, printing, foundries, machine construction	+	+	+	+	
Large-scale manufacture of consumer goods	Cars, household appliances, paints	+	+	+	+	+

international competition. It is precisely because production is on a large scale that the amount of waste can be reduced through changes in the production process and through in-process recycling. This is frequently offset by the large economic benefits resulting from greater efficiency in the production process. Changing the production process, the product or the raw materials to be used is not easy. It is costly and risky and it can therefore not be expected that industry will embark on such a course so as to meet the environmental standards set by the government.

If innovation or expansion of the production process is required for market reasons, then the opportunity arises to introduce such changes. The other factors will then also begin to play a role. Accordingly, a relationship between government and industry should be sought, so that the opportunity is not missed.

To measure the potential for prevention or limitation at company level, making a mass balance, however approximate it may be, is a first step. By doing such a mass balance at the level of individual process steps, the greatest "leaks" in the process leading to environmental pollution become visible. Here lie the starting points for the preventive actions outlined above.

The amount of waste reduction resulting from these actions should be clearly established. This can best be done in relation to the production output of a company or a part thereof. For all kinds of reasons both the volume and the nature of the production of a company can change over time. This argues in favour of measuring waste reduction in a relatively disaggregated manner. In doing so, however, a problem is whether firms are willing to supply such, often detailed, confidential business information. Determining the cost of or returns from preventing the generation of waste is likewise often difficult. Process-integrated measures can lead to improvement in the manufacturing process and at the same time limit the generation of waste. The best measure is probably the total cost price per unit of production, including the costs for final disposal of waste products. As in the case of waste reduction, here too a disaggregated approach will be required so that it is again uncertain whether companies will want to reveal such information. These problems can possibly be avoided by publishing disaggregated data only in a form that does not damage business interests. This requires co-operation between government and industry on a basis of mutual trust.

Priorities and How to Continue

An outline of the current environmental pollution load resulting from industrial activities, and the opportunities that are in principle available for prevention is given above. Several industrial activities deserve particular attention. These activities can now be compared with the prevention options which depend upon the characteristics of the industrial activities. This results in the following picture of the Dutch industrial activities deserving closer inspection.

For the production of medicines the only possibility is changes in the production process. For the manufacture of ethene and propene, of phosphoric acid and the other commodity chemicals, changes in the manufacturing process and in-process recycling are among the feasible options. An additional feasible option for the base and nonferrous metals industries is process modification. The metalworking, paint, dye and printing ink, and wood preservation industries have moreover the possibility of input substitution, but they can probably not change their end product.

In addition, several activities that cannot be characterized as true production processes demand further attention: garages, petrol stations, storage and transfer and waste disposal. A more specific approach for each of the industries mentioned, grafted on to the actual business activities taking place within them, is now required. Such a specific elaboration of the formulated prevention strategy has only been done in part. The electroplating industry is an example.[3] In collaboration with the industry concerned a systematic survey was made of the available prevention options, depending on the type and size of the activity. The information obtained was subsequently put in the form of a model with which, on the basis of mass balances, the effects of measures (including costs) can be studied. Added technologies were also included in this model because prevention does not always lead to the desired result. This model system proved to be a powerful tool for underpinning the establishment of standards regarding the emission of heavy metals and cyanides.

A general picture of the results of the research performed is outlined in Figure 10.2. This is a schematic representation of the relationship between the

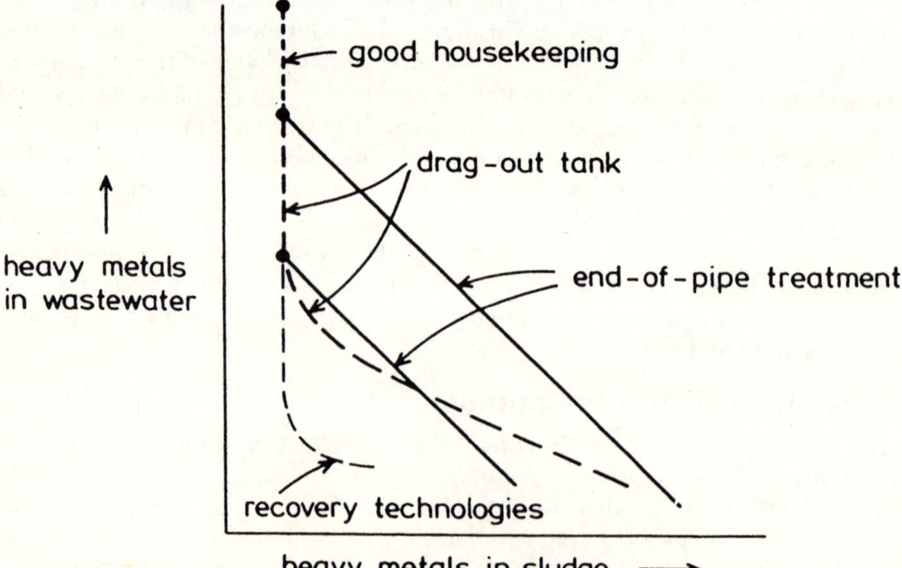

Figure 10.2 Relationship Between Heavy Metal in Waste Water and Sludge for the Plating Industry

discharge of heavy metals into wastewater and the discharge from waste sludge, depending upon process-integrated and added technological measures to be taken. The precise position and scale of these relationships depend on specific company characteristics. Measures aimed at good housekeeping (draining for a longer period above a bath and one drip-recovery bath) are very efficient because they are usually profitable. The addition of a second drip-recovery bath, or recovery techniques such as electrolysis, evaporation and membrane technology, become attractive if a further reduction is needed. Instead of, or in combination with, these measures additional technology can also be applied directly. However, additional technology alone is generally a slightly more expensive solution than the above-mentioned process-integrated measures.

The course outline for the electroplating industry appears to be also feasible in principle for the other types of industry mentioned. It can be initiated by the development of a more general model describing the importance of the role of added technology aiming at the treatment or reprocessing of waste streams. The potential for prevention in the industries concerned is used as a general starting point. It has been shown that prevention is, from the standpoint of environmental health and in many cases also of economy, an attractive solution.

References and Notes

1. Environmental Program of the Netherlands 1985–1989. 1985. Ministry of Housing, Physical Planning and Environment. Central Department for Information.
2. Congress of the United States, Office of Technology Assessment. *Serious Reduction of Hazardous Waste.* Washington, DC: US Government Printing Office.
3. Ros, J.P. Forthcoming. Modelling the Metal Plating Industry for Integral Control Strategies. *Recent Advances in the Management of Hazardous and Toxic Wastes in the Process Industries.* Elsevier.
4. National Research Council. 1985. *Reducing Hazardous Waste Generation: An Evaluation and a Call for Action.* Washington, DC: National Academy Press.
5. EPA, Office of Solid Waste and Emergency Response. October 1986. Minimization of Hazardous Waste. Washington, DC.
6. Huisingh, D. and Aberth, J. 1986. Hazardous Wastes – Some Simple Solutions. *Management Review*, June.
7. Information about the Proposed Technology Transfer Activities and Assistance in the Development of a Strategy on Clean Technologies. Ministry of Housing, Physical Planning and Environment.
8. Netherlands Indicative Multi Year Programme for Chemical Waste. 1985. Ministry of Housing, Physical Planning and Environment. Central Department for Information.

11

Risk Taking and Environmental Issues: A Survey of Issues and Problems from an Economic Perspective

Hanns Abele

People adopt very different strategies to cope with uncertainty. There are some who never board an aeroplane because there is a slight chance that it may crash. They rank the idea of this event as far more unpleasant than the chance of being involved in a car accident which from a statistical point of view is far more likely. There are others who despite their knowledge of the high risk of cancer and death continue to smoke or even start smoking. Some are fanatically and aggressively demonstrating against the theoretical risk of radiation from a nuclear plant while their actual dose of radiation because of smoking is several times the magnitude of the former. Obviously we are in need of a theory to explain the inconsistent observations.

Complex modern societies make much more intensive use of their natural environment, but experience shows that not all technological developments are beneficial. By means of industrial production techniques and exponential population growth, however, the transformation of the environment progresses on a large and accelerating scale. For a long time it had been possible to proceed as if the *ceteris paribus* model were correct, assuming that all negative side effects of an activity were local only and the ecological systems were elastic enough to cope with these "small" disturbances. The stability of the system was *not* thought to be in danger. Hence there was no immediate need for any containment of possibly detrimental activities like pollution or the production and use of hazardous materials. Experience has changed attitudes and people are now aware of environmental risks in a very general sense. Therefore we have to ask: Do we have policies to handle the uncertainties and risks posed by the environment which we cannot dispense with and are forced to live with?

Addressing the individual decision-maker first, this paper gives an overview of the problems involved. A widely used model is sketched and its critical points are discussed. This leads to a deliberation of important features of uncertain decision situations. Finally a non-technical discussion of the allocation mechanisms and the design problems is presented. This discussion helps to explain the difficulties involved in environmental issues – hazardous waste disposal being one example only – at both the theoretical and political level.

Individual Decisions in Risky Situations

Since the work of von Neumann and Morgenstern (1944) the expected utility (EU) model has been widely used in economics, management science and psychology to explain and describe decisions in uncertain environments. Although the first work on the EU model dates from the eighteenth century, it was not at the centre of theoretical interest before World War II. However, the following decades brought about a steadily increasing series of theoretical and empirical literature discussing and applying the EU model. The fact that decisions under certainty can be viewed as a limiting case of decisions under uncertainty contributed to the shift of interest in favour of the EU model.[1]

Economists as well as psychologists soon started to exploit the possibilities of the EU model and contributed theoretical and applied work. Both disciplines are well suited to use the EU model because they deal with decision making. The simplicity and formal elegance of the EU model made it a prime candidate for further research and development. In addition it rests on an axiomatic basis which induces its use at least as a benchmark for pertinent work. These axioms serve to generate a joint preference structure over outcomes and probabilities paving the way for the conclusions that individuals choose the action offering the *maximum expected utility*. The latter sometimes is formed by allowing transformations of the probabilities and using different concepts of utility. It is possible to give measures of the risk attitude of individuals.

Economists in particular have been attracted by the possibility of using the EU model at the individual level and thus of generating inputs for aggregate predictions. A number of applications, for example, in portfolio theory, capital market theory or insurance markets, were produced. These successful exercises added to the satisfaction of having a model at hand which explains behaviour and the allocation of risk within a society.

Since the very beginning, however, there has been a continuous critique of the axioms and the implications of the EU model. A vast literature has been accumulated showing behaviour even by highly sophisticated adherents of the EU model violating the axioms. On the other hand the basic comprehensive character of the orientation of the EU model has been shown in various situations to be descriptively misleading. The model seems to be very demanding in the way decision-makers are reaching decisions. Instead of giving an exhaustive review of the evidence let me concentrate on some points of special relevance.[2]

The EU model treats probabilities of possible consequences as exogenous, and therefore, any decision in an uncertain environment is heavily dependent on the input for this decision, even if the decision-maker behaves in conformity with the EU model. If decision-makers are unable to have correct estimates of the probabilities their decisions will be seriously distorted in general; the result may be even more misleading with respect to the aggregate outcome in society and may lead to inefficient hazard management.

Psychologists have produced a long list of judgemental biases in risk perception. This phenomenon is especially found in the cases of very low probabilities or high damages. Slovic *et al.* (1982) report the findings of several studies.[3] Similar results with respect to the ability to process probability information were found in other studies as well.[4]

In so far as these biases in the judgement of the probability of events lead to wrong conclusions, they may provoke alarmist attitudes and almost impede efficient provisions for hazards. On the other hand, extreme governmental regulation may produce a lulling effect which is clearly counterproductive, because people are induced by their underestimation of the probability of damage to change their behaviour and start being careless.[5]

A necessary ingredient for assessing probabilities and hence for the inclination to take risks is sufficient information. It is conceivable that the problems in processing probability information stem from the fact that the decision-makers do not have enough information. The more complex are the questions involved, as is the case for ecological or environmental problems, the less likely (lay)people have enough technical knowledge to interpret the situation and to draw (statistically) correct conclusions. This does not mean that enough information is a sufficient condition to generate responsible behaviour as we know from the campaigns with warnings about the risks of smoking. The last example underlines the difficulties in informing people about risky situations. One possible remedy could be to start training programmes to educate people to deal with risky situations.

People in general want to take a more active part in the management of the risks they have to face now, because of the increased awareness of their hazardous environment. The readiness to accept information, contrary to established beliefs, is another obstacle to an improvement in the management of hazardous environments. There is a tendency to interpret new factual evidence which should lead to a modification of beliefs as a strengthening of one's position. Similarly, very often beliefs are retained despite the fact that their evidential base has evaporated.[6]

It seems to be a good idea to look for experts in order to get guidance in difficult technical matters. Experience has shown, however, that it is possible that the opinions of experts are widely divided on central points of the problem at hand. Consider the case of the long-run effects of carbon dioxide on the climate or the problem of the emission of fluorocarbons and their eventual ozone-depleting effect.[7] Lack of consensus among experts contributes to the uncertainty of the public and generates emotional and sometimes bitter confrontations.

On the other hand overconfidence by experts is equally feeding public scepticism against relying on expert judgement. Studies show that experts themselves are not immune to be overly optimistic and can be severely misguided as a consequence.[8]

Environmental Problems and Risk Allocation in an Economy

The problems at the individual level concerning the decision model and the perception of risk are not encouraging if one would like to consider the aggregate level and derive conclusions about collective decisions in situations with uncertainty and risk respectively.

There is, however, another aspect enhancing the complexities of the problem which is characteristic for environmental questions. Decisions and actions by economic agents can have effects on the wellbeing of other individuals. As long as these spillover effects are reflected in the decisions of the agents, the overall result can be regarded as desirable. Very often this is not the case, because the decision-makers do not take into account the side effects of their actions. Thus the costs to the individual decision-maker differ from the costs to the society, that is, private and social costs diverge. Since an economy relies on signals to steer behaviour, prices being important signals cease to convey correct information leading to a distortion of decisions and probably to a misallocation of resources.

For almost a century economists have been trying to find solutions to reduce the (detrimental) effects of an agent's action on other agents because of the probable distortion of the allocation of resources and the distributive consequences. Whether one likes it or not, society cannot abstain from allocating risks in an uncertain world. Two extreme solutions have been proposed. One is to rely on market forces, the other is government regulation. The analytical problem looks rather simple at first glance. An *externality* has to be dealt with meaning that not all consequences of one's decision are included in the objective function of the decision-maker, because others are mainly effected by the decision. These effects can be positive or negative. Concerning the welfare aspects of an economy such phenomena lead to a departure from the social optimum, because a full consideration of the effects, an internalization of all costs, would bring about a different, more beneficial outcome.

It is rather straightforward to show that both extreme concepts can fail to achieve a satisfactory solution. Government regulation may be impossible since the regulating agency may not have enough information to fulfill its task. Very often the distribution of information is asymmetrical thus rendering efficient supervision extremely difficult and costly. Even worse a regulatory action may induce an expansion of the distortionary behaviour as a consequence of rational decisions of economic agents. Market-oriented solutions may break down too for several reasons. Risk attitude and asymmetric distribution of information can lead to the non-existence of an equilibrium. A large mismatch in the numbers of participants on the sides of the opponent interests can prevent the pursuit of their interests because it is impossible to organize a bargaining process. *A fortiori* these arguments are valid in the case of allocation of risks. Therefore designing appropriate decision mechanisms is a formidable task for economic theory in particular and social science in general.

Consider as an example the decision about the site of a hazardous waste facility. Everyone is ready to accept the abstract need for such a facility, however, nobody wants to have the site in his neighbourhood. The NIMBY (not-in-my-backyard) attitude does not in itself lead to a solution. *The core of the problem is the divergence of individual and collective rationality.*

The attitude of people to environmental problems has changed considerably during the last decades. On the one hand the awareness of long-run dangers as a consequence of careless behaviour has increased, on the other hand, it is not sufficiently widely recognized that there are costs necessitating changes of individual habits and behaviour involved in preserving the environment in an acceptable shape. Information about environmental issues is increasing, although in most cases very complex matters are involved dividing even experts on the likely consequences. It may be helpful to get an impression of the difficulties faced by environmental policies by reviewing public opinion on environmental issues.

The Council of Environmental Quality (1980) provides an excellent opportunity to obtain an impression of what the public thinks about environment and environmental policy. The reported opinions are the results of a nationwide poll done in January–February 1980. In order to put them into the right perspective as well as allowing the detection of intertemporal changes and trends, the outcomes of similar previous opinion surveys are summarized.

The basic conclusion emerging from the material is the fact that environmental protection has the support of a large part of the population. Environmental issues have established themselves as an important field of social concern in the same way as health or education. People are ready to pay for measures of environmental protection. With respect to industrial hazardous chemical waste, almost two-thirds expressed great concern in 1980 – a percentage higher than for air and water pollution before.

One question of the poll was devoted to revealing the preferences concerning the siting of five facilities, a ten-storey office building, a large factory, a coal-fired plant, a nuclear power plant, and a hazardous waste disposal site, near to one's home. Figure 11.1 depicts the responses.[9]

The acceptability among the five facilities differs enormously. It appears that 31% would not object to a ten-storey office building irrespective of how close it would be. On the other hand only 5% did not mind how close to them a hazardous waste dump would be located. A majority of the respondents accepted the office building as close as one mile, whereas the necessary distance for the same rate of acceptance for a hazardous waste disposal site had to be at least 100 miles. The disposal site for hazardous waste chemicals and the nuclear

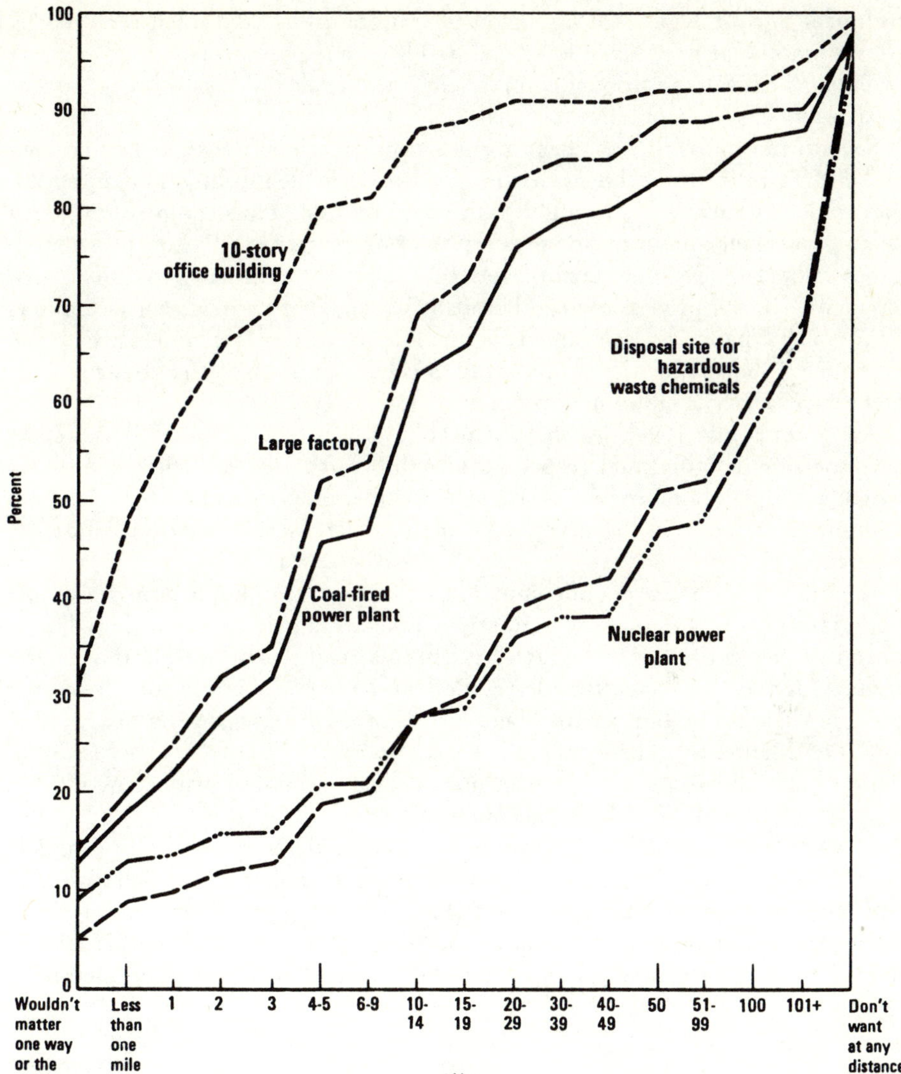

Figure 11.1 Cumulative Percentage of People Willing to Accept New Industrial Installations at Various Distances from their Homes

power plant seem to have a similar aversion profile as that of a large factory and a coal-fired power plant. In contrast the ten-storey office building is significantly less rejected in the near neighbourhood.[10]

These individual attitudes do not easily contribute to the solution of the problems at the societal level. Because of limited information on the complex technical and biological effects involved, rationality is very often replaced by strong emotional reactions. Some factors like the regional concentration of the problems generate additional complications for a social decision process, since it is often easy to mobilize protest.

Several mechanisms have been proposed to overcome the stalemate created by NIMBY attitudes.[11] The common idea is to provide incentives for accepting the facility by means of sharing the costs and burden. Thus compensations are offered and their amount leads to acceptance of the facility. The others who are happy not to have the facility nearby share the burden providing such payments. In the process of negotiations it is clear that strategic considerations by the participants are relevant. It is possible to misrepresent one's preference to improve one's position! Thus any mechanism design has to deal with preference and distributional problems.

An interesting issue in this respect was raised. Some studies report asymmetries in individual responses to gains versus losses. The evidence on whether people are more willing to pay to avoid an increase in risk or to gain an amount of money for the acceptance of risk increases is not decisive at the moment.[12]

Experience with automobile emission and safety regulation provides ample material to study the effects of regulation. There is ongoing research and debate to clarify the issues, some of which have been already mentioned in the general context above. The fact that legislation and governmental action not only provide regulations but set the stage for negotiation processes by distributing initial entitlements and specifying rules is important. Finally, it is necessary to point to the ethical questions concerning fairness, equity, or justice, involved in any distributional or risk management decision.

A number of related issues is under active study. Economists are trying to clarify the role of experts; they have started to analyse the delegation of decisions, they are continuing to search for allocation mechanisms, the effects of informational asymmetries. Several promising results are available. However, further work is required to integrate the results into a coherent picture.

Concluding remarks

From an economic perspective one may summarize the state of the art in the analytical handling of risk management of environmental hazards. Economic theory is not able to provide general solutions given the present state of knowledge. Economists are groping for solutions of special cases and trying to overcome the difficulties involved, some of which have been mentioned above. It is conceivable that there is no general solution to the design problem in the face of uncertainty. At least some general principles have been established by theoretical research. Without incentive compatibility it is not possible to reach stable solutions in the allocation of risks. In addition all decision mechanisms have to be informationally manageable and therefore implementable.

It is clear that society urgently needs solutions for serious environmental problems.[13] Environmental policies top the list of research agendas.[14] Despite sometimes shaky foundations and the complexity of the problems, it is necessary to formulate a comprehensive and coherent policy.[15] Partial approaches cannot provide a robust and stable basis for a satisfactory solution. Limited knowledge and uncertainty about the long-run consequences of behaviour make it doubtful that traditional institutional arrangements will suffice to secure an optimal development.

It is often thought that if the appropriate technology were developed the problem of say hazardous chemical waste would be solved. Numerous examples show, however, that even in this situation there has to be social consent about the application of technical knowledge, because some may have to take higher risks in order to relieve all others from detrimental influences. Therefore the critical social dimensions of environmental issues should not be underestimated.

That environmental issues are now well established in public opinion is an indication that there has been a dramatic change in the awareness of the long-run dangers facing mankind. Since pollution is a transboundary problem, an international response is necessary if it is to be dealt with effectively. It would be irresponsible to neglect the environmental problems of developing countries because currently they have no environmental movements to exert pressure on their policymakers. It should be possible to assist them in avoiding the mistakes which have occurred in industrialized countries' to the benefit of all nations.

Notes

1. For an excellent survey of the EU model, see Shoemaker, 1982.
2. See Arrow, 1982.
3. Slovic *et al.*, 1982, 466 "In general, rare causes of death were overestimated and common causes of death were underestimated. In addition to this general bias, sizeable specific biases were evident. For example, accidents were judged to cause as many deaths as diseases, whereas diseases actually take about 16 times as many lives. Homicides were incorrectly judged more frequent than diabetes and stomach cancer deaths. Homicides were also judged to be about as frequent as death by stroke, although the latter actually claims about 11 times as many lives. Frequencies of death from botulism, tornadoes, and pregnancy (including childbirth and abortion) were also greatly overestimated. . . In keeping with availability considerations, overestimated causes of death were dramatic and sensational, whereas underestimated causes tended to be unspectacular events, which claim one victim at a time and are common in nonfatal form."
4. See e.g. Jones-Lee *et al.*, 1985, 65.
5. Only in passing I want to point out that several generalizations of and alternatives to the EU model have been developed in order to deal with these violations of the basic axioms; see Kahneman and Tversky, 1979; Machina 1982, 1984; Loomes and Sudgen, 1982.
6. See Ross and Anderson, 1982.

7. Several contributions in the *American Economic Review,* Papers and Proceedings section, offer an introductory sample of opinions; see Bailey, 1982; d'Arge *et al.*, 1982; Dowing and Kates, 1982; Lave, 1982.
8. Slovic *et al.*, 1982, 475 ff. and with respect to nuclear power 485 ff.
9. The question asked was: "want to move to another place or to actively protest, or wouldn't it matter to you one way or another how close it was?" Council on Environmental Quality, 1980, 30.
10. The results are stated more extensively in Council on Environmental Quality, 1980, 30, 32: "People objected the least to the office building and the most to the nuclear power plant and hazardous waste disposal site. For the 10-storey office building, 31% said that it would not matter how near the building was located. Another 17% said that they would not be inclined to move or to protest if the building was located within one mile of their homes. As Figure 11.1 shows, at a distance of 14 miles, 88% were willing to accept the construction of a 10-storey office building; at the other extreme, 4% did not want it at any distance.

    In contrast, only 5% and 10%, respectively, of respondents did not care how near a hazardous waste dump or a nuclear power plant was located. Aversion to the siting of a disposal site for hazardous waste chemicals is very strong despite the fact that respondents were given special assurances about this kind of installation – that 'disposal could be done safely and that the site would be inspected regularly for possible problems.' A lower percentage of people accepted a disposal site within 5 miles of their homes than any of the other installations, although beyond 5 miles reactions to the disposal site and the nuclear power plant were almost identical. Neither reached majority acceptance until the distance exceeded 100 miles.

    People are much more averse to the siting of a nuclear power plant near their homes than they are to the siting of a coal-fired plant. Although only 28% would accept a nuclear power plant within 14 miles, 63% would accept a coal-fired plant within that distance.Taking into account the difference in wording of the question, the percentage of people who would accept a coal-fired plant within 5 miles is near that found by Harris in 1978. The acceptance profile for the coal-fired plant is very similar to that for a large factory, a fact which suggests that people evaluate them similarly as potential neighbours."
11. See Kunreuther and Kleindorfer, 1986; Mitchell and Carson, 1986.
12. See Smith and Desvousges, 1986; Knetsch and Sinden, 1984. An interesting methodological discussion is provided by Cummings *et al.*, 1986.
13. What was once regarded as beneficial and a great achievement, for example the durability of plastic, is now back to be a haunting headache; see Browne, 1987.
14. See Blomquist, 1982, for a survey of important theoretical issues.
15. A good introduction to both the technical and economic questions may be found in Tolley *et al.*, 1981.

References

Arrow, K.J. 1982. Risk Perception in Psychology and Economics. *Economic Inquiry,* January, 20, 1:1–9.

Bailey, M.J. 1982. Risk, Costs, and Benefits of Fluorocarbon Regulation. *Amercian Economic Review,* May, 72, 2:247–50.

Blomquist, G. 1982. Estimating the Value of Life and Safety: Recent Developments. *The Value of Life and Safety*. Jones-Lee, M.W. ed. Amsterdam e.a., 27–40.

Browne, M.W. 1987. World Threat of Plastic Trash Defies Technological Solution. *The New York Times,* September 6.

Council on Environmental Quality. 1980. Public Opinion on Environmental Issues. Results of a National Opinion Survey. Washington, DC.

Cummings, R.G., Brookshire, D.S. and Schulze, W.D. 1986. *Valuing Environmental Goods. An Assessment of the Contingent Valuation Method.* Totowa, NJ.

d'Arge, R.C., Schulze, W.D. and Brookshire, D.S. 1982. Carbon Dioxide and Intergenerational Choice *American Economic Review,* May, 72, 2:252–6.

Dowing, T.E. and Kates, R.W. 1982. The International Response to the Threat of Chlorofluorocarbons to Atmospheric Ozone. *American Economic Review,* May, 72, 2:267–72.

Jones-Lee, M.W., Hammerto, M. and Philips, P.R. 1985. The Value of Safety: Results of a National Sample Survey. *Economic Journal,* March, 95:49–72.

Kahneman, D., Slovic, P. and Tversky, A., eds. 1982. *Judgement under Uncertainty: Heuristics and Biases.* Cambridge e.a.

Kahneman, D. and Tversky, A. 1979. Prospect Theory: An Analysis of Decision under Risk. *Econometrica,* March, 47, 2:263–91.

Knetsch, J.L. and Sinden, J.A. 1984. Willingness to Pay and Compensation Demanded: Experimental Evidence of an Unexpected Disparity in Measures of Value. *The Quarterly Journal of Economics,* 99:507–21.

Kunreuther, H. and Kleindorfer, P.R. 1986. A Sealed-bid Auction Mechanism for Siting Noxious Facilities. *American Economic Review,* May, 76, 2:295–9.

Lave, L.B. 1982. The Global Commons II: Institutions and Decisions. Mitigating Strategies for Carbon Dioxide Problems. *American Economic Review,* May, 72, 2:257–61.

Loomes, G. and Sugden, R. 1982. Regret Theory: An Alternative Theory of Rational Choice under Uncertainty. *The Economic Journal,* December, 92:805–24.

Machina, M.J. 1982. "Expected Utility" Analysis without the Independence Axiom. *Econometrica,* March, 50, 2:277–323.

Machina, M.J. 1984. Temporal Risk and the Nature of Induced Preferences. *Journal of Economic Theory,* 33:199–231.

Mitchell, R.C. and Carson, R.T. 1986. Siting of Hazardous Facilities. Property Rights, Protest, and the Siting of Hazardous Waste Facilities. *American Economic Review,* May, 76, 2:285–90.

von Neumann, J. and Morgenstern, O. 1944. *Theory of Games and Economic Behaviour.* (2nd edn 1947) Princeton, NJ: Princeton University Press.

Ross, L. and Anderson, C.A. 1982. Shortcomings in the Attribution Process: On the Argues and Maintenance of Erroneous Social Assessments. *Judgement under Uncertainty: Heuristics and Biases.* Kahnemann, D., Slovic, P. and Tversky, A., eds. Cambridge e.a., 463–89.

Shoemaker, P.J.H. 1982. The Expected Utility Model: Its Variants, Purposes, Evidence and Limitations. *Journal of Economic Literature,* June, 20:529–63.

Slovic, P., Fischoff, B. and Liechtenstein, S. 1982. Facts versus Fears: Understanding Perceived Risk. *Judgement under Uncertainty: Heuristics and Biases.* Kahneman, D., Slovic, P. and Tversky, A., eds. Cambridge e.a., 463–89.

Smith, V.K. and Desvousges, W.H. 1986. Asymmetries in the Valuation of Risk and the Siting of Hazardous Waste Disposal Facilities. *American Economic Review,* May, 76, 2:291–4.

Tolley, G.S., Graves, P.E. and Blomquist, G.C. 1981. *Environmental Policy. Elements of Environmental Analysis.* Vol. I. Cambridge, Mass.

12
Costs and Benefits of Recycling

Sonia P. Maltezou

Recycling extends the life of resources and minimizes waste entering the environment. As a productive activity, it generates its own wastes. Since secondary resources are seldom homogeneous, recycling will rarely be a homogeneous process. The incentive to recycle secondary resources varies with the price of corresponding virgin resources; in other words, if these prices are high, the incentive to recycle is strong. If, however, a virgin resource input is cheap, the incentive to recycle will be less since recovering and recycling a secondary material may exceed the cost of the virgin material.

Other factors affecting recyling are the diversion between public and private interests in free-enterprise societies and the extent to which these interests influence legislated policy promoting or discouraging this economic activity. Although private production goals are relatively short-term economic objectives, the optimal allocation and conservation of resources according to social needs may require that public decision-makers plan these activities in a systematic, long-term manner. Consequently, the private and public sectors may diverge between short-term and long-term economic pursuits. Misallocation of resources and environmental externalities generated by microplanning may require microsolutions both for resource conservation and environmental damage problems.

As recycling is not necessarily an optimal economic activity in unregulated markets (from the private viewpoint), deliberate governmental planning may be needed in order to make it optimal and to ensure that sustained recycling activities are promoted. In essence, resources would be shifted from the present to the future in gradual steps, thus ensuring that the needs of future generations will, to a greater degree, be fulfilled.

Traditionally, environmental damage and resource waste have been dealt with as externalities. In treating them, various writers have suggested policy options to reduce or alter externalities at source (such as changes in the durability of products, adoption of technological processes producing less residues) or by recovering and recycling residues, or both. Recycling should not be confused with treatment of residual wastes, which alters the form of the materials but not the quantity.

The reduction of waste and pollution as well as the curtailment of waste, pollution, and the conservation of resources depends on the efficacy of solutions aimed at waste reduction in the production process and the efficacy of recycling activities aimed at reducing waste after waste generation. There are trade-offs between production processes generating less waste and reduction after the generation of waste. Assessment of the effectiveness of recycling as a component in a waste-management system requires consideration of costs and benefits from two different viewpoints: public and private.

From the private viewpoint, recycling will be undertaken to the extent to which it constitutes the most efficient allocation of privately owned resources. For a market system that has no imperfections, incentives or disincentives affecting the private decision to recycle will be purely economic. If regulations, fiscal tools, and other interferences alter the criteria in individual cases, then the motivation to recycle will be influenced accordingly.

On the other hand, the public decision to encourage recycling will depend not only on a cost–benefit evaluation of privately owned resources, but also on the social welfare function of a given country or region. The maximization of the public welfare function in terms of recycling a secondary resource in a given administrative area will be based on the cost–benefit analysis of both the public and private goods. These include net additions to the economy from rechannelling the wastes back to production and consumption; costs of converting the waste material into useful products; environmental damage if the waste materials were instead disposed of in the environment. Economic systems with regulated markets will, therefore, stimulate recycling activities to the extent that the benefits outweigh costs. A more detailed analysis of the variables that enter private and public decisions to recycle is presented below.

The Private Decision to Recycle

Recycling as the Main Productive Activity

Where recycling is the main productive activity, the use of virgin materials in production is auxiliary to the processing of secondary materials. In this situation an entrepreneur will not undertake recycling unless his expected benefits are greater than predictable costs. Analysis of the following factors would aid him in decision-making.

1. Recycling costs

   Recovery costs (including waste separation and decontamination, intermediate storage, and transportion); energy; virgin materials; reprocessing and other operating costs (including disposal of residues); capital cost

2. Available technology
3. Steadiness of flow of waste materials
4. Credit conditions
5. Government policy

Several factors govern the cost of recovery of waste materials. These are frequently nonhomogeneous and require costly separation and upgrading activities. The waste materials have to be stored until sufficient bulk or weight is attained to make transportation economical. Further, transportation costs will vary, increasing or decreasing in relation to the concentration or dispersion pattern of the waste materials.

Energy costs are an important variable. If the recyling of a given waste material requires large inputs of energy, this cost may become crucial in the decision whether or not to recycle. However, if the cost of energy is high but the amount required is small in comparison with that required to process virgin resources, then recycling will offer the additional incentive of larger profit margins. It should be noted that recycled products are usually priced lower than their virgin resource substitutes in order to compete in the market. The collection of secondary materials also requires energy. The crucial factor here is whether the market price of a unit of untreated waste material is higher than the costs of energy required for its collection.

Though waste materials will constitute the main input to the production process of a recycling plant, certain virgin materials will be needed. For example, wasteoil re-refiners often mix virgin oil with recycled oil to establish required or desired properties in the final product. The extent to which the prices of necessary virgin resources are within or beyond the entrepreneurs cost–benefit calculus will militate for or against recycling. The demand elasticity of the recycling unit, the existence of substitutes or near-substitutes, and the recycler's profit margins (including his ability to pass on increased costs) determine how crucial the prices of virgin resources as auxiliary inputs will be.

The cost of processing waste materials is particularly important among the operating costs of the recycling industry. This cost has an endogenous component (derived from whatever production process is selected), and an exogenous component (e.g. additives in lubricating oils have led to increases in recycling costs). Other operating costs (labour, materials, energy) normally increase or decrease in relation to general cost levels, the price of the particular inputs, labour market, union demands, and to the extent that certain recovery activities – such as separation of waste material – are performed in-plant. Substitutability and complementarity of materials and flexible margins of operating capacity will be relevant in the case of firms with rising operating costs.

Total operating costs also include the cost of disposing of residues generated during the recycling process. This cost depends on existing technological ability to re-channel residues back into internal (in-plant) or external recycling and on other available alternatives for disposal. For example, the cost of using residual waste for landfill is relatively lower than the cost of incinerating or neutralizing it. Again, the extent to which government regulations monitor the disposal of residues will increase or decrease the cost of the process. (In this case, the environmental factor *per se* is neglected: it is assumed that the entrepreneur is not concerned with the environment unless there are explicit penalties associated with its misuse.)

Available technology affects the decision to recycle. If it can be applied with reasonable capital investment and does not involve prohibitive operating costs, then there will be a greater incentive to recycle. The application of appropriate technology might significantly reduce such costs as in-plant separation, decontamination and other pretreatments of secondary materials, restoration of desirable qualities, reduction of residues or alteration of their composition to decrease their environmental harmfulness. Finally, technology, by improving the quality of the final product, could narrow the price gap between new and recycled materials (a gap may still exist – if only psychological) while simultaneously widening its market.

The steadiness of the flow of secondary materials into the production process is a desideratum the recycler cannot control or guarantee. In fact, producers of recycled products have been shown to possess much less control of the principal inputs of the recycling production process than do producers of products manufactured from virgin materials.

Credit conditions available to recyclers are affected by the general political and economic situation, by the individual institutional ratings of given recycling plants and by regulatory and incentive systems imposed by government. These conditions significantly influence the decision to start or expand recycling activities.

Government interference or noninterference with market forces can be detrimental, beneficial, or of mixed effect. If government chose to support producers and collectors of secondary materials, as well as suitable markets for recycled products, then several economic preconditions necessary for a variable recyling industry could be achieved.

The Private Decision to Recycle In-Plant by the Economic Unit Generating the Waste

Consideration of the feasibility of in-plant management of residual wastes involves an examination of factors influencing the quantity of production-generated residues and the quantity of residues recycled. It should be noted, however, that in-plant reuse of production-generated residues does not always presuppose recycling. The residues may be used for applications other than as product inputs.

The minimization to the technological limit of production-generated residuals depends primarily on the relative prices of production inputs, on the existence of government regulation and the degree of its enforcement contributing to the direct or indirect cost (nuisance value) of the disposal of residues. If the generation and disposal of residual wastes has zero cost there is no incentive for expanding the plant's technological capabilities to reduce it. However, if virgin materials and energy are costly and there is an explicit cost associated with disposal into the environment, the generation of residues will diminish to the technologically achievable minimum. Furthermore, there will be an incentive to improve technology through additional research and development.

In considering the factors influencing the quantity of residues recycled in a plant, it must be noted that although large volumes of reused residues imply low unit cost, in-plant reuse of residues depends on several factors other than the quantities generated. These factors include the costs of adapting existing equipment (for the reuse of residues) or of purchasing new equipment, separating or treating residues (if treatment prior to reuse is necessary), and purchasing virgin inputs and energy (if energy consumption in-plant increases with the reuse of residues). In addition, consideration must be given to the technology available, the type of residues, and government regulatory or incentive systems.

The availability of easily adapted existing equipment through which residues can be incorporated into the production mix is a strong incentive to reusing them: no additional funds or new investment are required. However, this benefit may be offset by an increase in maintenance costs.

The costs of recovering residues depends in part upon the type and degree of contamination they have sustained. Furthermore, the type of resource (whether it is exhaustible and nonrenewable, or replenishable) will influence the decision as to reuse. If the price of new materials or energy is expensive in relation to other inputs, any increase of in-plant energy consumption caused by reuse of residues becomes an important factor. Technology, in determining the use and adaptation of existing equipment, the use of residues, and the extent to which certain types of residual contamination can be dealt with, is an additional factor.

Finally, effluent charges, fines, taxes, or regulations about residues not recycled but discharged into the environment can directly increase the operating costs of the firm that does not control the disposal of residues. On the other hand, if government rewards the reuse of residues with tax rebates, then for a firm to forgo this benefit (reward + marketing return + savings in inputs are assumed to be greater than other costs of reusing materials) amounts to operating with higher cost.

The private decision to buy secondary or recycled materials and products will depend on:

1. Price differential between virgin and recycled materials and products
2. Consistency of supply of inputs of secondary materials and products and authoritative recommendation for their use

3. Adequacy of supply of inputs of secondary materials and sufficient output of recycled products to meet effective demand
4. Adequacy of available technology – whether or not recycled products will be of good quality; whether available technology requires costly replacement or new investment: and whether additional maintenance costs and higher rates of obsolescence are avoided
5. Government regulations and incentives systems: prohibition of or penalties on wasteful and environmentally hazardous disposal of waste resources; quotas for the purchase of new materials, higher depreciation allowance for firms using recycled materials and products (when use increases maintenance or investment needs); tax rebates

If available technology can produce substitutes (or near-substitutes) for virgin products, then the price differential will be the primary criterion for purchasers. It will become of secondary importance if not consistently sustained. Sustained price differential, sustained volume quantity and consistent quality of the recycled inputs or products are particularly relevant if their purchasing involves replacement or new investment, if large-quantity purchasers are involved or if marketing conditions favour long-term contracts. Potential buyers of recycled products might be influenced adversely by the fear of imperilling existing relationships with long-term suppliers. In fact, purchasers might forgo the benefits of price differentials between virgin and recycled materials and products for this reason.

However, all these considerations and others that come readily to mind can be positively influenced. If government decides to apply rigorous policy tools, there is a substantial likelihood that the positive configuration of conditions required for expanding recycling could be created.

The Public Decision to Recycle

The new waste materials as well as the energy consumed in the production process of recycling will generate their own residues. There are, in fact, costs and benefits of varying significance from the public and private viewpoints. Public decisions to encourage recycling will occur when public benefits outweigh costs. The most readily identified public benefits of recycling are:

1. Reduction in the demand for new resources
2. Reduction in the flow of residues to the environment; reduction in the use of additional natural resources – air, water, land – for waste assimilation (the use of the environment for disposal of residues competes with alternative goods, mainly recreation and health)
3. Reduction in the dependency on imported materials
4. Net additions to the national product

As noted earlier, the production processes of recycling produce residual wastes that cannot in all cases be recycled. This may cause environmental damage which in itself constitutes a public or social cost of recycling. In addition, the collection of waste materials may have an environmental or resource cost that can be expressed as a social cost.

Assessment of the socially optimal rate of recycling requires data on both virgin and secondary material flows. Environment damage functions (and also energy consumption) for the production of virgin and recycled materials must be assessed. However, the feasibility for properly assessing the cost of environmental damage can be questioned.

Traditional cost–benefit analysis fails to estimate adequately the damage that pollution can work upon the world's ecological systems and upon the health and lives of present and future generations. With the clear understanding that the orthodox cost–benefit approach cannot be a guideline for the formulation of the social decision of promoting recycling, traditional economic criteria should be revised to reflect environmental and resource considerations, thus, more equitably distributing the costs and benefits between present and future generations.

Finally, public decisions to use recycled products in government services will depend primarily on the quality of the recycled products and government directives promoting their marketing through appropriate legislation. In some instances, public decisions have been expressed through procurement guidelines issued to government agencies regarding given categories of products and including product specifications.

The public or social decision to recycle, to promote recycling and to buy recycled materials and products depends not only on the identified social costs and benefits produced through recycling, but also on the basic factors the private decision-makers, as described earlier, take into account.

Recycling is gaining additional support as recent studies have established that the external environmental cost of virgin materials are very frequently higher than those of recycled materials. Furthermore, the cost of depleting natural resources and the external cost of depriving future generations of the use of the materials cannot be neglected.

This analysis of the public and private considerations in deciding whether to recycle and whether to use recycled materials and products, indicates that the benefits involved in formulating the public decision to recycle are probably greater (since external costs are taken into consideration) than the benefits entering the private decision.

As already stressed, a key incentive for recycling secondary materials is the relative prices of resources entering the production process. If a virgin resource is readily available and reasonably priced in relation to other resources, little economic incentive exists for conservation and less for recycling.

The cases in which virgin materials are cheaper than secondary recovered materials should be understood to depend sometimes upon special circumstances. Virgin materials are possibly not taxed, or priced at a level sufficient to reflect the social cost of their extraction and the residues subsequently generated as well as the social cost of their depletion. Artificial incentives may lower their cost. Royalties for exhaustible resources may be abnormally low because of imperfect economic knowledge, and there may be negative user costs due to high rates of extraction or low labour costs.

Each resource constitutes a special case with different economic variables and physical and biological limitations: all require individual consideration. To prescribe arbitrarily a universal or comprehensive policy for all secondary materials and all recycled products is impossible. Government, therefore, in considering the tools and scope of possible policies should aim to create and maintain a balance between the costs of new and used materials. The tools employed can be both direct controls and fiscal regulation.

Clearly there is interaction between the private and public sectors. Private practices and decisions influence public policies and strategies, which alternatively alter the decisions and practices of the private sector. These interdependencies and interactions between the two sectors indicate the necessity for an integrated approach to pollution control problems incorporating both private and public interests.

13

Cost-Effective Management of Hazardous Chemicals

Jorg Aadahl

Societal Importance of Chemicals

Chemicals represent the main key to an improved standard of living for the majority of the world's population. Our dependence upon chemicals is manifested in such areas as health, sustenance, shelter, apparel, transportation, communications and publishing, recreation and music, to name but a few areas where great advances in technology have been made over the last few years. However, our dependency on chemicals includes a long list of hazardous ones, that threaten people's health and endanger animal life and the environment. There are short-term risks as well as longer term dangers. Some health hazards may not be apparent until after decades of exposure. The fast-growing high-technology electronics industry is a prime example of the multifaceted threat from chemicals so essential to the various production processes. We gain little from an increased standard of living unless goods and services can be produced safely.

Management Responsibility

As industrialists, managers and supervisors and as engineers and designers, we have no right to employ potentially dangerous materials and methods unless every possible step has been taken to protect people, facilities and the environment. We are responsible for *all* the resources that we employ, as well as to society for permitting us to engage in the various activities we have chosen. Many mistakes have been made throughout the history of industrialization and they are still occurring with alarming frequency. We read and hear about ecological disasters, chemical fires with disastrous effects, and chemically induced illnesses. Although we now have considerable knowledge about the risks associated with the chemicals we use, much is still unknown and will remain so for a long time. New chemical products are constantly being developed, and it takes time before the various risk factors are discovered and documented.

However, we are currently sufficiently well informed to protect people and the environment. We do not always need the scientific details that years of research will eventually bring. It is a matter of acknowledging our responsibility and employing what we know. And, there is absolutely no reason why developing countries should make the same mistakes that have already been made several times over by the more "developed" nations. Safety ought to be adopted as a prerequisite for the privilege to produce. Industrialization should, by definition, include the element of resource protection: people, facilities, environment, and the society at large. The more industrialized countries should not be permitted to move their production to less industrialized areas because the safety regulations are weaker.

We do not have the right to neglect proper safety precautions because those we employ are ignorant of the dangers. It is *our* responsibility to inform. It is our responsibility to find the best methods and tools available and put them to effective use. It may seem more costly in the short term to be safety conscious. However, in the longer term production is more economical if it can be done safely. It is also the *only* ethical way to conduct business.

Chemical Management System

SAFECHEM II has been developed as an administrative tool for those who deal with hazardous chemicals. It is a cost-effective solution to the problem of obtaining the right chemical information in a timely manner. The system is delivered as a self-contained database package that runs on personal computers (such as the IBM PC), it is menu-driven and user friendly (Figure 13.1). The system is preloaded with data for 1000 chemicals, which are described by means of 90 data categories. The user of the system can add his own inventory data and data for any chemicals or products not already covered by the system. Thus, the system can be tailored to each specific situation.

The system will keep track of the available chemicals, where they are employed and stored, their properties, how to store them correctly, how to protect the employees, and what to do in an emergency. Also, the system permits various analyses to be made and is geared to respond to the various questions about the management of the chemicals (Table 13.1).

Data Directory

The SAFECHEM system is a database divided into eight major sections: Chemical ID data, General data, Health data, Fire and Emergency Response data, Protection data, Reactivity data, Physical data, Site data (user added).

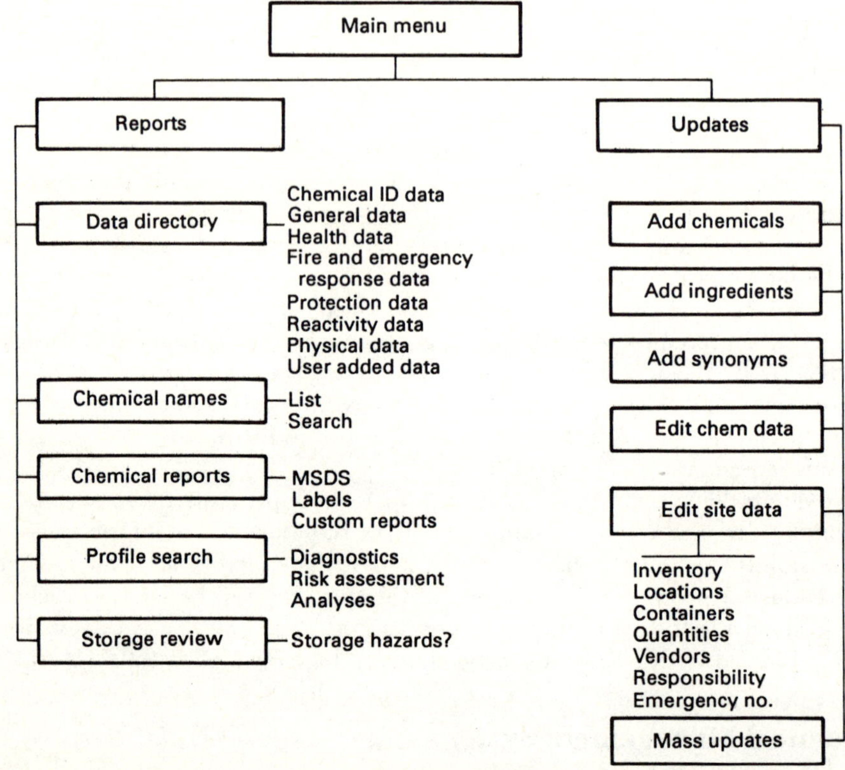

Figure 13.1 Safachem II Menu/Report Hierarchy (Copyright Safeware, Inc., 1987)

Each section consists of a number of related data categories, most of which consist of a series of preselected standard phrases or sentences. Internal references to appropriate characteristics exist in accordance with the nature of each individual chemical.

Planning and Emergency Response

The SAFECHEM system is designed for both planning and response. The information content permits careful planning of activities involving chemicals, so that personnel, facilities and the environment can be protected and risk and damage minimized, or ideally, avoided. However, when incidents occur, the system provides information about the proper response in terms of first aid, fire fighting, and spillage cleansing.

Chemical Reports

For individual chemicals covered by the system, including any added by the

Table 13.1 SAFECHEM II Data Categories (copyright Software, Inc., 4/87)

Chemical ID Data	*Health Data*	*Reactivity Data*	*Protection Data*
Chemical Name	Toxicity Index	Incompatible Chemicals	Osha Pel
Synonyms, Trade Names Family	Toxic Dose (LD50)	Incompatible Mixtures	ACGIH TWA
SAFECHEM #	Target Organs	Incompatible Classes	ACGIH STEL
EPA #	Routes of Entry	Incompatible Element Fragments	Niosh PEL
CAS #	Medical Warnings	Incompatible Radicals	Niosh IDCH
Niosh/RTECS #	Medical Symptoms	Hazardous Polymerization	Eye Protection
DOT/UN/NA #	Acute and Chronic Health Hazards	Polymerization Description	Protective Gloves
Ingredients	Medical Considerations	Stability	Ventilation
Formula	First Aid		Respiratory Protection
Chemical Classes			Resp. Threshold Levels
Active Element Fragments			Other Protection
Active Radicals			Hygienic Work Practices
User Part #		*Physical Data*	
		Appearance	
		Color	
		Odor	
General Data	*Fire and Emergency Response Data*	Boiling Point	*User Defined Data*
Hazard Classes	OSHA Flammability Class	Vapor Pressure (mm/Hg)	Status
DOT Classification	Flammability Index (NFPA)	Vapor Pressure Temp	Usage Locations
Storage Criteria	Reactivity Index (NFPA)	Vapor Density (AIR=1)	Storage Locations
Danger Level	Health Hazard (NFPA)	Solubility Description	Responsibility
Spill Cleanup Methods	Extinguishing Media	Solubility Value	Vendors
Disposal Methods	Special Fire Fighting Procedures	MOL Weight	Quantity
Environmental Listings	Unusual Fire & Explosion Hazards	Volatile by Volume (%)	Quantity Type
Reportable Quantities	Evacuation Description	Evaporation Rate (Butyl Acetate=1)	Container Type
Threshold Quantities	Evacuation Distance	Melting Point	Emergency Phone
Storage Precautions	Nature of Hazard	Flash Point	Part Number
Usage Precautions	Hazardous Decomp. Products	Flash Point Method	Internal Group Code
Regulatory References	Decontamination Procedures	Auto Ignition Temp	Revision Date
Warning Labels		Flammability Limit–Lower (%)	Verification Date
Literature References		Flammable Limit–Upper (%)	Revision Names
Special Notes		Specific Gravity	Verification Names
			User Notes
			Applications

user, one has a choice among standard data sheets, in-process label data, and customized reports.

Customized reports are useful for special needs where only selected data categories are of interest, for example a medical report, physical characteristics, emergency response data.

Profile Search

A unique feature of the SAFECHEM system is the profile search capability, where the system will find all chemicals with certain common properties or characteristics specified by the user. This could be a listing of all flammable chemicals in a particular building, or all carcinogenic, corrosive liquids with a potential impact on the human central nervous system, or the chemicals that result in the display of a certain set of medical symptoms.

Storage Review

SAFECHEM's storage review capability will identify all reactive material combinations in every storage facility, including laboratory room, or cabinet, if necessary. This feature can reduce drastically the risk of accidental mixing of incompatible chemicals.

Diagnostics

The profile search mentioned above can also be used to diagnose unknown materials, provided that the unknown is a chemical covered by the system. This would normally be the case in an industrial setting, where a spillage of an unknown substance is from the company's known inventory of chemicals. The system will then facilitate the narrowing down of potential chemicals by responding to each input of observed characteristics.

14
Legislative Measures to Prevent and Respond to Chemical Accidents

Will A. Irwin

Many kinds of legislation have been enacted since toxic and hazardous materials have become a focus of environmental policy in the last fifteen years. Emission limits or prohibitions on hazardous air pollutants and effluent limitations on toxic wastewater discharges have been adopted. Testing of new chemicals for short- and long-term human health and environmental effects before they may be placed into commerce is required, and programmes are underway to selectively test the most suspect existing chemicals already on the market. The generation, transportation, storage and disposal of hazardous wastes, sometimes called "special wastes", are regulated, and waste minimization and recycling are currently receiving special emphasis. Licensing requirements for pesticides and other intentionally toxic products have been made more rigorous, and notification of their characteristics to importing nations is now common. Occupational health laws regulate toxic substances in the work environment.

The use, storage, and transportation of poisonous, explosive, or flammable substances have long been regulated. Recently, prompted by serious accidents in Flixborough, Seveso, Beek, Velbert, Mexico City, Bhopal and Schweizerhalle, and hundreds of less spectacular accidents around the world annually, there is a new generation of laws designed to assess the risks posed by certain chemicals and other hazardous substances, prevent accidents from occurring where they are manufactured, stored, or used, and plan the response to such accidents if they do occur. This paper focuses on these new laws. It discusses their general characteristics, gives some examples of both preventive and responsive regulatory approaches, and suggests guidelines for developing nations that may wish to consider adapting such legislation to their circumstances.

Defining the Scope of the Legislation: General Characteristics

The focus of recent legislation has been on plants and depots that handle chemicals that pose risks or significant damage to public health – both inside the workplace and outside it – and to environmental resources. The consequences such legislation is designed to avoid are accidents that cause serious health effects, for example, death or disability, or contamination of the environment. Such accidents are distinguished from so-called "routine releases" and may be defined to limit them even further, for example to "major" accidents (i.e., those causing extensive damage or many casualties). Such an accident may also be defined as one ensuing from a disruption of normal operations and involving a fire, explosion or other dramatic release of a hazardous chemical.

Such laws may apply to a "facility", that is, the building(s), equipment, and contiguous area that generate, store, or handle one or more hazardous substances. (For certain purposes it may include means of transportation.) Or they may apply to an "industrial activity", that is, an operation in an industrial installation that involves a hazardous chemical. "Involving" such a substance may include its storage, use, or production as a product, byproduct, or waste. The operations included may be defined in terms of the industrial process employed at the plant, for example, oxidation or halogenation, or the substances processed or produced, for example, pesticides. Delivery of a chemical to a plant or transportation of it within the installation or storage of it in association with the installation may also be defined as a regulated industrial activity. Separate storage or temporary storage (e.g., for less than thirty days) or storage for certain purposes (e.g., for recycling) may be excluded, although storage in certain kinds of areas (e.g., floodplains, highly populated districts), may be included whether or not it is separate or temporary.

Usually whether a facility or activity is regulated is defined by whether it deals with more than minimum amounts of specified substances. Sometimes these substances and their threshold amounts are named in the laws themselves. More often they are generally defined, for example, material that, because of quantity, concentration, or physical or chemical characteristics, poses a significant present or potential hazard to human health and safety or to the environment if released. They are then specified in regulations promulgated under the authority of the law. The number and amounts of substances listed vary depending on the legal or scientific criteria for selecting them, or on the comprehensiveness of the regulatory programme undertaken. An initial programme may include only the most acutely toxic chemicals. Typically, toxic substances are included on the basis of the relative lethality of short-term exposure to them; explosive, reactive, and flammable substances are also usually included.

Some kinds of facilities that use, process, produce or store such substances may be exempted from regulation under these laws. Laboratories or specially designated areas used primarily for research, development, and testing activities involving hazardous substances rather than for production of goods for commercial sale are sometimes excluded from the definition of facility. Military installations, nuclear plants, explosives manufacturing or storage, mining operations, and hazardous waste disposal facilities are also often excluded.

Once such legislation goes into effect, existing facilities that have the substances listed are required to provide government authorities with information about them. This information must be kept up to date and must also be submitted by a previously unregulated facility if it is modified in a way that brings it within the ambit of the law or if it acquires a listed substance for the first time.

The nature of the requirements imposed by various regulatory programmes is elaborated below.

Preventive and Responsive Legislation: the 1982 "Seveso" Directive of the Council of the European Communities

The Provisions of the Directive

In June 1982 the Council of the European Communities adopted Directive 82/501/EEC concerned with "the prevention of major accidents which might result from certain industrial activities and with the limitation of their consequences for man and the environment".[1] Industrial activities are defined as "any operation carried out in an industrial installation referred to in Annex I [of the Directive] involving, or possibly involving, one or more dangerous substances and capable of presenting major-accident hazards, and also transport carried out within the establishment for internal reasons and the storage associated with this operation within the establishment".[2] Storage at installations other than those covered by Annex I – so-called "isolated storage" – is also an industrial activity if it occurs under conditions specified in Annex II of the Directive.

Annex I refers to six kinds of installations:

1. For the production or processing of organic or inorganic chemicals using any one of 14 specified processes as well as for the manufacture of phosphorus-containing compounds and for the formulation of pesticides and pharmaceutical products.[3] Installations that use distillation, extraction, solvation or mixing to process organic and inorganic chemical substances are also included in this category
2. For distillation, refining or other processing of petroleum or petroleum products
3. For total or partial disposal of solid or liquid substances by incineration or chemical decomposition
4. For the production or processing of energy gases (e.g., LPG, LNG and SNG)[4]
5. For dry distillation of coal or lignite and
6. For the production of metals or non-metals by the wet process or by means of electrical energy

Annex II covers storage of more than specified quantities of several named substances (e.g., flammable gases, chlorine, liquid oxygen) in groups of installations belonging to the same manufacturer "where the distance between the installations is less than approximately 500 m[eters]" and to each installation or group of installations belonging to the same manufacturer "where the distance between the installations is not sufficient to avoid, in foreseeable circumstances, any aggravation of major-accident hazards".[5] The amounts specified (in tons) are lower for the application of the obligations of Articles 3 and 4 of the Directive than for those of Article 5 (discussed below).

Excluded from the coverage of the Directive are nuclear installations and plants for the processing of radioactive substances and material, military installations, the manufacture and separate storage of explosives, gunpowder and munitions, extraction and other mining operations, and installations for disposal of toxic and dangerous wastes that are covered by other Community Acts to the extent where their purpose is to prevent major accidents.[6]

A major accident is defined as an occurrence such as a major emission, fire or explosion resulting from uncontrolled developments in the course of an industrial activity, leading to a serious danger to man, immediate or delayed, inside or outside the establishment, and/or to the environment, and involving one or more dangerous substances.[7]

Dangerous substances are also defined differently for the purposes of Articles 3 and 4 and Article 5. For Articles 3 and 4 they are those that are generally considered to fulfill criteria set forth in Annex IV. These criteria vary for "very toxic" and other toxic substances;[8] criteria are also set forth for flammable and explosive substances.[9] For Article 5, there are two lists of substances: the list of substances in Annex II (i.e., those whose isolated storage is defined as an industrial activity, see note 5, *supra*) and a list of 178 substances contained in Annex III. For each list, amounts are specified for each substance.

Article 3 requires each member nation of the Community to adopt provisions ensuring that any person in charge of an industrial activity is "obliged to take all the measures necessary to prevent major accidents and to limit their consequences for man and the environment". Article 4 requires member nations to take the measures necessary to ensure that any person in charge of an industrial activity is required to prove to competent authorities at any time that he has:

1. Identified existing major-accident hazards
2. Adopted the appropriate safety measures, and
3. Provided persons working on the site with information, training and equipment to ensure their safety

Article 5 requires member nations to introduce measures that require a person in charge of an industrial activity that involves or may involve any of the substances listed in Annex III in the quantity specified there or in charge of an isolated storage installation with an amount of a substance specified in Annex

II that is greater than the quantity listed there for the purposes of Article 5 to provide notification to competent authorities appointed for the purposes of receiving the notifications and for other purposes set forth in Article 7 of the Directive, described below.[10] The notification must provide information about: the substances listed, the installations, and possible major-accident situations.

Information about the substances that must be contained in a notification includes:

1. The data and information listed in Annex V of the Directive, that is, their chemical name, CAS number, name according to IUPAC nomenclature, other names, empirical formula, composition, degree of purity, main impurities and relative percentages, detection and determination methods available to the installation, description of the methods used or references to scientific literature, methods and precautions laid down by the person in charge of the installation in connection with handling, storage and fire, emergency measures laid down by the person in charge in the event of accidental dispersion, and methods available to the person in charge for rendering the substances harmless.[11,12] Annex V also calls for brief indications of hazards, immediate and delayed, for man and the environment
2. The stage of the activity in which the substances are involved or may be involved
3. The quantity (order of magnitude)
4. The chemical and/or physical behaviour under normal conditions of use during the process
5. The forms in which the substances may occur or into which they may be transformed in the case of foreseeable abnormal conditions; and, if necessary
6. Other dangerous substances whose presence could have an effect on the potential hazard presented by the relevant industrial activity

The information about an installation that must be contained in a notification includes:

1. The geographical location of the installations and predominant meteorological conditions and sources of danger arising from the location of the site
2. The maximum number of persons working on the site of the establishment and particularly of those persons exposed to hazard
3. A general description of the technological processes
4. A description of the sections of the establishment which are important from the safety point of view, the sources of hazard and the conditions under which a major accident could occur, together with a description of the preventive measures planned
5. The arrangements made to ensure that the technical means necessary for the safe operation of the plant and to deal with any malfunctions that arise are available at all times

The information about a possible major-accident situation that must be provided in a notification includes:

1. Emergency plans, including safety equipment, alarm systems and resources available for use inside the establishment in dealing with a major accident
2. Any information necessary to the competent authorities to enable them to prepare emergency plans for use outside the establishment, as required by Article 7(1) of the Directive and

3. The names of the person and his deputies or the qualified body responsible for safety and authorized to set the emergency plans in motion and to alert the competent authorities

A notification for a new installation must be submitted a reasonable length of time before its activity begins. Notifications must be updated periodically to reflect new technical knowledge relative to safety and the development of knowledge concerning the assessment of hazards.[13]

Member nations are to take appropriate measures to ensure that if an industrial activity is modified in a way that could have significant consequences for major-accident hazards, the person in charge must revise the measures taken in accordance with Articles 3 and 4 and must inform the competent authorities about the modification in so far as it affects the information contained in the notification submitted in accordance with Article 5.[14]

Article 7 requires member nations of the Community to establish or appoint competent authorities who are responsible for receiving and examining the notifications submitted in accordance with Article 5 and the information about modifications relevant to such notifications. The authorities are responsible also for ensuring that an emergency plan is prepared for action outside an establishment for whose industrial activity a notification has been submitted. If necessary they are to request supplementary information and ascertain that the person in charge of an industrial activity takes the most appropriate measures to prevent major accidents, given the various operations involved, and to provide the means for limiting the consequences of such an accident. Competent authorities are also to organize inspections or other means of control suitable to the kind of industrial activity involved.

Member nations are to ensure that persons who may be affected by a major accident from an industrial activity that must submit a notification in accordance with Article 5 are informed of the safety measures and of correct behaviour in the event of an accident. This same information must be provided to other member nations concerned, as a basis for consultations within the framework of their bilateral relations.[15]

Article 10 of the Directive requires member nations to ensure that as soon as a major accident occurs the person in charge of an industrial activity must immediately inform the competent authorities and must provide information, as soon as it becomes available, on the circumstances of the accident, the dangerous substances involved, the data available for assessing the effects of the accident on man and the environment, and the emergency measures taken. The person in charge must also inform the competent authorities of the steps planned to alleviate medium- and long-term effects of the accident and to prevent any recurrence of such an accident.

Competent authorities are required to ensure that necessary emergency, medium- and long-term measures are taken and, where possible, to collect the information necessary for a full analysis of the accident and, possibly, to make recommendations.[16]

Article 11 of the Directive requires member nations to inform the Commission of the European Communities as soon as possible of major accidents which have occurred within their territories. For each accident a member nation is to provide the Commission with information, as soon as it becomes available, about where and when it happened, the type of industrial activity, the type of accident, the substances emitted, a description of the circumstances, the emergency measures taken, the causes (these may be supplied later if they are not yet known), the nature and extent of the damage both inside and outside the establishment (including casualties, persons exposed, material damage, and whether or not the danger still exists), and the medium- and long-term measures taken, especially those aimed at preventing the recurrence of similar major accidents.[17]

The Commission is to establish and maintain at the disposal of the member nations a register containing a summary of the major accidents which have occurred, including an analysis of their causes, the experience gained and measures taken, so that member nations may use this information for prevention purposes.[18]

A member nation is also to inform the Commission of the name of an organization that might have information on major accidents that could advise competent authorities of other member nations which have to intervene in the event of such an accident. Member nations may also notify the Commission of any substance which in their view should be added to Annexes II and III and of any measures they may have taken concerning such substances. Such information is to be forwarded to other member nations by the Commission.[19]

Information received by the competent authorities in accordance with the provisions of Articles 5, 6, 7, 9, 10, and 12 of the Directive may not be used for any purpose other than that for which it was requested and the same prohibition applies to information obtained by the Commission under Article 11. [20] Commission employees may not divulge information obtained under the Directive, and employees of competent authorities may not divulge any information they obtain from the Commission, except in connection with the register provided for by Article 12 and in connection with the provisions of Article 18. [21] (Article 18 provides that member states and the Commission are to exchange information on the experience acquired with regard to prevention of major accidents and the limitation of their consequences, especially the functioning of measures provided for by the Directive. This information is to be the basis of a report by the Commission in 1987 on the application of the Directive.) These prohibitions do not preclude publication by the Commission of general statistical data or information on matters of safety that contains no specific details regarding particular undertakings or groups of them and does not jeopardize industrial secrecy.[22] Nor do they apply to the provisions of Article 20, which requires member nations to communicate to the Commission the provisions of national law which they adopt in the field covered by the Directive.

An Example of Directive Implementation and Proposed Improvements: The Federal Republic of Germany

EEC Directives are binding on member nations as to the results to be achieved; they are implemented by national legislation or regulations. One example of the implementation of the Seveso Directive – and of some ways its effectiveness could be enhanced – is the Twelfth Ordinance Implementing the Federal Emission Protection Act in the Federal Republic of Germany, known as the "Störfall-Verordnung".[23] The Federal Emission Protection Act, enacted in 1975, requires industrial facilities, as well as properties on which substances can be stored that could cause emissions, to be licensed.[24,25] Installations subject to licensing shall be established and operated so that 1) harmful environmental effects and other dangers, considerable disadvantages and burdens for the general public and the neighbourhood cannot be caused, 2) precautions against harmful environmental effects are taken, and 3) residues created by the operation of the facility are properly and harmlessly used or, if this is not technically possible or economically feasible, properly disposed of as waste.[26] The Act authorizes the Federal government to prescribe ordinances that the establishment, condition and operation of installations subject to licensing must comply with in order to fulfill these three general obligations, especially that the facilities meet specific technical requirements, that emissions from the facilities do not exceed specified limits, and that operators of the facilities must conduct or allow measurements of emissions from the facilities.[27] The Twelfth Ordinance was promulgated on the basis of this authorization.

The Ordinance applies to facilities subject to licensing under the Act that are listed in Appendix I of the Ordinance in which substances listed in Appendix II are present during normal operations or can arise during abnormal operations. The Ordinance does not apply to facilities in which the amount of these substances is so small that a public danger as a result of a disturbance of normal operations is apparently impossible.[28] The Bundesrat of the Federal Republic of Germany has recommended three changes to the applicability of the Störfall-Verordnung as necessary for improved precautions for environmental protection: 1) expansion of its applicability to all facilities subject to licensing under the Act in which Appendix II substances are present or can arise; 2) extending the list of facilities in Appendix I of the Ordinance; and 3) extending the list of substances in Appendix II of the Ordinance, especially to include substances that endanger water and soil.[29]

The Ordinance requires that the operator of a facility take the measures to prevent a *Störfall* that are necessary in view of the nature and extent of possible dangers. In fulfilling this obligation, he is to consider operational sources of danger, sources of danger from the surroundings, for example, earthquakes or floods, and acts of unauthorized persons.[30] In addition, precautions are to be taken to keep the effects of *Störfall* as small as possible.[31] To fulfill these obligations, in addition to complying with the requirements of sections 4 and 5 of the Ordinance, the operator of a facility must 1) continuously check it from a safety viewpoint and regularly maintain it, 2) conduct maintenance and repair work in accordance with generally recognized rules of technology, 3) take the required safety measures to avoid errors in operations, 4) prevent, by suitable operational and safety guidance and by training of personnel, mistaken conduct, and 5) instruct affected employees concerning the rules for their conduct provided in the facility's emergency plan in case of a *Störfall*.[32,33] The Bundesrat has recommended promulgation of administrative regulations to make the duties and requirements of sections 3–6 of the Ordinance concrete, especially those for facilities' emergency plans referred to in sections 5 and 6. [34]

The Ordinance provides that the operator of a facility must prepare a safety analysis that contains 1) a description of the facility and its process, including indicative process conditions during normal operations; 2) a description of the important parts of the facility from the safety technology standpoint, the sources of danger, and the circumstances under which a *Störfall* could occur; 3) the chemical name, state, and amount of Appendix II substances present in the facility in normal operations or that could arise during a disturbance of normal operations; 4) information on the effects that could arise from a *Störfall*; and 5) a presentation on how the requirements of sections 3–6 are being fulfilled.[35] According to the Ordinance, the operator of a facility is to keep such a safety analysis available and to present it to a competent authority upon request.[36] The Bundesrat has recommended that safety analyses as well as important information about safety, substances, and analytical procedures and references, be filed with the competent authorities of the *Länder*. It has also recommended authorization of evaluations and safety analyses by expert third parties at the cost of the facility as well as the introduction of a requirement for regular inspections by experts (without limiting the authority of government officials).[37]

The Ordinance requires an operator to inform the competent authority without delay of the occurrence of a *Störfall* or a disturbance of normal operations which causes the release or explosion of an Appendix II substance, and to provide a written report to the competent authority within a week. The report of a *Störfall* must describe the incident and its causes and effects in such a way that they can be evaluated from a safety technology point of view and the measures to control the *Störfall*, limit its effects, and prevent its recurrence. The written report of a disturbance of normal operations must also enable an evaluation of the circumstances from the safety technology point of view and must relate the measures taken to prevent a *Störfall*.[38] The Bundesrat has recommended that this duty to report be expanded to include disturbances of normal operations that pose more limited risks of danger.[39]

The Bundesrat has also recommended imposing a duty on warehouses to keep a current list of the amounts of dangerous substances on hand and their location and of the means to be employed to extinguish a fire. In addition, it recommends investigating the possibilities for improving the exchange of experience and knowledge about the field, for example, via a faciilty safety commission.[40]

Other suggested amendments to the Directive call for applying its provisions to a wider scope of isolated storage facilities and adopting common requirements for the safety reports prepared in accordance with Article 7. Measures for compensating for damage due to pollution are also being discussed.

Emergency Response and Planning Legislation

The U.S. "Emergency Planning and Community Right-To-Know Act of 1986" became law in October 1986.[41] It requires the U.S. Environmental Protection Agency to publish within thirty days of enactment an existing list of extremely hazardous substances, and to establish a "threshold planning quantity" for each substance, taking into account its toxicity, reactivity, volatility, dispersability, combustibility, or flammability.[42,43]

The Act requires the governor of each state to appoint a state emergency response commission.[44] The commission is to designate emergency planning districts to facilitate preparation and implementation of emergency plans, and to appoint members of a local emergency planning committee for each district.[45] Each committee must include representatives from elected state and local officials; law enforcement, civic defence, firefighting, first aid, health, local environmental, hospital, and transportation personnel; broadcast and print media; community groups; and owners and operators of facilitites subject to the Act.[46]

The owner or operator of a facility at which a substance on the list of extremely hazardous substances is present in an amount greater than the threshold planning quantity must inform the commission of the state in which the facility is located within seven months of enactment of the Act.[47] If the list is revised or if the facility acquires a listed substance for the first time, the owner or operator must notify the commission within sixty days that the facility is subject to the Act.[48] A commission (or governor) may designate additional facilities, after public notice and an opportunity for comment, and must notify the Administrator of the U.S. Environmental Protection Agency of each notification it receives from a facility and each designation it makes.[49]

Each facility subject to the Act must notify the local emergency planning committee for the district in which it is located of a facility representative who will participate in the emergency planning process as a facility emergency co-ordinator. [50] The owner or operator of a facility must provide the committee, upon request, with information necessary for developing and implementing the emergency plan, and must inform the committee promptly of any relevant changes at the facility.[51]

The local emergency planning committee must complete preparation of a plan no later than two years after enactment of the Act and must review it once a year, or more often if changed circumstances in the community or at a facility require.[52] Each plan must include:

1. Identification of facilities subject to the requirements of the Act that are within the emergency planning district, of routes likely to be used for the transportation of substances on the list of extremely hazardous substances, and of additional facilities contributing or subjected to additional risk due to their proximity to facilities subject to the requirements of the Act, such as hospitals or natural gas facilities.
2. Methods and procedures to be followed by facility owners and operators and local emergency and medical personnel to respond to any release of such substances
3. Designation of a community emergency co-ordinator and facility emergency co-ordinators, who shall make determinations necessary to implement the plan
4. Procedures providing reliable, effective, and timely notification by the facility emergency co-ordinators and the community emergency co-ordinator to persons designated in the emergency plan, and to the public, that a release has occurred
5. Methods for determining the occurrence of a release, and the area or population likely to be affected by such a release
6. A description of emergency equipment and facilities in the community and at each facility in the community subject to the requirements of the Act, and an identification of the persons responsible for such equipment and facilities
7. Evacuation plans, including provisions for a precautionary evacuation and alternative traffic routes
8. Training programmes, including schedules for training of local emergency response and medical personnel and
9. Methods and schedules for exercising the emergency plan[53]

A federal guidance document is available for the preparation and implementation of emergency plans.[54] The committee must forward its plan to the state emergency response commission for review and recommendations to the committee or revisions that may be necessary to co-ordinate the plan with those of other districts.[55]

Related to the information that must be submitted for emergency planning purposes, each facility that is required to prepare or have available a material safety data sheet (MSDS) for a hazardous chemical under the Occupational Safety and Health Act of 1970 (OSHA) must submit an MSDS for each such chemical *or* a list of such chemicals to the local emergency planning committee, the state emergency response commission, and the fire department with jurisdiction over the facility within one year of enactment of the Act. [56, 57, 58] (Currently, only manufacturers and importers in Standard Industrial Code categories 20–39 are subject to these OSHA requirements). If a list is submitted, it must include 1) the list of chemicals, grouped in categories of health and physical hazards, 2) the chemical name or common name of each chemical as provided on the MSDS, and 3) any hazardous component of each chemical.[59] Because the facilities that are subject to the emergency planning provisions decribed above may not include facilities subject to the OSHA requirements, the information available from MSDS may be useful to committees preparing plans. Committees, commissions, and fire departments may find it more practical to work with lists than the MSDSs until the latter become useful. The U.S. Environmental Protection Agency may establish threshold quantities for such hazardous chemicals below which no facility is subject to the requirement to submit an MSDS or list (or an inventory form, described below).[60]

In addition, beginning March 1, 1988, and annually thereafter, such facilities must also submit to the same entities an emergency and hazardous chemical inventory form containing information for the preceding year on 1) estimates of the maximum amount of chemicals in each category of health and physical hazards present at the facility at any time, 2) estimates of the average daily amounts of such chemicals, by category, present at the facility, and 3) the general location of hazardous chemicals in each category.[61] Upon request by a committee, commission, or fire department with respect to a specific facility, the owner or operator must provide, in addition to this information, the chemical name or common name of the chemical as provided on the MSDS, a brief description of the manner of storage of the chemical, the location at the facility of the chemical, and an indication of whether he elects to withhold information on its location from the public.[62] The fire department with jurisdiction over a facility that submits an inventory form may, upon request, conduct an on-site inspection of the facility and may require the owner or operator to provide it with specific location information of hazardous chemicals at the facility.[63] Information available from inventory forms, like MSDS data, may be useful in preparing emergency plans that cover facilities not subject to emergency planning provisions.

Another provision of the Act that requires information useful in preparing emergency plans, especially in listing methods for determining if a release has occurred and identifying the area and population most likely to be affected, provides that if a release of an extremely hazardous substance occurs from a facility at which a hazardous substance is produced, used, or stored, notification must be given by the facility owner to the community emergency co-ordinator for the local emergency planning committees for the districts likely to be affected and to the commission of any state that is likely to be affected.[64, 65] The notice shall include the name of the chemical released, whether it is on the list of substances to which emergency planning procedures apply, an estimate of the quantity that was released, the time and duration of the release, the medium or media into which the release occurred, any known or anticipated health risks, along with advice for medical attention to individuals exposed, precautions to be taken as a result of the release, and the name and phone number of the person providing the information.[66] A written follow-up notice must be submitted as soon as practicable after the release providing this information and any additional information about actions taken to respond to and contain the release.[67]

Preventive Legislation: The New Jersey Example

The State of New Jersey's "Toxic Catastrophe Prevention Act" came into effect in January 1986. It applies to "extraordinarily hazardous substances". These are defined as any substance or chemical compound that is used, manufactured, stored, or capable of being produced from on-site components in sufficient quantities at a single site so that its release to the environment would produce a significant likelihood that persons exposed will suffer acute health effects resulting in death or permanent disability.[68] The New Jersey legislature named eleven of them and required the Department of Environmental Protection (Department) to adopt a regulation with a list of them within 18 months.[69, 70]

Each facility that generates, stores, or handles any of the substances on the initial list (named by the legislature) was required to submit a registration form that provided:

1. An inventory of the substance(s) and the quantity(ies); the inventory is to identify whether the substances are end products, intermediate products, byproducts, or waste products
2. A general description of the processes and principal equipment involved in the management of the substance(s)
3. A profile of the area in which the facility is situated, including its proximity to population and water supplies
4. The extent to which the risks and hazards of the processes, equipment, and operations have been identified, evaluated, and abated – and the expertise and affiliation of the evaluators and any direct or indirect relationship between the evaluators and the owner or operator of the facility and
5. The names of all insurance carriers underwriting the facility's environmental liability and workers' compensation insurance policies, and the scope of these policies, including any limitations and exclusions[71]

Within three months of the adoption of the list of additional substances, facilities generating, storing, or handling them are to submit a registration form for these substances.[72]

If the owner or operator of any facility that has submitted a registration form has an established "risk management programme", he may submit it to the Department for review. A risk management programme means all programmes to minimize extraordinarily hazardous accident risks. Such a risk is defined as a potential for release of an extraordinarily hazardous substance into the environment which could produce a significant likelihood that persons exposed may suffer acute health effects resulting in death or permanent disability.[73] A risk management programme may include requirements for safety review of the design for new and existing equipment, for standard operating procedures, for preventive maintenance programmes, for operator training and accident investigation procedures, for risk assessment for specific pieces of equipment or operating alternatives, for emergency response planning, and internal or external audit procedures to ensure that programmes are being executed as planned.[74]

If the Department finds the risk management programme has any material deficiencies or omissions, it must recommend changes or additions to the owner or operator, and the owner or operator must submit his proposed revisions to the plan within sixty days. If the Department and the owner agree on the changes needed, they enter into a consent agreement. If they do not, the Department must conduct a hearing, prepare written findings of fact, and issue an administrative order requiring changes or additions needed to correct the deficiencies. Cost effectiveness, risk reduction effectiveness, and technical feasibility of any changes or additions recommended by the Department or owner must be considered.[75]

For each facility which does not have a risk management programme as a result of these procedures, the Department, in co-operation with the owner or operator, shall develop an "extraordinarily hazardous risk reduction work plan". In this plan must be reported:

1. The identity and quantity of all substances generated, stored, handled, or that could unwittingly be produced in the event of an equipment breakdown, human error, design defect, procedural failure, or the imposition of an external force
2. The nature, age, and condition of all equipment and instruments involved in the handling and management of the extraordinarily hazardous substance(s) at the facility, and the schedules for their testing and maintenance
3. The measures and precautions designed to protect against the intrusion of external forces and events, or to control or contain discharges within the facility
4. The circumstances that would have to exist in order for the discharge of a substance to occur, and the practices, procedures, and equipment designed to forestall such an event
5. Any alternative processes, procedures, or equipment which might reduce the risk of a release of a substance while yielding the same or commensurate results, and the specific reasons they are not employed
6. Any training or management practices in place which impart knowledge to personnel regarding dangers posed by a release of a substance and the training provided to prepare them for the safe operation of the equipment and for unanticipated occurrences and
7. Any other preventive maintenance measures or on-site emergency response capability or other internal mechanism developed to safeguard against the occurrence of an accidental release of a substance[76]

On the basis of this work plan, any facility without an approved risk management programme shall be subject to an "accident risk assessment" either by the Department or by a consultant selected by the Department with the advice of the owner or operator of the facility.[77] Such an assessment is a review and safety evaluation of those operations at a facility which involve the generation, storage, or handling of an extraordinarily hazardous substance.[78] Thus, the work plan sets forth the scope and detail of the accident risk assessment.[79] The owner or operator of the facility pays a fee for the assessment based on its costs.[80]

After reviewing the risk assessment, the Department, if appropriate, must order the owner or operator to undertake a risk reduction plan. The order must identify for risk(s) that must be abated and a reasonable timetable for the implementation of the plan. (The Department must, by regulation, establish criteria or quantative standards for determining risk that reflect, *inter alia*, the size of the potentially exposed population and the gravity of the consequences of a release.) The Department may order a facility that has not abated a risk on schedule to cease operations until the risk reduction plan has been implemented.[81] The owner or operator may request a hearing on such an order, during which the time for compliance will be suspended.[82]

The Department may enter a facility at any time to verify compliance with the law.[83] It must develop regulations requiring the owner or operator to report to the Department on all risk assessment and risk reduction efforts made in accordance with the law, all ongoing maintenance measures taken, all unanticipated and unusual events, and any other information the Department deems appropriate. These records are to be part of an information system that does not permit their destruction or alteration.[84]

The Department shall also, by regulation, adopt principles and procedures governing the internal management of confidential information supplied to the Department under the Act. The regulations must provide that such information may be disclosed only to its employees or agents to assist in enforcing the Act, or for use in a civil or criminal proceeding.[85] In addition, an owner or operator may petition to withhold information on the grounds that it contains or relates to a trade secret or constitutes security information.[86]

International Initiatives in Accident Prevention, Preparedness, and Response

Several initiatives by international organizations concerning prevention or response to accidents involving hazardous materials are in various stages of development.[87]

The United Nations Industrial Development Organization (UNIDO) has been implementing several technical assistance projects incorporating hazard assessment and hazard control technologies and methodologies. It has also undertaken studies and simulation exercises in emergency planning. The Director-General of UNIDO has appointed earlier this year an international Task Force to work out a plan of action for UNIDO aiming at fostering chemical safety in developing countries through *ad hoc* UNIDO assistance and established programmes. The Workshop for which this paper is being written is one of the many activities related to industrial safety that UNIDO has planned.

The Executive Director of the United Nations Environment Programme (UNEP) has proposed the preparation of an international treaty under which governments would inform each other of the timing and type of chemical releases that might have harmful transboundary effects, the location of the facility involved, the likely transport and fate of the chemical released, and protective measures taken or planned. A second convention would provide for measures for prompt assistance among governments after an accident in order to minimize consequences and protect life, property, and the environment. These conventions would parallel those concerning nuclear accidents recently signed in Vienna under the auspices of the International Atomic Energy Agency. UNEP also has proposed a programme under which governments, in co-operation with industry, would identify acutely toxic chemicals present in their communities and the existing and needed measures to control and limit possible accidental releases of these chemicals, for example, via emergency planning and additional training and equipment. Guidance for developing such emergency plans would be provided.[88] UNEP has been consulting with governments, industry, and international organizations about these proposals and will discuss them at the June 1987 meeting of its Governing Council. It has suggested that the International Register of Potential Toxic Chemicals (IRPTC) in Geneva might assist in implementing these proposals. The IRPTC presently provides, on request, information on characteristics of hazardous chemicals and advice on how to handle an accidental release.

The International Labour Organisation (ILO) is establishing and putting into operation major hazard control systems funded by the Federal Republic of Germany in India and by the United Nations Development Programme in Thailand. It is also preparing a manual providing guidance on assessment and prevention of major hazards in industrial processes, emergency planning, prevention systems, and management of major accidents.

The World Health Organization's European Regional Office will update its guidelines on contingency planning for emergencies and on preventive measures and responses to chemical accidents that it developed in 1981. It is preparing a report on emergency response and accident prevention and is performing an analysis of the causes and circumstances of chemical accidents. It has established a task force of experts in management, toxicology, and environmental health who are sent on request to help when a major accident has occurred. In collaboration with the International Programme on Chemical Safety (IPCS), the European Office organized an international conference on chemical accidents jointly with the Instituto Superiore di Sanita in July 1987 in Rome.

The IPCS was jointly created by UNEP, ILO, and WHO. It provides assistance to member states in improving safety practices in the chemical industry. It is preparing a checklist for identifying hazards during the manufacture, storage, and transport of chemicals and drafting guidelines on preventing chemical accidents.

The World Bank's Office of Environmental and Scientific Affairs has adapted its "Guidelines for Identifying, Analyzing, and Controlling Major Hazard Installations in Developing Countries" from the provisions of the Seveso Directive as part of its function of appraising and supervising industrial development by evaluating the adequacy of measures to control major hazards from developments contained in proposals for financial support. It published a Manual of Industrial Hazard Assessment Techniques in October 1985 that was prepared in co-operation with Technica, Ltd., of London, for the hazard analysis required for proposed installations. The methodologies are also applicable to existing operations and to rehabilitation or expansion projects. The World Bank is currently supporting the development of an integrated safety audit system for major accident prevention and is proposing a programme of research on important management and organizational factors in preventing major accidents.

The Organisation for Economic Co-operation and Development's Environment Committee will appoint an *ad hoc* group of policy and technical experts to prepare a workplan and timetable for its implementation concerning the prevention of and response to unintended releases of hazardous substances to the environment, including transfrontier releases, with potentially significant environmental effects. The *ad hoc* group is to take full account of relevant activities in other bodies and arrange for the exchange of information on experience gained by government and industry concerning prevention of and response to accidents. It is to prepare an outline of common principles and procedures to improve prevention of and preparedness for such accidents, prepare outlines for guidelines that set out common policies for preventing major releases, and make all possible efforts to develop criteria to identify the priority chemicals and processes which should be the focus of international co-operative efforts.

Policy-Making, Legislation, and Technical Assistance

In any field – in this case, the field of protecting public health and natural resources from the risks posed by hazardous materials – the purpose of legislation is to implement policy. Defining a social problem and formulating a policy or set of policies to alleviate the problem must therefore precede drafting laws that will institutionalize the policy. But it is essential to establish the policies in laws; policies do not implement themselves. Without laws, thoughtful policy statements are merely well-meaning rhetoric.

Preventing and responding to major accidents may not be a nation's highest priority problem involving hazardous materials, either because the risks are not present or because they are already adequately controlled by the implementation of existing legislation. As indicated in the introduction to this paper, there are many aspects to comprehensive control of the risks from chemicals in society, including industrial siting and risk assessment, discharge, limits, transportation, occupational health, and emergency planning. And there are trade-offs among these aspects, so they must be co-ordinated so that problems are actually dealt with, not simply transferred. If they are not co-ordinated, reducing allowable discharges of hazardous pollutants may just result in increased amounts of hazardous wastes, for example. Similarly, reducing inventories in order to reduce the risks of accidents involving chemicals in storage may simply result in increased risks of accidents from more frequent transport or delivery-related events.

Because circumstances and legal systems differ among nations, drafting "model" legislation to address a problem is often of only limited utility. A more promising approach is to prepare guidelines for legislative drafting based on an examination of a representative selection of up-to-date laws from jurisdictions with different problems and legal systems, and a presentation of the alternative approaches they reflect. Checklists of common components of such laws, as well as selected legislative language with brief commentaries explaining its origins and purpose (i.e. its legislative history) and any experience gained under it are also useful. In other words, the process of defining a problem and formulating a policy can be performed by analysing existing laws involving similar problems in other jurisdictions. Adapting these laws to local circumstances involves judgements about how the problems differ, what technical and administrative resources, public and private, may reasonably be allocated to addressing the problem, and what is politically acceptable.

As this paper indicates, the new generation of laws concerning the prevention of and response to accidents involving hazardous materials poses several policy issues that must be resolved. Experience under existing laws regulating such materials as well as national and international initiatives must be examined before adopting such new laws so that policies are as co-ordinated and forward-looking as possible.

Hazardous materials policy is pre-eminently a subject for co-operation among disciplines and between those who handle the materials and those who would govern how they are handled. Information must be shared, plans developed with a view to who will implement them, and controls imposed that address the most important risks to health and environment first. UNIDO's role in providing technical assistance to developing nations could perhaps most effectively be to provide persons experienced in hazardous materials regulation as members of interdisciplinary teams that would work with requesting nations in defining the problems to be alleviated, analysing existing national policies that bear on those problems, identifying alternative approaches to their alleviation, and preparing draft legislation and accompanying rationale that will establish the policies the nation chooses and their purposes into the nation's institutions and customs in an integrated fashion.

Notes

1. Council Directive of 24 June 1982 on the major-accident hazards of certain industrial activities (82/501/EEC), *Official Journal of the European Communities*, 5 August 1982, Article 1.1.
2. *Id.*, Article 1.2(a).
3. The 14 processes are: alkylation; amination by ammonolysis; carbonylation; condensation; dehydrogenation; esterification; halogenation and manufacture of halogens; hydrogenation; hydrolysis; oxidation; polymerization; sulphonation; desulphurization, manufacture and transformation of sulphur-containing compounds; and nitration and manufacture of nitrogen-containing compounds.
4. "LPG" is liquefied petroleum gas.
5. The full list in Annex II is: flammable gases as defined in Annex IV (c)(i) [substances which in the gaseous state at normal pressure and mixed with air become flammable and the boiling point of which at normal pressure is 20°C or below]; highly flammable

liquids as defined in Annex IV (c)(ii) [substances which have a flash point lower than 21°C and the boiling point of which at normal pressure is above 20°C]; acrylonitrile; ammonia; chlorine; sulphur dioxide; ammonium nitrate; sodium chlorate; and liquid oxygen.

6. Council Directive, *supra* note 1, Article 2.
7. *Id.*, Article 1.2(c).
8. Both "very toxic" and "other toxic" substances are defined in tables setting forth lethal doses to 50% of a population of rats from oral administration, of rats or rabbits from cutaneous application, and of rats from inhalation for four hours.
9. The criteria for flammable gases and highly flammable liquids are provided in note 5, *supra*. Flammable liquids are defined as substances which have a flash point lower than 55°C and which remain liquid under pressure, where particular processing conditions, such as high pressure and high temperature, may create major-accident hazards. Explosive substances are those which may explode under the effect of flame or which are more sensitive to shocks or friction than dinitrobenzene.
10. The kinds of involvement named in Article 5 are as substances stored or used in connection with the industrial activity concerned, as products of manufacture, as by-products, or as residues.
11. CAS is the Chemical Abstracts Service.
12. IUPAC is the International Union of Pure and Applied Chemistry.
13. Council Directive, *supra* note 1, Article 5.2 and 5.3.
14. *Id.*, Article 6.
15. *Id.*, Article 8.
16. *Id.*, Article 10.2 (a)–(b).
17. *Id.*, Article 11.1, Annex VI.
18. *Id.*, Article 12.
19. *Id.*, 11.2–11.3.
20. *Id.*, Article 13.1.
21. *Id.*, Article 13.3.
22. *Id.*, Article 13.4.
23. Zwölfte Verordnung zur Durchführung des Bundes-Immissionsschutzgesetz (Störfall-Verordnung), 12. BImSchV, 27 June 1980, Bundesgesetzblatt I, page 772. A "Störfall" is defined as a disturbance of normal operations in which a substance listed in Appendix II of the Ordinance is released or explodes and causes a public danger. *Id.*, section 2. For another example of implementation of the Seveso Directive, see *A Guide to the Control of Industrial Major Accident Hazards Regulations 1984*, Health and safety services booklet HS(R)21, Health and Safety Executive 1985. London: HMSO.
24. Gesetz zum Schutz vor schädlichen Umwelteinwirkungen durch Luftverunreinigungen, Geräusche, Erschütterungen, und ähnliche Vorgänge (Bundes-Immissionsschutzgesetz), BImSchG, 15 March 1974, Bundesgesetzblatt I, pages 721, 1193.
25. BImSchG, Articles 3(5) and 4(1).
26. *Id.*, Article 5.
27. *Id.*, Article 7.
28. BImSchV, *supra*, note 23, section 1.
29. Beschluss des Bundesrates zur Entschliessung des Bundesrates zur Vorsorge gegen Schadensfälle in der chemischen Industrie, Drucksache 622/86, 13 March 1987, pages 4–5. The Bundesrat also recommended expanding the scope of facilities subject to licensing under the Act in accordance with the Fourth Ordinance Implementing the Federal Immission Protection Act. *Id.*, page 4.
30. 12. BImSchV, *supra*, note 23, section 3(1)–(2). To fulfill this obligation, the operator shall 1) lay out the facility to withstand foreseeable stresses; 2) provide sufficient

protection of the facility against fire and explosion; 3) equip the facility with sufficient warning, alarm, and safety devices; 4) furnish the facility with sufficient reliable measuring, control and regulatory equipment that are of different kinds, independent, and existing in several places; and 5) protect the parts of the facility that are important from a safety technology standpoint from the acts of unauthorized persons. *Id.*, section 4.

31. *Id.*, section 3(3). To fulfill this obligation the operator shall 1) ensure that the nature of the foundations and load-bearing portions of the building cannot cause any additional dangers during disturbances of normal operations; 2) equip the facility with the necessary safety equipment and take other protective measures; and 3) establish emergency plans that co-ordinate with local catastrophe control planning. *Id.*, section 5.
32. See notes 30 and 31, *supra*.
33. 12. BImSchV, *supra*, note 23, section 6.
34. Beschluss des Bundesrates, *supra*, note 29, page 5.
35. 12. BImSchV, *supra*, note 23, section 7. For the obligations of sections 3–6, see text at notes 30–33, *supra*.
36. *Id.*, section 9.
37. Beschluss des Bundesrates, *supra*, note 29, page 5.
38. 12. BImSchV, *supra*, note 23, section 11.
39. Beschluss des Bundesrates, *supra*, note 29, page 5.
40. *Id.*
41. Sections 300–330, Public Law 99–499, 100 Stat. 1728–1758, 42 U.S.C. 11001–11050, October 17, 1986. This Act is Title III of the Superfund Amendments and Reauthorization Act of 1986 ("SARA") and is often referred to as "Title III"
42. *Id.*, section 302(a)(2).
43. *Id.*, section 302(a)(3)(A)(i). The list of 402 substances and accompanying explanations appears at 51 *Federal Register* 41570–41592 (Nov. 17, 1986).
44. *Id.*, section 301(a). If the governor does not appoint a commission within six months of enactment of the law, he serves in its role until he does.
45. *Id.*, section 301(b). The districts are to be designated within nine months of enactment of the Act.
46. *Id.*, section 301(c). Members are to be appointed not later than thirty days after districts are designated. Committees are to establish rules for public meetings to discuss the emergency plan, public comments, response to the comments, and distribution of the plan.
47. *Id.*, section 302(c).
48. *Id.* Section 302(a)(4) authorizes the Administrator to revise the list and thresholds from time to time.
49. Id., section 302(b)(2), 302(d).
50. *Id.*, section 303(d)(1).
51. *Id.*, section 303(d)(2)–(3). This should enable a committee to obtain information required for its plan on a facility's equipment, emergency response capabilities, emergency response personnel, and evacuation plans, subject to the trade secret restrictions in section 322.
52. *Id.*, section 303(a).
53. *Id.*, section 303(c).
54. See *id.*, section 303(f). The *Hazardous Materials Emergency Planning Guide*, issued in March 1987, is available from HAZMAT Planning Guide (WH-562 A), US Environmental Protection Agency, 401 M Street, S.W., Washington, D.C. 20460.
55. *Id.*, section 303(e).
56. Title III of SARA defines a hazardous chemical by reference to the definition of 29 *Code of Federal Regulations* 1910.1200(c) but excluding 1) any food, food additive,

colour additive, drug, or cosmetic regulated by the Food and Drug Administration, 2) any substance present as a solid in any manufactured item to the extent exposure to the substance does not occur under normal conditions of use, 3) any substance to the extent it is used for personal, family, or household purposes, or is present in the same form and concentration as a product packaged for distribution and use by the general public, 4) any substance to the extent it is used in a research laboratory or a hospital or other medical facility under direct supervision of a technically qualified individual, and 5) any substance to the extent it is used in routine agricultural operations or is a fertilizer held for sale by a retailer to the ultimate consumer. *Id.*, section 311(e).

29 CFR 1910.1200(c) defines a hazardous chemical simply as "any chemical which is a physical hazard or a health hazard". These terms are defined in the same regulation, respectively, as follows: "Physical hazard" means a chemical for which there is scientifically valid evidence that it is a combustible liquid, a compressed gas, explosive, flammable, an organic peroxide, an oxidizer, pyrophoric, unstable (reactive) or water reactive. "Health hazard" means a chemical for which there is statistically significant evidence based on at least one study conducted in accordance with established scientific principles that acute or chronic health effects may occur in exposed employees. The term "health hazard" includes chemicals which are carcinogens, toxic or highly toxic agents, reproductive toxins, irritants, corrosives, sensitizers, hepatotoxins, nephrotoxins, neurotoxins, agents which act on the hematopoietic system, and agents which damage the lungs, skin, eyes or mucous membranes. Appendix A [of the regulation] provides further definitions and explanations of the scope of health hazards covered by this section, and Appendix B describes the criteria to be used to determine whether or not a chemical is to be considered hazardous for purposes of this standard.

57. 29 U.S.C. 651 *et seq.* (1982).
58. Section 311(a)(1), Public Law 99–499, *supra* note 41. If the facility later becomes subject to the OSHA requirement to prepare or have available an MSDS, he must submit the MSDS or list within three months. A revised sheet must be submitted by any owner or operator within three months of discovering significant new information concerning an aspect of a hazardous chemical for which an MSDS was previously submitted. *Id.*, section 311(d). An MSDS provides 1) the chemical name of a chemical and its characteristics, including 2) toxicity, corrosivity, and reactivity, 3) known health effects, including chronic effects from exposure, 4) basic precautions in handling, storage, and use, 5) basic counter-measures to take in the event of a fire, leak, or explosion, and 6) basic protective equipment to minimize exposure.
59. *Id.*, section 311(a)(2). If a list is submitted, the facility must submit an MSDS upon request of a committee. A member of the public may obtain an MSDS from the committee. *Id.*, section 311(c).
60. *Id.*, section 311(b). Proposed rules implementing this authority were published January 27, 1987. See 52 *Federal Register* 2836–2855 (Jan. 27, 1987).
61. *Id.*, section 312(a), (d)(1).
62. *Id.*, section 312(d)(2), (e)(1). Any other state or local official acting in his official capacity may have access to this information from a committee or commission. A member of the public may have access to it for chemicals stored in quantities exceeding 10,000 pounds at a facility in the preceding year. For chemicals stored in smaller amounts, the committee or commission may provide the information at its discretion. *Id.*, section 312(e)(2)–(3). If an owner or operator requests it, the committee or commission shall not disclose the location of any specific chemical. *Id.*, section 324(a).
63. *Id.*, section 312(f).
64. See text at note 53, *supra*.

65. Section 304(a),(b)(1), Public Law 99–499, *supra* note 41.
66. *Id.*, section 304(b)(2).
67. *Id.*, section 304(c).
68. Section 13:1K--21 e., New Jersey Code.
69. The initial extraodinarily hazardous substance list was: hydrogen chloride (HCl), and allyl chloride in quantities of 2000 pounds or more; hydrogen cyanide (HCN), hydrogen fluoride (HF), chlorine (Cl_2), phosphorous trichloride, and hydrogen sulphide (H_2S) in quantities of 500 pounds or more; and phosgene, bromine, methyl isocyanate (MIC), and toluene-2, 4-diisocyanate (TDS) in quantities of 100 pounds or more. Section 13:1K-22 a., New Jersey Code.
70. Section 13:1K-22 c., New Jersey Code. The list shall correlate the substances or compounds with the quantities thereof required to produce the potentially catastrophic circumstance. The Department shall have the power to amend, by regulation, the extraordinarily hazardous substance list to accommodate new chemical compounds that may be developed or reflect new information or scientific data that may become available to the Department. *Id.*
71. Section 13:1K-22 b., New Jersey Code.
72. Section 13:1K-22 d., New Jersey Code.
73. Section 13:1K-21 a., New Jersey Code.
74. Section 13:1K-21 i., New Jersey Code.
75. Section 13:1K-23 a.–c., New Jersey Code
76. Section 13:1K-24, New Jersey Code.
77. Section 13:1K-25, New Jersey Code.
78. Section 13:1K-21 d., New Jersey Code.
79. Section 13:1K-21 g., New Jersey Code.
80. Section 13:1K-25, New Jersey Code.
81. Section 13:1K-26 a., New Jersey Code.
82. Section 13:1K-26 b., New Jersey Code.
83. Section 13:1K-27 a., New Jersey Code.
84. Section 13:1K-27 b., New Jersey Code.
85. Section 13:1K-29 a., New Jersey Code. The regulations shall include requirements that the facility owner must label all confidential information as such, that receipt of such labelled information be acknowledged in writing by an authorized employee of the Department, that only specifically authorized personnel have access to the information and then only on a "need-to-know" basis, and that the Department establish secure areas for the storage of such information. *Id.*
86. Section 13:1K-29 b., New Jersey Code.
87. Unless otherwise noted, the information in this section is based on "Activities of Intergovernmental Organizations in the Area of 'Chemical Accidents' During 1987–1988", Organization for Economic Co-operation and Development, Environment Directorate, Room Document No. 1 for the Third High Level Meeting of the Chemicals Group, 17–18 March 1987.
88. "International Measures to Deal with Chemical Emergencies – Proposals by the Executive Director," Committee of Permanent Representatives to UNEP, UNEP/CPR.6/8, 3 December 1986.

15
Effects of Changing Regulations on Hazardous Wastes Flow in Ohio

W.B. Clapham, Jr.

The patterns of industrial development in the United States are too varied to be categorized easily. This is doubly true of its patterns of industrial waste flow. Plants in the same general category produce different configurations of wastes, since they operate in the slightly different ways. Different parts of the country have different industrial mixes, different problems imposed by the geological situation, and different ways of managing wastes.

Industries in the United States, like those all over the world, are faced with a changing regulatory situation. One set of regulations is set by the Federal Government, but each State implements those regulations and otherwise affects industrial activity in its own way. To attempt to forecast the effects of changing regulations on a national scale is very difficult. On one hand, we must deal with the overall complexity imposed on the industrial situation by the country's geography and federal system of government. On the other, we must determine a meaningful "typical" industrial facility to simulate waste generation in the industry.

It is reasonable, in this light, to consider an *area* that is typical in many ways of the industrial United States and consider it as a surrogate for the country as a whole. The State of Ohio located in the industrial Midwest, just south of the Canadian border, on Lake Erie is an appropriate area. Depending on how "hazardous wastes" are defined, it varies from second to fifth, by states, in generation of hazardous wastes, according to various US government agencies.[1] Its industrial mix includes all of the industries typical of most of the United States, including automotive, chemical, steel, and manufacturing, and it is quite typical of the industrial patterns that one generally perceives as the American norm.

Another reason to use Ohio as a surrogate for the country as a whole is that the data base for hazardous waste generation and flow for the State are fairly good – indeed a recent survey by the National Governors' Association suggests that the annual report data for Ohio are as good as those for any State and considerably better than most. It is possible to use this data base to characterize the patterns of hazardous waste generation and management, and then to simulate the effects of changes imposed on the system.

Industrial Waste Generation in Ohio

Ohio has a broad spectrum of industrial waste generation (Table 15.1). A majority of the industrial groups produce less than 10,000 metric tons of hazardous waste. In these cases, the overwhelming majority of the wastes are shipped by the generator to commercial facilities for treatment or disposal. Industries within the chemical, petroleum, and related industries grouping generated much more wastes, with a majority being managed on the site of generation by the generator itself. Nevertheless, a substantial amount of waste was sent off-site for commercial treatment. Most of the largest waste streams

Table 15.1 Industrial Waste Generation and On-Site Management in Ohio (in metric tons)

SIC Code	*Industrial Grouping*	*Commercial Facility*	*On-Site Management*
12–20	Mining, Construction, and Food Products	73.2	0.0
22–23	Textile Mills, Apparel, and Textile Products	1,848.0	595.4
24–25	Lumber, Wood Products, Furniture, and Fixtures	511.9	351.7
26	Paper and Allied Products	2340.1	5.7
27	Printing and Publishing	1404.4	0.0
28	Chemicals and Allied Products	364,787.8	1,043,290.5
281	Industrial Inorganic Chemicals	299,429.4	6367.5
282	Plastics Materials and Synthetics	4271.7	16,622.8
285	Paints and Allied Products	19,530.5	3491.0
286	Industrial Organic Chemicals	33,276.7	996,408.1
287	Agricultural Chemicals	251.4	15,172.4
289	Miscellaneous Chemical Products	6110.1	5214.7
29	Petroleum and Coal Products	16,242.7	38,406.1
30	Rubber and Miscellaneous Plastics Products	5757.8	110,782.2
31	Leather and Leather Products	62.0	0.0
32	Stone, Clay, and Glass Products	9091.2	38.3
33	Primary Metal Industries	359,881.7	200,659.4
331	Blast Furnaces and Basic Steel	331,403.8	174,825.7
332	Iron and Steel Foundries	4974.5	21,130.0
335	Nonferrous Rolling and Drawing	6247.2	4702.8
339	Miscellaneous Primary Metal Products	13,673.4	0.0
34	Fabricated Metal Products	30,745.7	133,053.1
341	Metal Cans and Shipping Containers	1511.2	47.4
344	Fabricated Structural Metal Products	2630.2	86.9
346	Metal Forgings and Stampings	2297.0	132,276.0

347	Metal Services	12,212.5	40.8
349	Miscellaneous Fabricated Metal Products	8626.8	581.5
35	Machinery, except Electrical	16,072.4	83,723.3
36	Electric and Electronic Equipment	6438.5	355.4
37	Transportation Equipment	49,352.9	18,667.7
371	Motor Vehicles and Equipment	42,447.4	18,504.8
372	Aircraft and Parts	3705.2	108.5
374	Railroad Equipment	226.8	54.4
379	Miscellaneous Transportation Equipment	2072.1	0.0
38	Instruments and Related Products	3538.4	12,598.5
39	Miscellaneous Manufacturing Industries	111.8	0.0
40	Railroad Transportation	26.5	0.0
42	Trucking and Warehousing	9631.4	354.4
44	Water Transportation	1385.9	0.0
45	Transportation by Air	17.2	3.6
46	Pipelines, except Natural Gas	27.9	0.0
47	Transportation Services	2457.5	34.6
49	Electric, Gas, and Sanitary Services	89,807.7	1,388,467.5
495	Sanitary Services	89,154.6	1,388,134.5
50	Wholesale Trade – Durable Goods	2095.5	40.8
51	Wholesale Trade – Nondurable Goods	1958.6	45.0
67	Holding and Other Investment Offices	228.6	0.0
72	Personal Services	85.8	0.0
73	Business Services	7807.2	54.9
76–89	Repair, Health, Educational, and Miscellaneous Services	1880.6	0.0
91–97	Government	304.0	48.8
	Total Statewide Waste Production	985,974.8	3,031,243.9

came from the metals and manufacturing sectors. Considerable quantities of these wastes are managed both on-site and at commercial facilities. The other very large waste stream comes from the poorly named "Sanitary Services" sector, which includes the hazardous waste management corporations (i.e. landfills, incinerators, and treatment facilities).

Hazardous Waste Flow from Industries in Ohio

Thousands of companies throughout the State generate hazardous wastes. In order to visualize the patterns of waste flow, it is most meaningful to aggregate these wastes by county. Ohio has 88 counties, with an average area of approximately 1215 sq. km each. The pattern of hazardous waste generation from all sources in the state is shown in Figure 15.1a. The state total is 3,823,195 metric tons of waste, of which the largest county generates 1,390,714 tons. The size of the "pies" in Figure 15.1a indicates, relative to this largest county, the amount of waste generation for each county. The pattern of the pies indicates disposition of the wastes among the three categories: commercial management off-site, but within the State; commercial management off-site, out of Ohio; and management on-site.

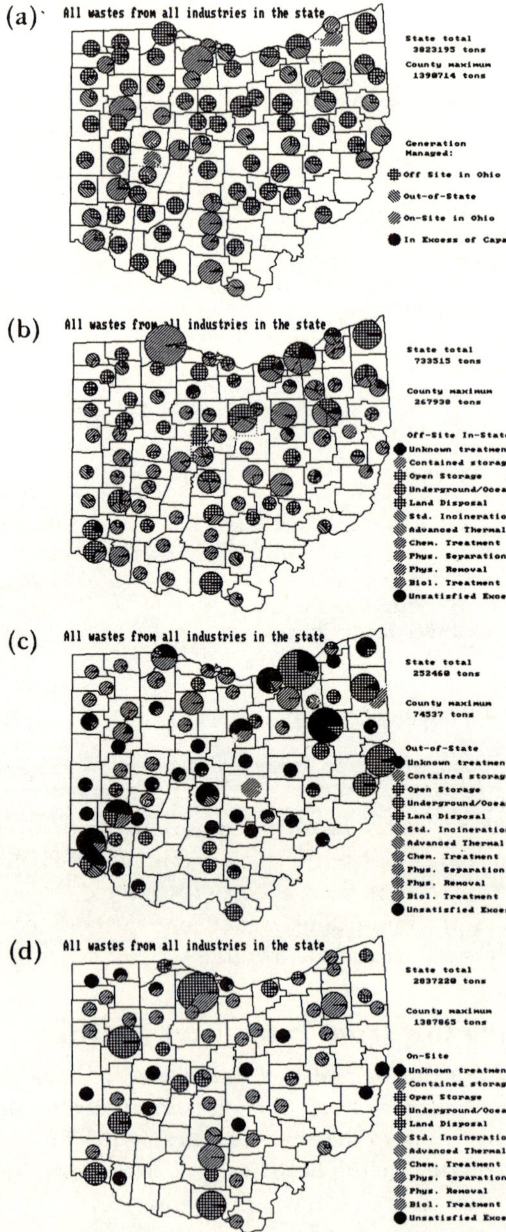

Figure 15.1 Waste Flow Patterns for all Hazardous Wastes Generated in the State of Ohio during 1984: (a) Disposition Off-site, Out-of-state or On-site; (b) Method of Treatement at Commercial Facilities in Ohio; (c) Method of Treatment at Out-of-state Commercial Facilities; (d) Method of Treatment On-site

We can see what is actually done with the wastes by looking at each of these three categories in greater detail. Figure 15.1b shows the disposition of wastes treated by commercial facilities within Ohio. Management is considered in ten different categories, plus "unknown treatment", if the treatment cannot be derived from the data base. Commercial facilities within the state manage 733,515 metric tons of material. The size of the pies in Figure 15.1b refer to the relative amount of hazardous waste *generated* in the county in question and managed by in-state commercial facilities. Figure 15.1c shows the disposition of wastes managed by commercial facilities in other states. Data are much worse for these facilities, so that a much larger proportion of the waste treatment appears as "unknown." The quantity of hazardous waste sent out-of-state for treatment totals 252,460 metric tons. Once again, the size of the pies in Figure 15.1c refer to the relative amount of hazardous waste *generated* in the county in question and managed by out-of-state commercial facilities. The treatment types considered are the same as those in Figure 15.1b. Finally, Figure 15.1d shows the disposition of wastes managed on-site. The quantity of waste that falls into this category totals 2,837,220 metric tons.

Figures such as these are not particularly meaningful because the information is highly aggregated. Let us also consider a more specific waste stream, such as combustible wastes and solvents from the plastics industry.[2] Figure 15.2a shows the generation of these wastes throughout Ohio; Figures 15.2b–d document their disposition on commercial facilities within the State, out-of-state, and on-site, respectively. Interpretation of the waste amounts and treatment types are the same as for Figure 15.1. Within the general waste type shown in these figures exists a wide variation. Nevertheless, the various subsets of the waste type will be treated in a similar way as the regulations governing hazardous wastes evolve, and the different units within the plastics industry will be affected in basically the same way by this regulatory evolution.

Changes Affecting Hazardous Waste Flows and Management

The pattern of hazardous waste flow is not static. It responds to a number of factors, including economic growth within specific industries, possibilities for waste reduction, closure of existing waste-management facilities, opening of new waste-management facilities, and regulatory changes dealing with classes of waste management. The first two affect the net generation of hazardous wastes; the last three are the factors that will determine changes in the available capacity for waste management.

The effect of economic growth in an area is to expand (or contract, if growth is negative) the gross generation of wastes. This is a reasonable simplistic assumption, but it is the best that will normally be possible. To assume that, *ceteris paribus*, a 5% increase in the economic status of a particular industrial area will result in a 5% increase in production, which will result in a 5% increase in waste generation, is a reasonable estimate, at least as long as it is

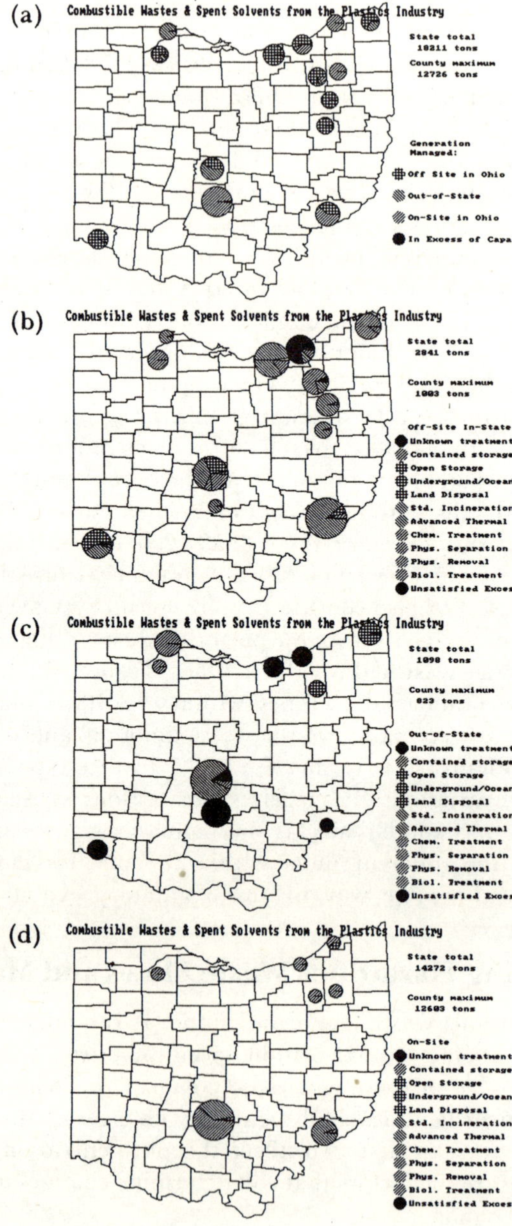

Figure 15.2 Waste Flow Patterns for all Combustible Wastes and Spent Solvents Generated by the Plastics Industry in Ohio during 1984: (a) Disposition Off-site, Out-of-state or On-site; (b) Method of Treatment at Commercial Facilities in Ohio; (c) Method of Treatment at Out-of-state Commercial Facilities; (d) Method of Treatment On-site

understood as an order of magnitude and not as an exact estimate. Of course, the *ceteris paribus* argument does not hold. A major effort is now underway to increase the efficiency of industrial processes in general, and the chemical industry has been one of the most effective in reducing the amounts of wastes generated in the production process. We can expect major reductions in the generation of some wastes, and minor reductions in others. As was pointed out in a recent study by the Office of Technology Assessment, it is impossible to estimate the precise levels of reduction that will take place in all basic industries, much less on a plant-by-plant level.[3] If it is at all reasonable to make quantitative estimates for future waste reductions, it must be done in a very rough way, and the estimator will have to exhibit a level of humility in asserting the credibility of his or her estimate that few of us are capable of.

We can calculate the net waste generation as follows: gross waste production = current waste production × (1 + growth factor), net waste production = gross waste production × (1−waste reduction factor). If it is difficult to estimate the net waste production for a particular waste within an industrial group, it is even more difficult to estimate the future available capacity for waste management. Few facility owners really want to close their facilities, but rather will try to expand, enlarge, or renew them as long as they can. Nevertheless, facility closing is the easiest factor to estimate, at least in the short to medium run. The opening of new facilities is difficult to estimate, since it involves both the capabilities and interests of the waste management industry and the dynamics of the siting process. In the United States, Federal law sets certain standards for hazardous waste management facilities, as well as minimum requirements for the siting process. Depending on the nature of State laws and the stage of negotiation between the US Environmental Protection Agency (USEPA) and the State environmental regulatory agency, installation and operation of a new facility may require only a Federal permit, only a State permit, or both a Federal and a State permit. A Federal permit is relatively easy to obtain. Most industrial states, however, require a State permit. Each state has its own requirements and procedures. Some states have been unable to follow through the siting process; other states (including Ohio) have been successful in siting new facilities even though the siting criteria are much more severe than those found in Federal law. As is shown in Figures 15.1c and 15.2c, a considerable amount of waste crosses State lines. Therefore, any additions to the hazardous waste management capacity available to one area depends greatly on the siting practices followed in nearby states. In general, it is very likely that few major new facilities will emerge in the industrial Midwest over the next few years. There will be some, but they will be fewer in number than the number that will close. Where and what kinds they will be is an open question that is, at best, an educated guess.

Regulatory changes are matters of policy under the control of the relatively small number of Federal and State environmental regulatory agencies. Examples of these changes are the banning of disposal of particular wastes on land, as called for in the Hazardous and Solid Waste Amendments of 1984, and possible restrictions on underground injection of hazardous wastes. Because these changes are a matter of policy, unlike all of the other factors that lead to changes in hazardous waste flow and management patterns, they can be predicted in at least a qualitative manner and incorporated into a scenario analysis. Indeed, one of the reasons one carries out an analysis of a system is to gauge the effect of different regulatory scenarios.

A Scenario for Change in Hazardous Waste Flow and Management

There are many reasons why one should not attempt to do a numerical analysis of changes in hazardous waste management. As with all analyses of this sort, the basic data base is suspect and may be misleading.[4] Scenarios of regulatory change are fairly straightforward, and are typically intended as qualitative, or at best semiquantitative indications of the results of policy change. The other four factors are virtually an open field. One cannot simulate the process by which facility closings and sitings take place. One can use, at best, simple projections of economic growth to characterize potential growth in gross industrial waste production, and estimates of waste reduction are all but impossible to project.

We shall make some very simple assumptions, recognizing that they are representative of the changes that *may* happen, but they that do not represent a likely specific scenario. We shall assume that over the next few years, the production in the Ohio chemical industry (including plastics) will grow by 10%. During the same interval of time, the productivity of other industries will grow by 7.5%. Waste reduction in the chemical industry during this period will be by a factor of 15%; waste reduction in all other industries will be by a factor of 10%. No new plastics plants will be opened; no existing plants will be closed. Two landfills accepting wastes in 1984 will close (one of which has already closed). A major incinerator will open up along the Ohio River near the Pennsylvania border and 40% of its capacity will be available to Ohio industries. Finally, USEPA will implement a land ban of materials going into land disposal, surface impoundments, and underground injection, so that 60% of the wastes now entering these facilities will be diverted to other facilities.

The results of this scenario are shown in Figures 15.3 and 15.4. Figure 15.3 shows the effect of the changes postulated on overall waste flow in the State as a whole. Figure 15.4 shows the effects of the same changes on the management of combustible wastes and spent solvents from the plastics industry.

The aggregated results of the scenario (Figure 15.3) indicate that the State can expect a major shortfall in management capacity under the conditions postulated. But it is difficult to draw any other conclusions, since so much of the information is aggregated. We can, however, draw meaningful results from Figure 15.4. For most of the generators in the State, the amounts treated by

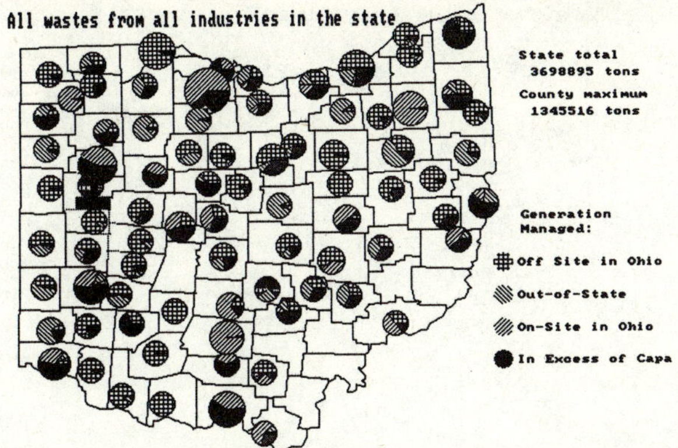

Figure 15.3 Waste Flow Patterns for all Hazardous Wastes Generated in the State of Ohio under a given Scenario

incineration increase, but only slightly. The exceptions are for generators near Cincinnati, in the southwestern part of the State, who send major tonnages of waste to the nearby Cecos landfill, and the one on-site underground injection well used for these wastes. The picture for the plastics industry, then, appears to be significantly different from that for industry as a whole. Evidently, current patterns of waste treatment will not be greatly disrupted by regulatory changes that have a major effect on other industries.

Let us briefly consider two of these other industries. The electroplating industry has a serious problem, since so much of the waste generated there currently goes to landfills (Figure 15.5).[5] Estimates have suggested that full compliance with regulations designed to control the hazardous wastes from this

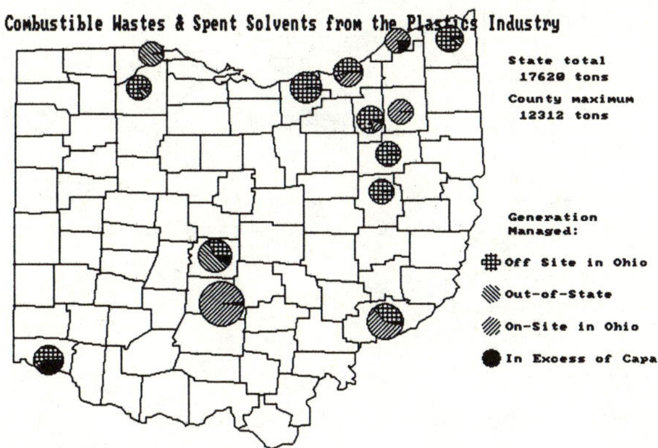

Figure 15.4 Waste Flow Patterns for Combustible Wastes and Spent Solvents Generated by the Plastics Industry in Ohio under a given Scenario

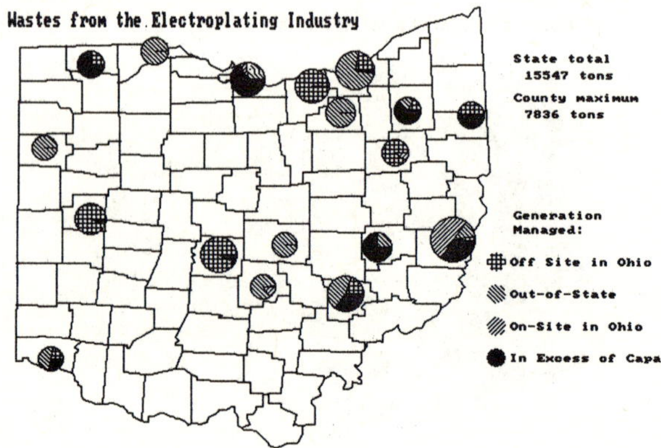

Figure 15.5 Waste Flow Patterns for all Hazardous Wastes Generated by the Electroplating Industry in Ohio under a given Scenario

industry will cause a large number of electroplating shops to go out of business. Yet most electroplating shops are fairly small, and they are concentrated in the State's larger cities. Relatively small specialized treatment facilities located in these cities may be capable of meeting the needs of most of the electroplaters even in the context of regulatory shifts of the sort envisaged in the scenario presented here. Citywide facilities of this sort are both feasible, and it should be possible to get them sited fairly quickly, since they would be small and related to a clear local need.

The situation is quite different with the automobile manufacturing industry.[6] Once again, Figure 15.6 indicates a major problem. Unlike the electroplating shops, however, each of which is fairly small and produce similar wastes, major

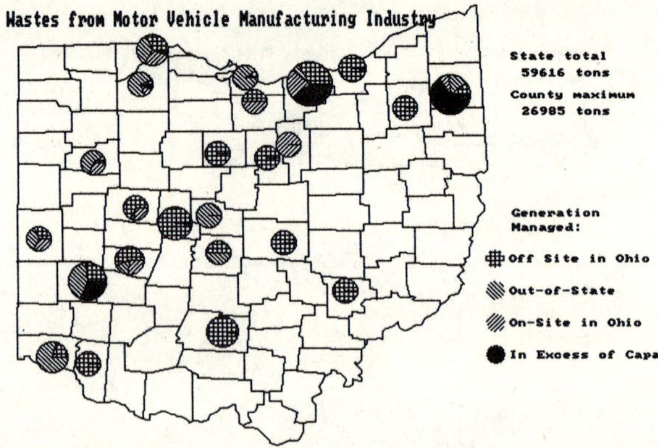

Figure 15.6 Waste Flow Patterns for all Hazardous Wastes Generated by the Automobile Manufacturing Industry in Ohio under a given Scenario

factories may be very large, and different plants produce very different wastes. The treatment plants required to meet the needs of manufacturing would likely to be one of two types. The first would have new on-site facilities; the second would have large regional facilities designed to service many different plants. On-site treatment may be a new venture for many companies, and one that they are not well equipped to handle. Regional treatment facilities are likely to be very controversial, given both their size and their being only indirectly related to perceived local needs. Both types of facility are likely to be more difficult to site than a citywide facility designed to serve the electroplating industry.

Discussion

We cannot *forecast* the effects on hazardous waste management of regulatory changes now being discussed in the United States. Many factors are so fluid that they cancel each other out and confer a level of uncertainty on an analysis of the hazardous waste management system tht makes precision inconceivable. But we do not need precise estimates in order to carry out the level of public policy debate that needs to be followed in order to plan adequately for hazardous waste management. We need an analytical tool that will enable us to predict the degree and location of changing facilities requirements. Even in its simplest incarnation, we can use this type of tool to gain insight into the interactions of regulatory change, economic development, the closure and opening of facilities, and improvements in technologies that allow waste reduction and greater efficiency of raw materials use. We can also gain insight into the interactions between the amounts of waste generated in one industry and the requirements for waste treatment capacity for others. Even where we do not know and cannot estimate the trajectory of technology development (as with waste reduction), we can use sensitivity analysis or Monte Carlo analysis to improve the usefulness and credibility of the analytical results.

The types of interpretation we can make for individual industries is very different from those we can make when the only information available is highly aggregated. As shown in the examples provided in this paper, the aggregate picture (Figure 15.3) suggests that the situation is very serious, and indeed it is. But it is much more serious for certain industries.

Some industries, for example plastics, will evidently not be greatly affected. Others, such as electroplaters, will have to suffer severe short-term dislocation as citywide treatment facilities are built, but the solution to their problem should be feasible. Still other industries, such as manufacturing, will find it more difficult to accommodate regulatory change.

For this type of analysis to take place, sufficient data to enable a characterization of hazardous waste flow, such as that collected by the Ohio Environmental Protection Agency, is required. However, the data is not easy to collect, and many states within the United States do not have data in a form that can easily be used. The data available to the USEPA in their Biennial

Reports are not yet adequate to characterize the flow of hazardous wastes, although this will change within the next year or two.

It would not be a trivial undertaking for Third World countries to collect the type of data on hazardous waste flow that are used here. But information technology is relatively inexpensive, and UNIDO or some other UN agency is ideally suited to act as the clearing house for data. Third World countries have been providing information to UN agencies for many years on phenomena that are more difficult to quantify – for example, census data and foodstuff production. Plant managers must already have data on waste at their disposal – the basic problems posed by wastes are the same in developed and Third World countries. As with many other lines of enquiry, the UN will be able to learn from the successes and failures of its member countries. For example, some variant of either the annual/biennial reporting system used in the United States might well serve as an adequate basis for estimating waste flows. But the 4-digit classification scheme used by USEPA is so aggregated in some places and disaggregated in others that it is virtually useless for any analytical purpose.

It is clear that Third World countries share many needs and problems of hazardous waste management with developed countries. They must be able to plan adequately for sufficient waste management capacity to serve their industrial base. Indeed the planning role is probably more acute in Third World countries, since there is less money for waste investment, and those funds must compete with other acute needs, such as health care, population control, industrial development. These countries must use their resources, including the ability to impose regulations, to develop new technologies and facilities, to use those facilities, and to plan, efficiently in order to ensure that expansion in their industrial development has the most effective waste management capability. Using current patterns of hazardous waste flow as a baseline against which to postulate changes resulting from the use of these resources is a fully feasible way to improve the system of hazardous waste management. UNIDO clearly has a role in helping to determine the most appropriate methods to collect and classify data.

Notes

1. For example, Carol, Daniel, and Kenneth Rubin. 1985. *Hazardous Waste Management: Recent Changes and Policy Alternatives.* Washington: US Government Printing Office, xxi+96pp.
2. For purposes of this paper, the plastics industry will be defined as all plants with the 3-digit SIC code 282. Combustible wastes and solvents will include all wastes with EPA codes D001, F001, F002, F003, F004, F005, K019, P022, U002, U007, U009, U004, U080, U111, U112, U122, U210, U213, U223, and U226.
3. Office of Technology Assessment. 1986. *Serious Reduction of Hazardous Waste.* Washington: US Government Printing Office. x + 254pp.
4. In the case of the data base used in this study, the data have been obtained from the Annual Reports of the Ohio Environmental Protection Agency. Data are provided by

the industries to the OEPA under provisions of State law. The OEPA does not, however, have the means to verify them independently.

5. For the purposes of this paper, the electroplating industry is defined as all plants with the 3-digit SIC code 347. All wastes produced by this sector are considered.
6. For the purposes of this paper, the automobile manufacturing industry is defined as all plants with the 3-digit SIC code 371. All wastes produced in this sector are considered.

16

Health Implications of Hazardous Wastes Disposal

C.R. Krishna Murti

Rapid industrialization has been accepted by the Less Developed Countries as a vital component in the strategy for progress. As of 1974 ten LDCs qualified to be included among the comity of industrialized nations on the criterion that manufacturing industries contribute more than 30% of their Gross National Product.

Waste generation and environmental pollution are visible consequences of any form of industrial or mining activity. The nature and diversity of health problems that have begun to surface in LDCs as a result of environmental pollution have been discussed elsewhere (Krishna Murti and Sandoval, 1983). Ideally environmental pollution is avoidable provided certain well recognized prerequisites for efficient captive recycling of nonrenewable materials of production and strict control of pollutant emissions into the environment are fulfilled. Since such ideal conditions do not prevail in many of the industrially advanced countries (Harthill, 1984), it is understandable that LDCs during the process of industrialization will encounter problems of pollution in a new setting and socio-political milieu and are called upon to manage issues related to the disposal of enormous quantities of wastes.

Specifically the following questions need to be addressed.

1. What is the state of awareness of the health problems associated with industrial wastes in LDCs?
2. Are there attempts in LDCs to inventorize wastes generated in manufacturing industries and make risk assessments?
3. Have these countries faced Love Canal-like episodes? (Paigan, 1984)
4. What is the action being initiated to deal with problems of industrial waste disposal in LDCs?

It is hoped that the background information provided in this paper will be useful in identifying problems related to the health aspects of industrial wastes and in developing adequate measures for their safe disposal without impairing public health or compromising environmental sanctity.

Categorization of Hazardous Wastes

A simple categorization of industrial wastes can be made on the basis of their hazardous nature including their toxicity, flammability, explosivity, infectivity and corrosivity. A model checklist is given in Table 16.1. A typical list of hazardous wastes is given in Table 16.2. The compilation has been based on regulations existing in industrialized countries. The inclusive listing system of hazardous industrial wastes used in Belgium, Denmark, France, Federal Republic of Germany, the Netherlands, Sweden, the United Kingdom and the United States envisages three distinct approaches:

1. Identification of characteristics by testing and declaration of hazard; also known as "Listing by Characteristics"
2. Assumption of the presence or absence of hazard; also known as "General Listing Approach"
3. Prior listing as hazardous wastes based on the profile of the potential health effects of the constituent chemicals of the waste site; also known as "Listing Approach"

The relative advantages and disadvantages of this type of inclusive listing have been discussed at length in an interim document of WHO (WHO/EURO, 1982). In an explorative study commissioned by the Department of Environment, Government of India, on the State of Art of Industrial Waste Management in India, the National Productivity Council (NPC) surveyed the situation in three industrialized states: Maharashtra, Gujarat and Tamil Nadu for which a classification system was adopted based on the physical state of the wastes and the degree of health hazard posed by them. This listing is shown in Table 16.3. A description of wastes generated in selected industries in India is given in Table 16.4.

Effects on Human Health

The impact of hazardous wastes on public health cannot be assessed without having more reliable information on the environmental health scenario of the Less Developed Countries. The depressing environmental status of LDCs has provided the foundation for the widely publicized World Bank model of global diseases, morbidity and mortality (World Bank, 1975). This model is based on the finding that environmental factors contribute to more than 80% of mortality and morbidity in LDCs and less than 20% in industrialized countries, where life style is the major factor.

Overcrowded slums and shanty towns nestling close to high-rise multi-storey buildings and enveloping huge industrial complexes, subhuman living conditions for a large section of the population who provide indispensable

Table 16.1 Category and Source of Hazardous Wastes

Category	*Source*
Toxic chemicals	Chemical industry, heavy industries Coal based thermal power plants, Pharmaceuticals, pesticides Plastics and polymers
Flammable	Oil sludges, solvents, plasticizers Light metal discards
Explosives	Ordnance factories, Oil tankers, safety match Pyrotechnics
Corrosives	Acid slurries
Infective/Biologicals	Hospital wastes, wastes from vaccine and serum institutes Fermentation industries Biotechnology

Table 16.2 Hazardous Wastes in Developing Countries

Waste	*Source*
Cyanide wastes	Electroplating, Metal processing, Chemicals
Metal finishing wastes	Cutting oils, Acid slurry
Solvents	Vegetable oil recovery, Chemical Industry
Mercury wastes	Chloralkali plants
Fluoride	Bauxite, Fertilizers
Arsenic	Fertilizers, Wood processing
Pesticides	Manufacture and formulation, Outdated disposal
Plastics, Monomers	Plastics
Phenols	Iron and Steel, Petrochemicals
Asbestos	Asbestos cement, Insulation industry, Building industry

human services, overburdened transport systems and community health facilities, and the prevalence of many communicable and noncommunicable diseases – this is the situation in any urban human settlement in LDCs in Asia and Latin America. In a study of world cities, Hardoy and Satterthwaite (1985) make the following apt observations:

> The Third World's cities are swelling as farmers and villagers hark to the siren promise of a better life. But up to half the inhabitants of most of these cities wind up in slums and shanty towns where drinking water supplies, sewage system, garbage collection, disease prevention measures, and primary health care are mere dreams while dysentery, typhoid, food poisoning, malnutrition, cramped conditions, and household accidents are daily realities.

Table 16.3 Physical Form and Degree of Hazard (Indian NPC study, 1985)

	Description	*Example*	*Degree of Hazard*
Solid	Very hard mass Filter cake Moisture up to 30%	Spent catalyst	Varied
Semi-solid	Pasty, sticky, viscous	Spent solvent Distillation Residue	Substantial
Sludge	Soft – High Moisture content up to 80%	Biomass Dyestuff Residues of wastewater treatment	Acute
Slurry	Suspensions of solids Moisture content up to 90%	Arsenic waste from petro-coke process	Chronic

Table 16.4 Generation of Solid Wastes in Industries

		Waste Generated TPA	
Industry	*No of Units**	*Total*	*Hazardous*
Chloralkali	9	9682	9022
Drugs and Pharmaceuticals	20	201,710	106,250
Dyes and Pigments	19	48,654	48,654
Engineering	4	57,871	49,873
Nitrogenous Fertilizer	4	5735	433
Phosphatic Fertilizer	5	413,320	—
Inorganic Chemicals	27	412,188	62,364
Organic Chemicals	30	34,445	19,974
Pesticides†	15	5088	5088

* Units covered in one survey in one state;
† only formulation plants.

Undoubtedly, biological pollution is a major problem and hence the emphasis on primary health care and sanitation. Nonetheless, according to a WHO Expert Committee on Environmental Pollution Control in Relation to Development, problems of environmental pollution arising out of developmental activities

> . . . have already begun to pose threats to human health; this is clear from the report on the world environment between 1972 and 1982 by the United Nations Environment Programme. (Holdgate *et al.*, 1982.) These problems can, therefore, no longer be ignored; to do so at this stage of development might prove to be a costly mistake. (WHO, 1985a)

The limited sample survey conducted under the Global Environment Monitoring System (GEMS) on biological monitoring for blood lead and kidney cortex cadmium (Valter, 1982) and for organochlorine residues in human breast milk (Storach and Vaz, 1983) has established beyond any doubt that exposure to toxic chemicals is on the increase in LDCs.

In assessing the risk of exposure to hazardous wastes in LDCs one has to keep in mind the important fact that the health and living standards of the people are not comparable to those enjoyed by the populations of the industrialized countries. It must also be remembered that exposure limits and other regulatory guidelines in existence today are derived mainly from the experience gained in industrialized countries. The health effects on LDC populations exposed to toxic chemicals have not been fully documented.

Epidemiology in the Study of Health Effects Associated with Hazardous Wastes

Industrial wastes often contain potentially toxic chemicals, therefore the health effects of industrial wastes cannot be evaluated without reliable information on the amounts and nature of the toxic chemicals present.

Two main approaches are used to assess the effects of toxic chemicals on human health: toxicity studies in animals or surrogates such as organs and mechanisms by use of bioreactive macromolecules, and epidemiological studies.

Principles and guidelines for the above are embodied in Environmental Health Criteria documents released by WHO (WHO, 1978; 1983; 1984; 1985b).

A profile of the biological effects of toxic chemicals is shown in Figure 16.1. Target organs and specific effects induced by toxic chemicals are summarized in Table 16.5. The concept of safety evaluation and its relevance to estimating societal risk are shown in Figure 16.2. The criteria used for establishing the cause–effect relationship in disease outcome associated with hazardous wastes are summarized in Table 16.6.

Epidemiological studies are very expensive and require highly skilled trained personnel. Certain priorities must be fixed before a decision can be taken to conduct epidemiological surveys on human populations living close to hazardous waste dumps or treatment sites. These priorities are summarized in Table 16.7. The role of epidemiology both in the investigative (diagnostic) phase and in the preventive phase is indicated in Table 16.8 and the issues to be considered in establishing preventive measures based on retrospective and prospective epidemiological studies are summarized in Table 16.9.

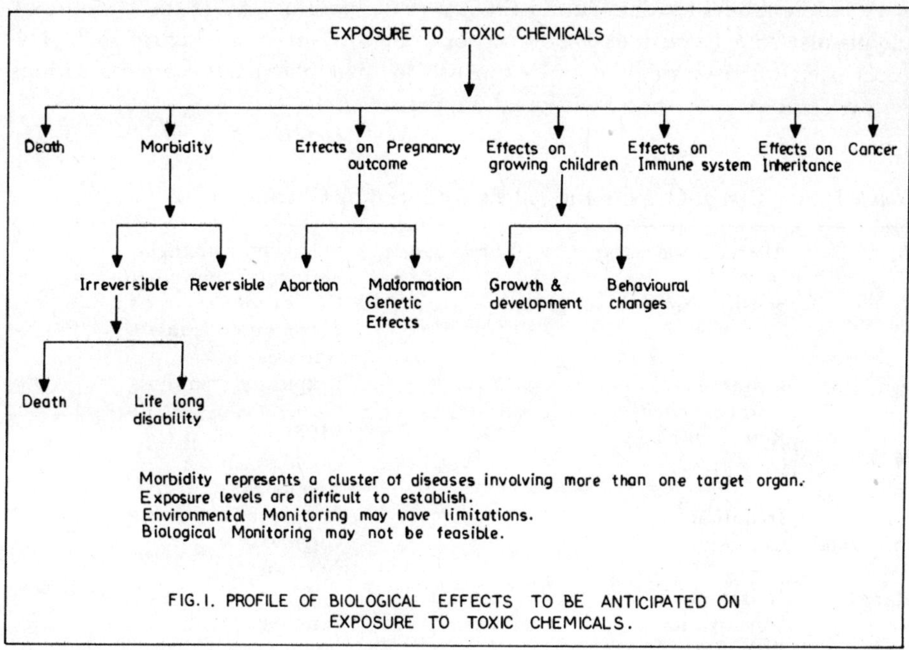

Figure 16.1 Profile of Biological Effects to be Anticipated on Exposure to Toxic Chemicals

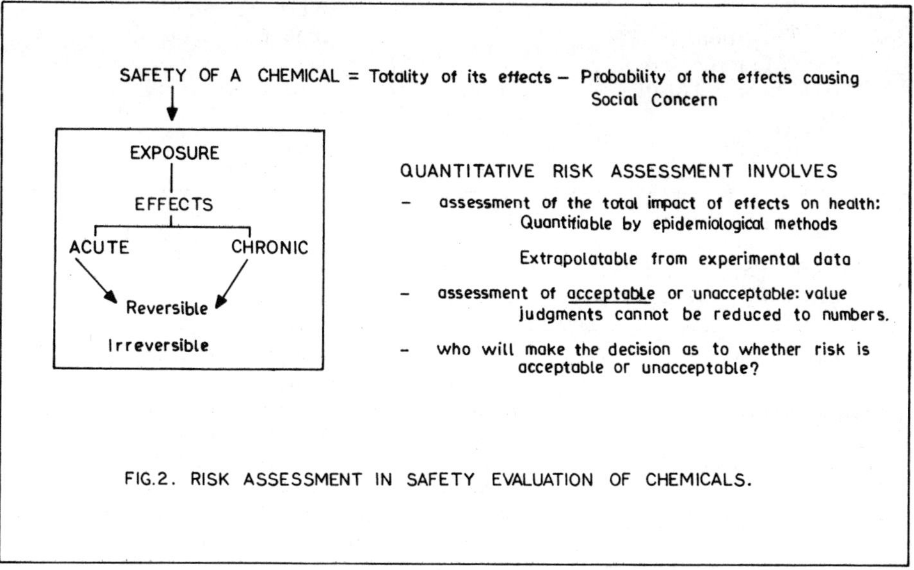

Figure 16.2 Risk Assessment in Safety Evaluation of Chemicals

A simple model for elucidating the routes of exposure and pathways of toxic chemicals from hazardous waste disposal sites is shown in Figure 16.3. The tools used for assessing human exposure to toxic chemicals from hazardous waste disposal sites are mentioned in Figure 16.4.

Table 16.5 Target Organs and Effects Induced by Chemicals

Skin	Altered appearance Irritation Sensitization Corrosion	*Nervous System*	Behavioural changes Peripheral neurophathy Depression Locomotor ataxia Narcosis Respiratory paralysis
Eye	Irritation Corneal opacity Retinal damage Corrosion	*Liver*	MFO induction Choleostasis Neoplasia Adenoma Carcinoma Cirrhosis Necrosis
Mucous Membrane	Irritation Corrosion		
Lung	Irritation Sensitization Pneumoconiosis Fibrosis Adenoma, Carcinoma Neoplasia Asphyxiation	*Kidney*	Aminoacidurea Uremia Lithiasis
		Haemapoeisis	Bone marrow depression Leukemia Aplastic anaemia Methamoglobinaemia
Foetus	Abortion Malformation. Neonatal death	*Musculoskeletal System*	Osteoporosis Corrosion
		Immune System	Suppression

Table 16.6 Criteria for Establishing Cause–Effect Relationship in Disease Outcome of Hazardous Wastes (from Grisham, 1986)

Consistency of Association	Observations can be reproduced using different methods and populations
Strength of Association	Incidence is greater in exposed population and increases with intensity of exposure
Temporal Relationship of Association	Disease appears after exposure allowing for required induction period
Specificity of Association	Does exposure lead to a specific health effect. Is an observed health condition the result of an exposure
Coherence of Association	Does past experience bear out present situation

Table 16.7 Priorities for Epidemiological Studies on Hazardous Wastes (adapted from Grisham, 1986)

Identify types, characteristics and concentrations present in leachates, ambient air, water or other media (food) of exposure.

Develop a relative exposure index based on actual ports of entry and absorption.

Select methods for objective assessment of health effects previously attributed to specific chemicals.

Evaluate potential for exposure at concentrations likely to cause adverse effects.

In case the effects are not agent specific, select control population from equivalent socio-economic standard.

Make detailed plans of investigations with design and protocols and establish communication with local authority, industry concerned and the community affected.

Table 16.8 Epidemiology to Assess Health Outcome of Hazardous Wastes

Epidemiology	*Action*
Investigative	
1. Health problem attributable to exposure to a chemical exists in a population	Find source of the chemical
2. A toxic chemical is known to be present at a site	Find evidence for excess of health problem in population exposed
3. A chemical present in a hazardous waste site previously shown to lead to a health problem	Test hypothesis in new or unknown site where similar exposure takes place
4. Preventive	Establish potential diffusion of chemical in exposed population. Anticipated health problems. Install preventive measures

Table 16.9 Issues to be Addressed in Establishing Preventive Epidemiology Services (USDHSS/CDC/CEH, 1984)

1. Standardized questionnaires. Base for minimal information to be solicited for survey of populations in exposed and unexposed areas.
2. Identification of sentinel health events similar to those known for occupational exposures. Include a spatial or temporal clustering factor.
3. Site specific census and health registries. Birth registries.
4. Identification of control communities for comparison prospectively

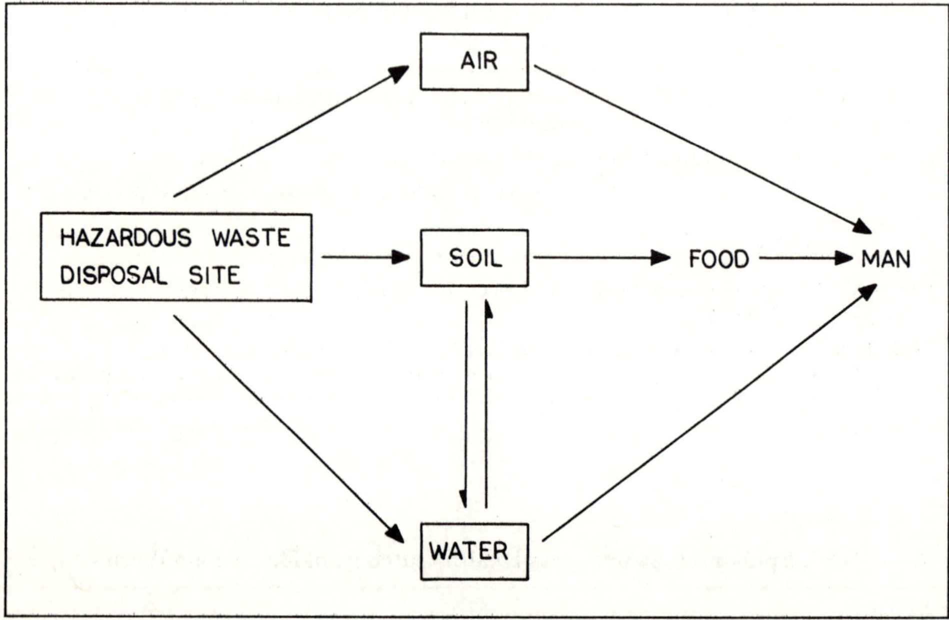

Figure 16.3 Exposure from Hazardous Wastes through Environment

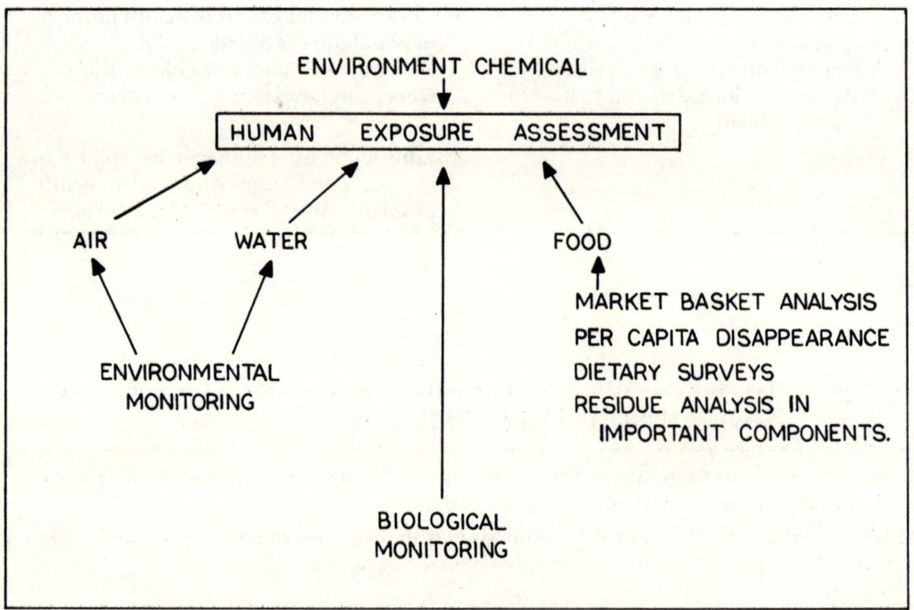

Figure 16.4 Tools Used for Assessing Human Exposure

Discussion

Historically, industrial wastes have been disposed of either on-site where the waste is generated or off-site through transportation by truck or rail. Site selection is dictated mostly by cost criteria and availability of land for disposal. Long-term environmental or human health considerations began to receive some attention in the United States after the passage of the Resource Conservation and Recovery Act of 21 October 1976. Generators of wastes did not analyze the chemical constituents of the wastes unless there was some material to be recovered or recycled (Fields and Lindsey, 1975). This is also true of other countries. The first task must therefore be to determine the chemicals at various disposal sites. It must be remembered in this content that all disposal sites have the potential for contaminating the environment by releasing toxic substances.

Any investigation undertaken on the health effects on human population living in the vicinity of hazardous waste disposal sites should envisage detailed analysis of the chemicals present at and escaping from the site, exploration of the routes of transport of these chemicals through the environment, concentration in the environment and estimates of the uptake and absorption by the exposed population.

Adverse health effects in the population living near hazardous waste disposal sites may involve any organ of the body or any of the vital physiological functions. The effects would depend upon the specific chemical, the characteristics of the individual such as age, sex and genetic make-up, the metabolism of the chemical and the operation on confounding variables such as personal habits and prevalence of other diseases. Of primary concern are cancer, genetic defects, congenital anomalies, reproductive abnormalities, alterations of immune status and disorders of the central nervous system and behaviour.

Epidemiological studies of populations living close to hazardous waste disposal sites include retrospective, prospective and case control studies. Comparisons are made with a population with similar characteristics. Epidemiological studies also enable us to establish preventive and anticipatory procedures to deal with future situations of toxic exposure. Monitoring and surveillance activities are essential components of any meaningful epidemiological programme to establish baseline data and provide a basis for registries. Such studies also help in generating valuable data on exposures which can be used for estimating the probability of exposure, absorption and the outcome as a disease cluster.

It is also important to evaluate ecological risks associated with hazardous waste disposal sites. Ecotoxic effects on terrestrial and aquatic flora and fauna will have to be evaluated with special emphasis on endangered species and sensitive ecosystems.

Maps of sites used for land disposal of hazardous wastes in all the industrial zones must be prepared giving details of the contiguous human habitations and ecosystems. Inventories of sites and nature of wastes disposed must be maintained and updated. Surveillance of the health of the population living in new townships constructed on erstwhile garbage and industrial waste dumps must be part of any regional health plan. Legislative mechanisms for exercising strict control must be activated. Plans for the training of manpower needed to handle various tasks involved in hazardous waste management must be evolved and implemented (Krishna Murti, 1984).

References

Fields, T., Jr. and Lindsey, A.L. 1975. Landfill Disposal of Hazardous Wastes: A Review of Literature and Known Approaches. EPA/530/SW-165. Washington DC.

Grisham, J.L., ed. 1986. *Health Aspects of the Disposal of Waste Chemicals.* A Report of the Executive Scientific Panel. New York: Pergamon.

Hardoy, J. and Satterthwaite, D. 1985. Third World Cities: The Environment of Poverty. *Journal 85.* Washington, DC: World Resources Institute, 45.

Harthill, M., ed. 1984. Hazardous Waste Management. *Whose Backyard?* AAAS Selected Symposium, 88. Colorado: Westview Press.

Holdgate, M.W., Kassas, M. and White, G.F., eds. 1982. *The World Environment 1972–1982.* UNEP Report. Dublin: Tycooly, 358–404.

Krishna Murti, C.R. 1984. Management of Toxic Wastes. *Productivity*, 24: 467–673.

Krishna Murti, C.R. and Sandoval, H. 1983. Environmental Pollution and Health Effects with Special Reference to Developing Countries. EPP/EC/WP/83.10. Background paper for WHO Expert Committee on Environmental Pollution Control in Relation to Development.

Paigan, B.J. 1984. Methods for Assessing Health. Hazardous Wastes Management. *Whose Backyard?* Harthill, ed.

Storach, A.S. and Vaz, R. 1983. Global Environmental Monitoring System (GEMS). Assessment of Human Exposure to Selected Organochlorine Compounds Through Biological Monitoring. Prepared for UNEP and WHO by the Swedish National Food Administration, Uppsala.

Valter, M. 1982. Assessment of Human Exposure to Lead and Cadmium Through Biological Monitoring. Prepared for UNEP and WHO by the Swedish National Institute of Environmental Medicine and Department of Environmental Hygiene, Stockholm.

World Bank. 1975. *The Assault on World Poverty. Problems of Rural Development; Education and Health.* Baltimore: Johns Hopkins University Press.

WHO. 1978. *Environmental Health Criteria, 6. Principles and Methods for Evaluating the Toxicity of Chemicals.* Part II. Joint sponsorship UNEP/WHO. Geneva: World Health Organization.

WHO/EURO. 1982. Hazardous Waste Management. Policy Guidelines and Practices. Report of a Working Group. Interim Document. European Regional Office of World Health Organization, Copenhagen.

WHO. 1983. *Environmental Health Criteria, 27. Guidelines on Studies of Environmental Epidemiology.* Joint Sponsorship UNEP/ILO/WHO. Geneva: World Health Organization.

WHO. 1984. *Environmental Health Criteria, 34. Principles for Evaluating Health Risks to Progeny Associated with Exposure to Chemicals During Pregnancy.* Geneva: World Health Organization.

WHO. 1985a. *Environmental Pollution Control in Relation to Development*. Report of a WHO Expert Committee. Technical Report Series, 718. Geneva: World Health Organization.

WHO. 1985b. *Environmental Health Criteria, 46. Guidelines for the Study of Genetic Effects in Human Population*. Joint sponsorship UNEP/ILO/WHO. Geneva: World Health Organization.

17

Risk Management in the Netherlands: A Quantitative Approach

C.J. van Kuijen

Reports of damage or threats to the environment are nowadays a regular occurrence. Sometimes the effects are only too clear, as in recent disasters in chemical industries. In other cases there is still no clear picture of the true seriousness or extent. Examples of these are possible climatic changes due to increasing CO_2 content in the air, the impoverishment of the environment in developing countries and pollution of the underwater soil in the large European rivers. It is being increasingly perceived that environmental problems are not just an irritating side-effect of economic and social development. The consequences of disrupting ecological processes and of carelessly managing natural resources are substantially threatening or even already undermining the conditions for wellbeing and welfare for present and future generations.

It has become clear that in order to manage our living environment adequately we must consider the physical environment as a whole, built up of ecosystems on various scales, from local/regional to global. These ecosystems, of which air, water, soil and organisms including humans being the constituent parts, are maintained by processes and cycles. Managing the risks of all kinds of pollution for humans, animals, plants and goods and being able to create the conditions for a lasting protection and development of our environment, therefore requires a more complete insight into the ecosystems and particularly into the processes and cycles that take place within these systems. This will enable us to formulate our environmental targets and to make the unavoidable trade-off between these targets and other societal constraints.

However, setting environmental standards only is not sufficient to solve environmental problems as has been clearly illustrated by Ruckelshaus.[1] In addition, a strategy is needed to make these standards implementable by governmental agencies and pollutors.

The formulation of these targets is very much hampered by the fact that our knowledge of the relations between pollutant discharges and ecosystem impacts is still very scanty. How should we establish the acceptable risk level of conceivable major hazards from chemical installations? Yet, decisions have to be made concerning emission and immission levels, the appropriateness of risk-reducing measures for hazardous installations and the siting of these installations in relation to their surroundings.

A significant fact is also that public concern about environmental problems has become an increasingly important political force. This demands a system of environmental policy making that enables the public to participate in the economic and social trade-offs required for deciding about the quality of its environment.

To summarize, the requirements of an effective environmental policy must be:

1. To cope with scientific uncertainty regarding essential toxicological and ecological relations in setting its quantitative environmental targets
2. To provide tools for authorities and industry to attain these targets
3. To enable true public involvement

Some years ago the Dutch government formulated the outlines of such a policy.[2] They were approved by Parliament in 1985. An explanation of this policy and some of its applications is given below.

Towards an Effective Environmental Policy

This "strategic" environmental policy is being formulated along two tracks: an effect-oriented and a source-oriented policy (Figure 17.1). The effect-oriented policy must make clear which objectives are being pursued with respect to the quality of the environment. Its ultimate goal is such a quality of the environment that no detrimental effects for people, animals, plants and goods of pollution or physical interference can be expected.

This goal is too abstract to give guidance for environmental management and especially to confront economic realities successfully. It is therefore necessary to make this "no risk" level clear by quantifying the concentrations of specific substances in water, soil and air, or the exposure of organisms or goods to noise, radiation, bad smell or danger, which are considered to be ultimately desirable from the environmental point of view. This quality level is designated as the target value.

For most environmental conditions the target value can only be reached in the long term. In such cases it must be pursued in several smaller steps, taken in an interim period. These "interim targets" are designated as environmental quality objectives, which reasonably guarantee that the risks will remain limited to an acceptable level. Unlike the target value, which is mainly based on ecological data, this quality objective is the outcome of a policy decision which

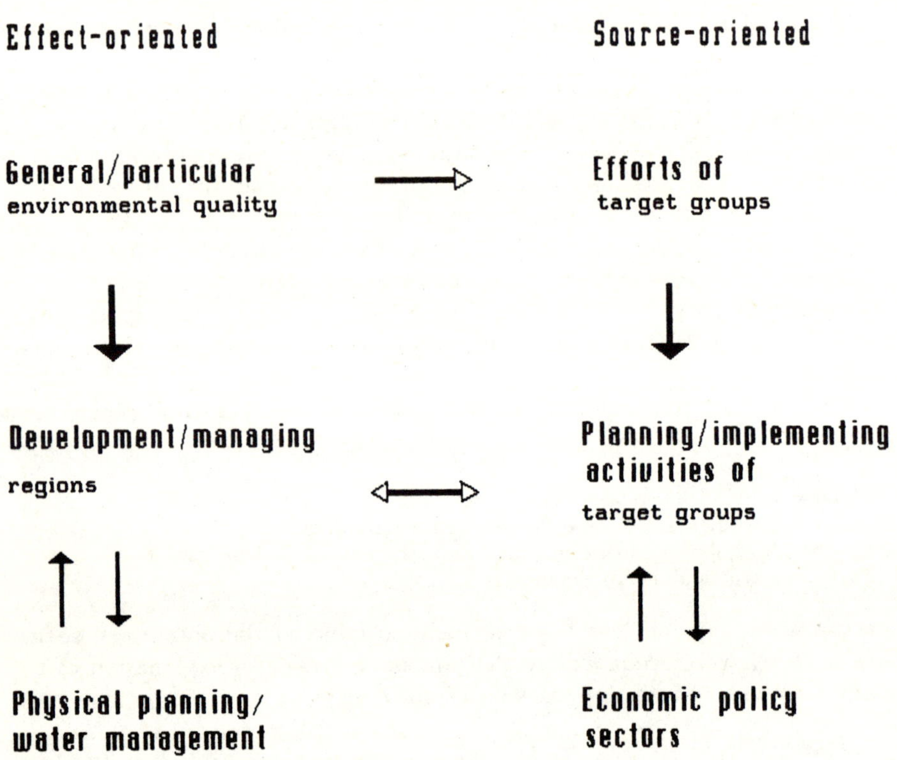

Figure 17.1 Two Track Policy

takes into account environmental interests as well as social and economic interests and technical possibilities. The space within which this trade-off has to take place (the "grey area") is bounded on one side by the target level and on the other side by the pollution or exposure level which is judged maximally tolerable. This approach is presented in Figure 17.2.

Environmental quality objectives are the link between the effect-oriented policy and the source-oriented policy, for they give an indication to what extent efforts must be made by industries and other sources of pollution. These are the target groups of environmental policy.

The source-oriented policy makes clear the way and the timescale in which the behaviour of target groups will be "corrected", with attention paid to the environmental quality objectives and tasks formulated in the framework of the effect-oriented policy. Depending on the target group in question, it will have to be determined what "mix" of regulatory and stimulatory instruments is most effective in achieving the necessary change in polluting behaviour. The choice regarding the package of measures will have to be made jointly by the environmental policy sectors and other policy sectors of the government.

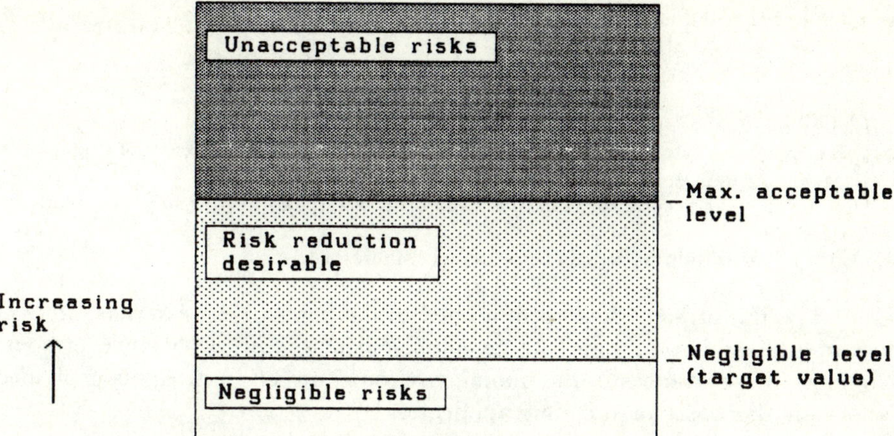

Figure 17.2 Dealing with Risks in Environmental Policy

Dealing with Risks

Within this environmental policy framework a risk management strategy has been elaborated. An essential element in this strategy is determining the two levels mentioned above (Figure 17.2). As a first step this has been done for people. Establishing these levels for ecological risks is an even more complicated task, especially because of the complex relationships that play a role in organisms and in ecosystems. Nevertheless it is expected that, as a result of research into ecosystems and ecological processes, more parts of the intended integral risk management system can be gradually completed.

Risk

Risk is defined as the chance of undesired events occurring in relation to the possible extent of the events' consequences to the population. The combination of chance and effect is the crucial element in the definition.

Risk to people can be expressed in two ways, as individual risk and as group risk. *Individual risk* may be defined as the chance per year that a person located at a specific position with respect to the risk source is affected by the undesired consequences of an event. For example, the risk of a random person in the Dutch population dying in a traffic accident is 1.4×10^{-4} per year. *Group risk* may be defined as the chance per year that a certain group of persons in a certain environment will simultaneously experience the detrimental consequences of an undesired event during an established period. This concept is used especially in determining the risks from accidents of which the societal consequences can be sizeable. Application of this concept makes it possible to take better account of the size of the group who can be simultaneously victims of an undesired event in the decision regarding risk acceptability. Accidents such as those in Mexico City and Bhopal illustrate the importance of this approach.

Risk Management

There are four sequential steps in the process of dealing with risks:

1. Identifying the dangers to people or the environment
2. Quantifying the extent of these dangers. Both the chance and the extent of exposure as well as an agent's detrimental properties play a role in this
3. Determining the acceptability of the risks of the activity and the desired risk reducing measures
4. Control – maintaining a situation of acceptable risk

Until now this strategy has been applied to the fields of the external safety of hazardous installations and of the management of the occurrence of some carcinogenic substances in the human environment. The strategy is explained below on the basis of the first application.

External Safety Policy

This policy is directed towards preventing unusual incidents with undesirable consequences for the surrounding area involving activities with dangerous substances and towards limiting those consequences as much as possible. Managing the so-called "small probability–large consequence" risks requires special attention.

Identification

In this step it is established whether a chemical plant or installation can cause a hazard to its surroundings and the nature of the hazard is defined. This is done on the basis of established or suspected detrimental effects, mainly on humans. The recent incident on the Rhine near Basel emphasized the urgent need to pay more attention to potential effects on the environment.

Quantification

Using methods of quantitative risk analysis the individual risk contour and the group risk curve of the activity under consideration are calculated. Quantification of the risk of "disaster type" incidents is currently completely feasible. In this field the progress of science aided by data processing developments has been remarkable. In 1982 the Ministry of Housing in the Netherlands decided to initiate a project to develop a computerized risk quantification tool using the most recent information available in the fields of the dispersion of heavy gases, unconfined vapour cloud explosions and fireball radiation, and the response of man to exposure to large concentrations of toxic chemicals during a relatively short period. The SAFETI computer package has resulted from this project.

This package consists of some 35 separate programmes that assist the risk analyst in the quantification of the risks of chemical plants and the associated transport. From a data base of vessels and process conditions for a particular plant, a set of potential failure scenarios is generated. The frequency of these accidents is derived from a data base which contains information on the frequency of failure of process plant components. The consequences are calculated using a built-in physical property data base for the substances involved. These consequences can result from outcomes of accidents such as fires, explosions (including BLEVES) and toxic vapour clouds. An elaborate set of models to calculate the various phenomena that occur as the initial accident scenario develops is available. The results of the consequence calculations are combined with data on the local weather conditions, population distribution and ignition source locations to calculate the final impact of the scenarios on the population. These results are finally aggregated into the individual risk contours around a plant or a site and into a group risk curve (Figure 17.3). The

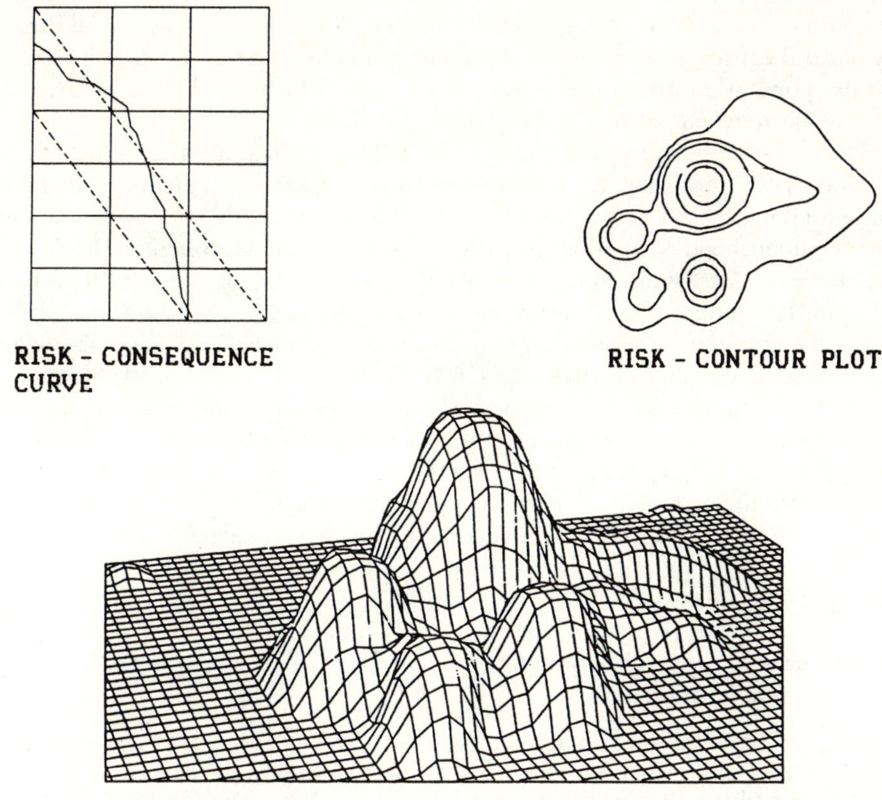

Figure 17.3 Quantifying the Risks associated with Chemical Plant Location

use of advanced computing techniques allowed an appropriate level of complexity to be incorporated into the modelling at each stage in the risk quantification procedure. This results in an enhanced precision in the calculation when compared with methods used previously.

The accuracy of this model has recently been systematically analysed.[3] The inherent uncertainty in the modelling has been found to be a factor of 3 and the uncertainty in the estimated frequencies about 10. Although this uncertainty is still large, the results of the model are considered to be sufficiently reliable and close enough to observed effects and frequencies to be usable in decision making.

Decision about Acceptability, Risk reduction

For this step in the decision-making process the calculated risk has to be evaluated against the levels mentioned above (Figure 17.2). For activities with dangerous goods in the Netherlands the maximum acceptable level for individual risk has been taken as the risk level which increases the risk of death by natural causes by a maximum 1%. The individual "natural death" risk run by the population group of 12- to 16-year-olds, which is 10^{-4} per year, has hereby been taken as the basis of risk.

The maximum acceptable individual risk has thus been established in general as 10^{-6} per year. In other words the risk of a fatal accident to which an individual is exposed because of his continuous presence (365 days per year) in the neighbourhood of an activity with dangerous goods shall be less than once in a million years. Risk exposure levels of less than 10^{-8} per year, or less than once in 100 million years, are considered negligible.

For the maximum acceptable group risk level a chance of 10^{-5} per year of an incident with a maximum 10 deaths has been chosen. A chance of 10^{-7} per year with 10 deaths is taken as the negligibility level for group risk. Further, a heavier weight must be assigned to the larger consequences of accidents. It has been decided in this connection that a consequence n times greater must correspond to a chance n^2 smaller, as it appears from literature that the seriousness of the societal consequences of an incident is judged to increase with the square of the number of people killed. These risk criteria are depicted in Figure 17.4.

A differentiating policy is desirable for both individual and group risks falling within the grey area. Risks from new industrial activities falling in this area are only acceptable after:

1. Adequate risk reduction measures have been instituted or safer alternatives have been chosen, aimed at reaching the target values for individual or group risk
2. The permitting authority has weighed the risks and disadvantages of the activity involved against its benefits and is convinced that the relationship between risks and benefits is acceptable
3. The interests and perceptions of the population liable to the risk of the activity have been considered by the permitting authority in a balanced and responsible way

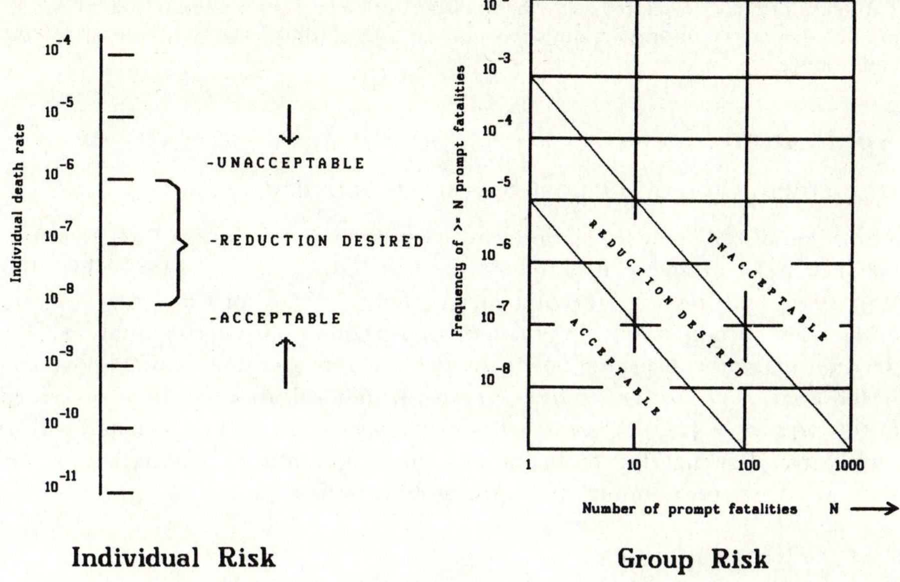

Figure 17.4 Risk Criteria for the Policy on External Safety in the Netherlands

Risk reduction can be achieved in two ways: first of all *in situ*, by measures such as the layout of the plant, the application of additional safety devices and the employment of a less hazardous activity and, secondly, by zoning. Often a combination of both types of reduction is needed. Limiting the size of the zones by measures at the installation and maintaining sufficient distance to sensitive areas promote both prudent space use and the good "fitting in" of installations where activities with dangerous substances take place. One of the major advantages of risk quantification is that it can provide information about the cost-effectiveness of different sets of risk-reducing measures, and it provides a tool for zoning.

Control

When an acceptable level of risk has been decided, decisions have to be made and implemented to safeguard this situation. Which specific measures have to be taken will depend on the type and scale of the activity involved.

Generally speaking the following actions will or may be required:

1. For stationary sources the licence will have to specify what safety provisions have to be taken and what procedures should be followed to test these safety devices
2. The municipal authorities will have to implement the required zoning measures and to maintain them. Where the necessary distances between the installation and the population cannot be achieved, removal of vulnerable dwellings or the hazardous installations will have to be considered

3. In case of risks associated with the transport of hazardous materials, action will be promoted either by improving the safety of the means of transport or by means of routing and zoning, or both

Applications

Implementation of the post-Seveso Directive

At the end of 1987 new legislation was in force to implement this EEC Directive on the major-accident hazards of certain industrial activities. This new legislation will require the industries concerned to provide the relevant authorities with a notification comprising a quantitative risk analysis. For decision making concerning new activities the approach mentioned above will be followed. For existing activities, about which notification has to be presented at the latest on 8 July 1989, further policy measures are to be taken and will be formulated after that date on the basis of the insight into the installation's safety, that will then be obtained from the notification.

Natural Gas Pipelines

Over the years there has been considerable argument about the distance that should be maintained between natural gas lines and housing developments. In the Netherlands land use has to be as optimal as possible, yet the safety of the population has to be guarded. To settle this dispute a risk analysis of the gas lines was performed on the basis of the risk criteria put forward above.[4]

It was concluded that the group risk criterion could not be applied for the very reason that the frequency of incidents grows with increasing length of the line, while on the other hand different persons are affected along the line (and not the same people as in the case of a plant). It was, therefore, decided that the safety distances should be based on individual risk alone. As a result, a directive concerning zoning along main transmission lines for natural gas was published by the Minister of Housing in 1984.[5]

Liquefied Petroleum Gas (LPG)

This approach was also adopted in LPG policy. In 1978 the large oil companies in the Netherlands expressed their expectation that the market for LPG as motor fuel and feedstock would see a spectacular growth. Within a few years import and transport were expected to grow from 1 million tons a year to 10 million tons a year. In addition this transportation would pass through densely populated areas and cities. The safety of these activities was a subject of general concern for the public and the authorities. The Netherlands government commissioned a study to examine the safety of all parts of the chain of activities from the importation through the sea harbours to the distribution of gas stations selling LPG.[6] The results of these studies formed the basis for the policy statement of the government to Parliament in 1984.[7]

In new legislation the risk contours are translated into safety distances to be applied to unloading facilities, depots and storage tanks, and measures are presented to increase the safety of barges and road tankers. Existing stations which are located within 15 metres of dwellings will be closed. For the other existing stations located at greater distances (up to 120 metres from dwellings) a programme of risk prevention measures will be executed. The cost of this (about 125 million guilders) will be borne by those companies selling LPG. The new legislation also specifies that new stations should be located along designated highways considered safe for LPG transport, or allocations which can be easily reached from these highways without passing residential areas.

General Electric Plastics

In 1980 General Electric Plastics applied for a permit to extend its plastics facilities in Bergen op Zoom in the south-west of the Netherlands. The facility imports chlorine (approximately 600 tons per week) and uses phosgene as an intermediate. The inhabitants of the nearest village, 600 metres from the premises, appealed against the granting of the permit on the grounds of it being hazardous. A risk analysis was performed. (This was before the computerized techniques became available and the costs were relatively high, approximately 250,000 guilders). The results were that the phosgene part of the process did not contribute greatly to the risk, but that as a result of the risks of the chlorine unloading facility the extension of the plant would bring the 10^{-6} individual risk contour over a large distance into the village. It was, therefore, decided that the permit for the extension could only be granted if General Electric Plastics agreed to build its own chlorine production facility. This would reduce the risk and bring the 10^{-6} contour outside the inhabited part of the village. General Electric Plastics agreed to this and the permits were granted.

Zoning around the Dutch State Mines (DSM) Site

In the south-east of the Netherlands, in the province of Limburg, is the main site of one of the larger chemical companies of the Netherlands, DSM. This company produces, among other products, over half a million tonnes of ammonia and just under half a million tonnes of acrylonitrile, most of which is further processed to produce a wide range of fertilizer, polymers and other products. The company employs about 10,000 people on this site. The installations are built on a site that up to 25 years ago housed a large coal mine. The housing developments around the site are still influenced by the previous planning of coal miners' housing, which means that residential areas are built in relatively close proximity to the site and thus to the chemical installations.

At this site in 1975 there occurred one of the largest explosions in the history of the process industy. In this accident 14 people were killed and 104 injured. The concern raised by this incident was reinforced by more recent, even more damaging incidents in Italy, Mexico and India. In this same period the awareness and concern about the environment in general and especially air pollution, noise, and waste problems led the government of the Netherlands to initiate an environmental action plan for the region.

The objectives of this plan are, on the one hand, to reduce the environmental impact by measures at the site and on the other hand to establish a zoning policy to ensure sufficient separation between the site and the surrounding population. The two determining factors for this zoning exercise are noise and risk (of hazards).

The methodology described above was applied to the DSM site, where there are some 35 separate installations. A first screening showed that some 10 of these posed a risk outside the site boundary. Each of these was then analyzed in detail (Figure 17.5). (This risk analysis, because of the availability of computerized methods could be performed for only just over 300,000 guilders.)

These risks proved to be lower than the authorities had initially (before the study) feared. Nevertheless, there were several problem areas for which measures had to be taken. An interesting aspect of the results was the breakdown of risks associated with the various process plants on which quantitative results were prepared. In general, the major production units (ethylene crackers and ammonia plants) were not the dominant features of the overall risk. Rather the ammonia distribution and storage system of the LPG handling system was more significant. This was not known prior to the study and insight of this sort in managing risks is often most valuable.

From this study it was decided that in areas where the risk is above 10^{-5} per year and no risk-reducing measures can be envisaged, no new housing developments will be allowed. In the zone between 10^{-5} and 10^{-6} only replacement of existing houses will be allowed. Furthermore, engineering studies will be undertaken to investigate the feasibility of reducing the risk. The costs of these risk-reducing measures will be compared with the costs of the zoning requirements and an appropriate course of action will be set out subsequently.

Transport of Dangerous Substances

Recently in the Rijnmond area the risks of the transport of chlorine and ammonia have been tackled. From the results, the iso-risk contours and F–N curves for the various routes and modes can be examined.

Plans for the Future

The development of these computer-assisted methods has also been a starting point for more advanced developments. The Ministry of Housing is presently engaged in a collaborative exercise with the Joint Research Centre of the EEC and the International Institute for Applied Systems Analysis to develop a decision support system. This system will allow data on the risks of different modes of import, production, storage, transport and use to be combined with socio-economic data to help the decision maker to seek an optimal solution to complex problems. Other lines of development are groundwater transportation models, long-term atmospheric models and the effects of incidents at nuclear installations.

The methods described above are aimed at dealing with risk problems only, not with problems connected with the *perception* of risk or with different views on

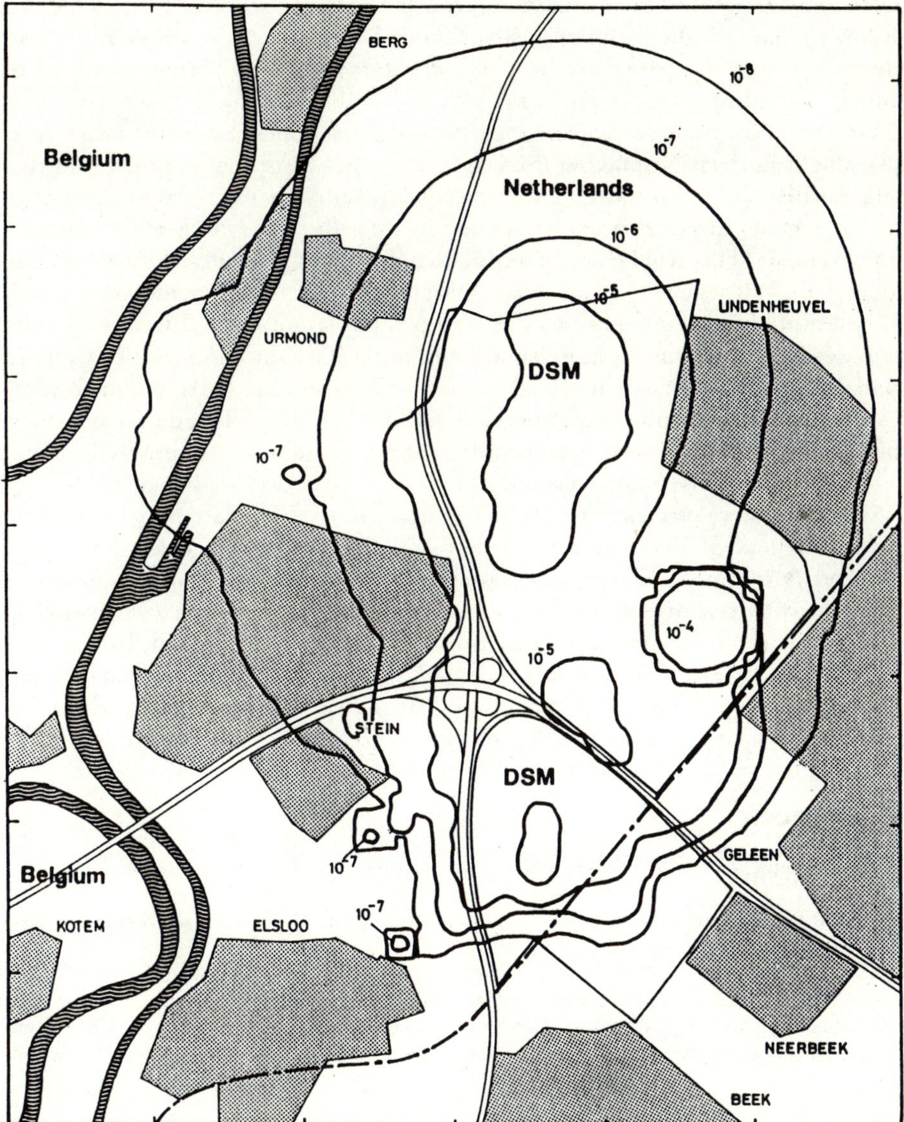

Figure 17.5 Risk Contour Plot for the Whole Site of DSM

the further development of our society. The nuclear power problem is an example of a still more complex issue. In these types of problem it is tempting to devote all efforts to solving the risk issue, only to find out later that solving the risk issue has not solved the problem.

Conclusion

As shown above this quantitative approach has been proven to be viable. It allows a balanced allocation of means to reduce the risk there, where it is most effective. It has also proved to be useful in solving physical planning problems and in the siting of new chemical plants.

Yet there is much scepticism regarding the usefulness of quantitative risk assessment in decision-making from industry. First, it is being argued that the data needed for quantitative risk analysis in many cases is too scattered to justify a generalized approach. We agree that the data base needs further improvement. This will be one of our main objectives for the near future. On the other hand the fact remains that decisions have to be made about the acceptability of risk, often in relation to very complex situations. In our opinion such decisions will have a more solid foundation if quantitative risk analysis is available, than when such decisions are taken on the basis of qualitative assessment only, despite the imperfections in some data. The decision maker should be aware of the uncertainties involved and accommodate these uncertainties in his final decision.

Secondly, there is a fear that the use of quantitative risk data may jeopardize the willingness of the public to accept additional hazards, even if the risk involved is marginal. It must of course be recognized that public opinion is determined by risk aversion, especially where non-voluntary risks are involved. However, it is for this reason that we prefer a normalized approach. In the view of the Dutch government it is only by means of politically established standards that endless disputes about the acceptability of a specific type of hazard can be avoided. Industry will eventually recognize that this is in its interest too.

References

1. Ruckelshaus, W.D. 1985. Risk, Science and Democracy. *Issues in Science and Technology*. Spring.
2. Indicatief Meerjaren Programma Milieuhygiëne 1986–1991. Tweede Kamer, 1985–1986, nr. 19204, 1 en 2.
3. AVIV. 1986. De onzekerheid van effectberekeningen in risico-studies. Enschede.
4. TNO. 1983. Risico-analyse van ondergrondse buisleidingen.
5. Circulaire van de Minister van Volkshuisvesting, Ruimtelijke Ordening en Milieubeheer inzake zonering langs hogedruk aardgastransportleidingen, November 1984.
6. TNO. 1983. LPG Integraal.
7. Integrale Nota LPG. Tweede Kamer, 1983–1984 nr. 18233, 1 en 2.

18

Risk Assessment: Danish Practice and Experience

Niels K. Vestergaard

In 1976 an accidental discharge in Italy containing dioxin, initiated the preparation of a directive to be issued by the European Communities to regulate hazardous activities. The resulting directive on "Major Accidents Hazards of Certain Industrial Activities" was passed as legislation in 1982 by the European Parliament. During the intervening period, legislation with a more or less similar purpose was introduced in the Danish Parliament on the basis of the national environmental and labour protection laws.

Danish Practice

Originally, in Denmark the administration of the regulations governing hazardous activities was split between three ministries; the Ministry of the Environment, the Ministry of Labour, and the Ministry of Justice. This created some co-ordination problems and on the basis of the EEC directive, the Ministry of the Environment and the Ministry of Labour decided to present a common procedure in which only one registration would be necessary for a hazardous activity application.

The basic philosophy of Danish practice is that:

1. Industries with a certain level of risk, because of storage, handling or production of chemicals, are obliged to document their safety precautions to the authorities
2. The same industries will, after possible approval, obtain a permit which will allow an operation without further interference from the authorities

Danish practice is hereby stated in a guideline issued by the National Agency of Environmental Protection. This guideline stipulates that it is only necessary for the applicant to complete one report (since it is automatically distributed to the other authorities) and that only one response will be given to the applicant. After an appeal of a number of weeks' duration, the response will form the basis for the issuing of a permit, valid for a given period. Under normal circumstances the permit will be valid indefinitely.

A risk analysis can be carried out at various levels of complexity. For some plants only a simple risk analysis is necessary proving that the probability of accidents or consequences of such accidents are minimal. For other plants it is possible to implement risk analyses performed at plants with an identical structure as reference analyses, thus bypassing the need to prepare a new analysis.

Moreover, some plants will only need to prepare a few pages stating the quantity, type and location of the hazardous materials, if they fulfill the following conditions:

1. The hazardous material must not be present in a way which can cause major hazards
2. The hazardous material is only present in quantities less than the minimum limit stipulated by the EEC.

On the whole, about 50 major plants in Denmark will need to undergo a risk analysis. The common concept is given in Figure 18.1.

Experience

Only a very limited number of "normal" industries in Denmark have to date prepared a risk analysis leading to a risk assessment by the authorities. Two companies have prepared risk analyses in accordance with the "old" national regulations. The first plant, a chlorine production plant, located in the heart of Copenhagen had a risk analysis carried out between 1979–80. In the spring of 1987, the Danish Parliament decided to overrule the local authorities and let the next parliamental session prepare a law for this specific plant. The second plant, the Kommunekemi facility in Nyborg, which treats hazardous wastes, had a risk analysis carried out in 1983 in response to public concern. Contrary to the results of the analysis carried out at the chlorine producing plant, this analysis resulted in the acceptance of Kommunekemi by the majority of the municipality.

These two analyses are, however, not representative of the new Danish procedure as they were brought about by public concern and national press coverage and not by a standard procedure to assess major hazards.

Between 1986 and 1987 Danish chemical plants began risk analysis and approval procedures in accordance with the directive on "Major Accidents Hazards of Certain Industrial Activities". The analyses carried out on the basis of the common concept illustrated in Figure 18.1 made it possible to deal with both minor and major hazards within the same framework of the analysis and legislation.

However, summer 1987 was too early to discuss the experience gained by this

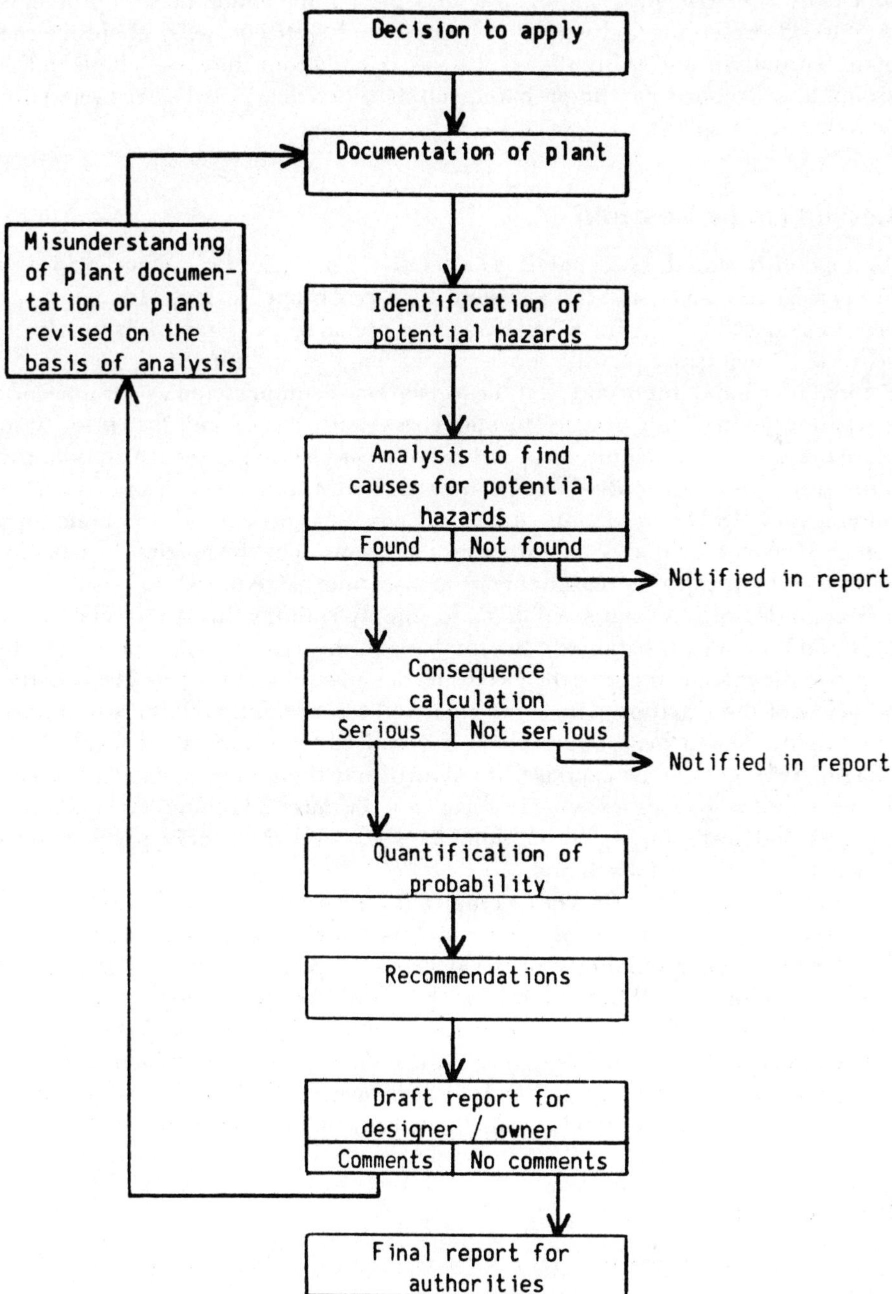

Figure 18.1 A Common Risk Analysis Concept

procedure. On the other hand, branches presenting minor hazard problems (e.g. wood preserving industry) are able to pool their efforts by choosing one plant for analysis and using the results as a reference for the other plants in the group. Short reports can be prepared stating differences between the reference and the actual plants.

Lessons to be Learned

As previously stated, risk analysis is a relatively new tool for the authorities to use to raise the safety standards in some high-risk branches. Even though it is a new tool, the first "chemical" risk analysis prepared in Denmark is nearly 10 years old, and consequently some lessons have been learned.

First, it is found important that the legislation is administered as a framework legislation giving high priority to major risk contributors and less priority to more normal risk contributors. If this is not clearly stated in the legislation, the authorities will not only have a new tool but they will also drown in bureaucracy. In Denmark only about 10 factories at present need to undergo a comprehensive risk analysis. At present, this seems to be the maximum capacity of the authorities, if a reasonable response time is required.

Second, factories which are obliged to comply with the directive seem – after a period of some resistance – to implement the risk analysis positively to improve the management of their contingency planning. This provides a better overview of the situation which can then be used in public relations to inform the employees and neighbours about the real dangers and arrest fears about imagined dangers. This exercise has shown that the money spent on the risk analysis, to a greater extent, is recovered because planning for imagined dangers and installing safety equipment, which provided only psychological benefits, can now be excluded.

Finally, on the basis of several completed minor analyses, it has been shown that the implementation of risk (and reliability) analysis has given a significantly higher reliability of the system and thereby generally improving industrial economy. This might be the most important aspect in newly industrialized and developing countries. It might therefore be important for newly industrialized countries to make use of the experience obtained elsewhere to support production reliability and safety by using a single tool, since it will be the higher production reliability which will bear the costs of safety precautions.

References

Major Accidents Hazards of Certain Industrial Activities, EEC Directive of 24 June 1982.

Pligter ved risikobetonede aktiviteter, vejledning fra Miljøstyrelsen nr. 3. 1985.

Redegørelse om risiko i forbindelse med Dansk Sojakagefabrik's kloralkalianlæg, Miljøklontrollen i København, 9. October 1980.

Schepper, L., Rasmussen, B., Vestergaard, N.K., Ott, S., and Grønberg, C. 1986. Risikovurdering af Kemisk Værk Køges herbaphenanlæg. February.

Taylor, J.R., Schepper, L., Kampmann, J., and Styr Petersen, H.J. 1983. Sikkerhedsanalyse af Kommunekemi's affaldsbehandlingsanlæg ved Nyborg. June.

Vestergaard, N.K., and Rolskov, H. 1987. The Use of Risk Analysis as a Design Basis. World Conference, Chemical Accidents, Rome 7–10 July, sponsored by WHO, International Programme on Chemical Safety and Istituto Superiore di Sanita-Roma.

Vestergaard, N.K., and Taylor J.R. 1986. Tanker Incidents Probability as a Basis for Environmental Emergency Planning. SRE-symposium, Otaniemi, Finland, 14–16 October.

19

Modelling of Uncertainty in Hazard Assessment

Niels C. Lind

Uncertainty is a major impediment to success in industrial development. A project that is apparently profitable may never come to be initiated just because the risk of failure is unacceptably large; if the project is undertaken, it may turn out to be unprofitable because the true conditions were not known; unforeseen circumstances may lead to accident, catastrophe or ruin; or the project may just turn unprofitable because too many resources must be allocated to implement safety measures that are insufficient. Uncertainty stands in the way of effective investment of resources. This is particularly true for countries, regions, or industries under development, because they must compete for capital against alternatives that are less risky.

It is therefore desirable to develop optimal ways to deal with uncertainty. Procedures must be rational rather than merely operational. They must be lucid and deliberate, clearly oriented towards a well-defined goal, whether it be just maximum payoff (within societal constraints) or some means to the public good. The methods must therefore be *documentable, replicable* and *defendable* in rational dispute among well-informed professionals ("experts"), or in public forum.

Because of this need for critical evaluation, it is not just idle philosophy to ask, even in the most practical contexts: "What is uncertainty, anyway?" We can restrict the discussion to a technological project: a system that is designed, built (more or less as designed) and has to perform well in a changing, unknown and unpredictable natural and social environment – in spite of an imperfectly monitored process of deterioration.

Uncertainty has been classified into "observational" uncertainty, that would remain even if the characteristics of the system were perfectly known, and "analytic" uncertainty, which is further divided into modelling uncertainty and parametric uncertainty (Paté-Cornell, 1987). Similar classifications are sometimes used with different labels, for example, "ambiguity". For the purposes of this paper it is convenient to consider four major categories of uncertainty (in logical order):

1. *Model uncertainty,* being an expression of our awareness of the difference between the *weltbild* of the design and the real world containing its practical realization and
2. *Vagueness* in the descriptions of the system states, the parameters, or in the criteria of performance
3. *Indeterminacy,* meaning the uncertainty in our knowledge about the parameters of the system and the environment
4. *Randomness,* reflected in the irreducible fluctuations from time to time and from place to place of the environment, or the variations in the parameters of the system between one realization and its replication

For example, consider the project of a chemical waste disposal site in the vicinity of the floodplain of a river. The river level is a random process in time. There are regular fluctuations, and superimposed random fluctuations over the projected lifetime of the site. If this lifetime is sufficiently short that climatic changes can be neglected, it is possible in principle to determine the distribution and the correlations of river level to any desired tolerance of detail. The river level can then be modelled as a stationary stochastic process of known type with known parameters; it is purely random – category 4.

In practice, we must rely on limited data and make "best" guesses about the type of process and the value of its parameters. These are the indeterminacy of the model – category 3.

Broadly speaking, there exists for this waste disposal as for every other designed system an optimum amount of data and, utilized in optimal fashion, this gives models that are "best" in some well defined sense. If, in unrelated projects, a series of reservoirs are built above the site, it is easy to see that the model ignores some fundamentally important variables; this exemplifies model uncertainty (category 1) in our model of the waste disposal project. Model uncertainty can be reduced by smart modelling but cannot be eliminated; every model has an uncertainty that can only be assessed by comparison with a better model.

Finally, should flooding occur, the consequences are imprecise. There will be some leaching – but how serious is it? There will be some environmental damage – but how much does it matter? And there will be *ripples* (Slovic 1987), that is, social and economic impacts beyond those directly affected. These impacts are often of greatest importance and they may involve costs to the company or agency that far exceeds the direct costs. The pollution of Love Canal, N.Y. is a familiar example. All this constitutes vagueness (category 2) in the problem.

Randomness and Indeterminacy

Many concepts and techniques have been developed to model randomness, indeterminacy and vagueness individually but not together. Historically, the classical frequentist probability theory has been employed to describe randomness and to summarize data for the estimation of model parameters. Many people believe that only frequentist probability is sufficient and necessary to decide if a technological risk is acceptable. However, all projects involve acts that affect the future, not just observation of the past; every project embodies *forecasts* and estimates of trends. Often, moreover, the available evidence is scanty. All probability used in the design and choice of a project is therefore subjective. It follows that frequentist probability is conceptually inadequate for the design or licensing of hazardous facilities. Finally, it is inappropriate in the assessment of risk associated with a *particular* industrial plant or process, as pointed out in the weatherman example of Jaynes (1976).

Bayesian probability concepts have been developed for this context and are naturally suited when randomness dominates the uncertainty. The Bayesian rationale, strictly speaking, is incomplete; it offers merely a recursive operation to update a probability assignment. It presupposes the existence of a valid *a priori* probability which is intrinsically subjective; it provides no suggestions as to how a legitimate prior assignment can be made.

In theory it is possible to reduce the indeterminacy in a project by collecting more data and updating probability in a Bayesian way. However, this is often not económical: the value of data is subject to a law of diminishing returns. Moreover, it is impossible to produce non-existing historical data on rare natural phenomena that affect the risk of hazardous waste or industrial plant, as, for example, earthquakes, cyclones or floods.

Decisions must nevertheless be made; when data is insufficient something else must be substituted. Methods based on the principle of maximum information-theoretic entropy (Kapur and Kesavan, 1987) can be defended as the most objective or least biased.

One way to produce surrogate data is to ask for the opinion of those that supposedly know best, the "experts". A formal procedure has been suggested by Cook and Unwin (1986) to produce probability assignments based on maximum entropy. Methods to reconcile expert opinion (when in conflict) are available, such as the Delphi method. Special methods for pooling or aggregating probability information have also been developed (Rosenblueth, *et al.*, 1987; Genest and Zidek, 1986; Lind and Novak 1988). When alternative methods are available, one must face the question as to whether they will yield the same result. It is not necessary in practical applications that the alternative methods yield identical probability output for the same input; it is sufficient that they lead to the same decision.

The method of Rosenblueth *et al.* (1987) is strictly Bayesian, processing expert opinion as if it were data. A method of pooling probabilities is said to have *Bayesian consistency* if the two operations of Bayesian processing and pooling are commutative. This means that if all experts are Bayesians and if they all receive new data, then their updated probabilities when pooled will be identical with the Bayesian update of their prior pooled probabilities. Lind and Novak (1988) have shown that certain ways to pool data are deficient, in that they lack this Bayesian consistency. Foremost among these pooling formulas is the obvious one: A weighted linear combination of individual probability distributions. On the other hand, the logarithmic pool, aggregating the weighted sum of the logarithms of probabilities, does possess Bayesian consistency. But unfortunately it yields zero probability wherever just one of the experts reports a zero probability. For example, if just one of many experts believes that the probability of harm from a certain chemical in concentrations less than x is equal to zero, then the logarithmic pool will reflect this view over all others. Clearly, "Bayesian consistency" and "non-dictatorship" are two desiderata that may be in conflict. Genest (1984) has shown that Bayesian consistency and "non-dictatorship" cannot be achieved simultaneously with another axiom which may be called "independence of irrelevant alternatives".

The importance of this result for the present context is that there does not exist a universally valid way to pool conflicting probability data. What is appropriate for one context may not be appropriate for another; it may be a subject for debate. These questions are probably best settled in context by professional consensus.

Thus, sufficient data are sometimes not available for a project and pooling opinions of experts may not be entirely free of controversy. Apart from this, the Bayesian paradigm offers a defensible (self-consistent) rationale for decision making in the presence of randomness and indeterminacy. However, one important aspect of indeterminacy remains: the extent to which its mere presence may affect the decisions. In part because of large risk uncertainties, for example, probabilistic safety analysis is not used in Canada in the process of licensing of nuclear facilities, and it is used only in the United States "as a supplement to the regulatory process". The agencies are clearly reticent about abandoning the frequentist belief of probabilities. They are also interested in more than the expected values of the probability and they are concerned that expert opinion may not be an acceptable basis in sensitive areas involving public safety. Paté-Cornell (1987) has described the notion of secondary probability as an aid to explaining such aversion to indeterminacy (as distinct from the well known "risk aversion").

Agencies that reject probabilistic safety analysis reflect the belief that you cannot make rational decisions if you use "soft" (i.e. subjective) probability data. This is wrong, of course – decisions, whether private or public, must be made in the real world on the basis of whatever data that can realistically be procured. We all use subjective probability when we cross the street. However, decision makers have a need for procedures that yield optimal decisions under conditions of strong indeterminacy, just as engineers need methods that optimally use the information in the data.

Summarizing, randomness and indeterminacy can, with some reservations, be treated in a unified way using subjective probability under the Bayesian paradigm. The essence of the reservations stems from the private nature of subjective probability.

Concerning subjective probability, French (1986) argues that if you are a subjectionist, another person's probability to you "is no longer part of his thought processes and therefore, not a probability. It belongs to no one". Whatever any averaging over individuals may produce, it is not a probability. However, we are led to ask: "If it is not a (subjective) probability, then what is it?" An agency that contemplates the use of such an average may take the view that it is at least a good *betting basis* because it must be assumed that there exists a true value and that each expert has an unknown associated error. It is like betting on experts as if betting on horses, and splitting the ante. If these errors are small, it is reasonable to approximate the payoff as a quadratic function of the errors; the average maximizes this payoff in the usual fashion of least squares theory. Subjective probabilities, when pooled, can be viewed as a rational basis for decision making and could be considered as a special form of decision makers' probability.

With randomness thus subsumed into indeterminacy, we proceed to consider the realm of decisions where indeterminacy and vagueness are both present.

Vagueness and Indeterminacy

Since its introduction in 1965, the theory of fuzzy sets has been rapidly explored as a basis for a calculus of indefinite attributes in a broad variety of fields (Dubois and Prade, 1980). The theory admits fractional measures of the membership of an element in a set. Some professionals have accepted this theory rather uncritically and apply the operations of fuzzy set calculus, like the stone of the Magi, to models in a variety of contexts where vagueness dominates. Others admit the mathematical integrity of this theory but are sceptical with respect to its application in engineering on the grounds that the fractional membership measures are not defined or that they, when they are, turn out to be substitutes for probability. Also, certain particular aspects of fuzzy set theory are simplistic representatives of the way real language works (e.g. the value of adverbial operators such as "very"). Ultimately, crisp decisions must be made on the basis of fuzzy numbers or memberships, and there is no known way to relate such decisions to rational behaviour in the sense of von Neumann and Morgenstern (1947). Decision rules at the end of fuzzy operators must always contain a crisping feature, overt or covert. You cannot approve the operation of a plant or decide that a process is acceptably safe "with membership" 0.9 or whatever.

Thus, in spite of its many attractive features, fuzzy set theory has shortcomings for professional or public decision making; many of these will undoubtedly be remedied in time. The theory is at once too permissive (on the input) and too restrictive ("linguistic" operators, for example). The output of complex fuzzy set operations on fuzzy sets is often too fuzzy to yield much information, and the problem of reproducibility, crucial to professional credibility, has not as yet received enough attention. Finally, many argue against fuzzy analysis on the basis of need: are there any contexts where it is necessary? Many Bayesian engineers hold that they can represent any uncertainty by means of subjective probability.

Blockley and Baldwin (1987) have acted upon some of these shortcomings, developing a theory using the notions of *necessity* and *possibility* of propositions. Their calculus amounts to a generalization of fuzzy set theory, with probability theory as a special case. The theory admits – indeed requires – the user to decide on the necessity and possibility of any compound event from those of the elementary events *ad hoc* according to the situation. This flexibility lends great strength to the theory, insofar as applicability is concerned, but the credibility may suffer correspondingly, because these decisions must be justifiable to the decision maker or to the public. As it stands, this theory cannot be utilized in politically sensitive contexts such as the management of hazardous substances.

Another difficulty with the calculus of Blockley and Baldwin (1987) arises when many sets are intersected and united: the range of possibility may become large and the answer to a problem therefore correspondingly indefinite. This is unfortunate because risk analysis often requires the convolution of random variables which numerically involves the sum of many intersections.

If fuzzy sets and their generalizations are not adequate for public risk assessment, it is necessary to devise a different method to handle vagueness. When the risks posed by physical and chemical systems are involved, a straightforward approach is to follow the paradigm of natural sciences in treating vagueness: operational definition. Everything that can be of interest with respect to such systems can be thought of as attributes that may be expressed in terms of adjectives. Each adjective and its negation characterize a scalar variable, and a numerical scale may be defined conceptually for this vague concept. Once a scale has been established, every object for which the attribute is meaningful will have an *image* on the scale. This image is a single value, possibly unknown, on the scale associated with it. Certainly, to establish this value is quite another matter, but it is a question of measurement rather than of processing fuzziness. Two examples follow.

1. In the emergency planning for a particular plant site, an expert has stated that evacuation speeds will be "quite low". The expert is next asked what "quite low" means, and answers: "Reasonable lower and upper bounds are 0 and 45 km/h, but a credible value is 2–20 km/h." Following the fuzzy paradigm, mathematics may be introduced into the processing of this observation by defining the membership function as, for example

$$m(x) = \begin{cases} 0.5 \text{ if } x < 2; \\ 1 \quad \text{if } 2 \leqslant x \leqslant 20 \\ 0.04 \text{ if } 20 \; x \leqslant 45. \end{cases}$$

But an alternative approach is to define a *credible value* as a value between 5% and 95% probability – which the expert may then judge to be between 2 and 20 km/h – and define an *incredible value* as a value of 0% or 100% probability of exceeding – which the expert may then judge to be outside the range 0 to 45 km/h. Average speed in evacuation of a site is, after all, a well defined but uncertain entity. The concept of "quite low speed" is fuzzy, but unnecessary. For a probabilistic discussion of this example, see Cook and Unwin (1986).
2. The water in Lake Ontario is safe"."Safe" could be defined to mean "safe with respect to each of the hazards on some list, taken individually". The list would specify a set of certain toxic or carcinogenic substances, selected bacteria, and specify for each item a safe limit of concentration calculated, perhaps, from LD(50) for the hazard or in some other way. It is debatable, of course, whether this means roughly the same as "safe water" in common parlance. The definition does reduce a vague problem to a problem of measurement, and establishes a practical way to administer the problem in the interest of the public.

The approach suggested here will retain much of the useful quality of fuzzy set methods, but will force the calculus to respect the laws of measurement, which makes it compatible with Bayesian probability.

Risk Analysis as a Profession

Probabilities do not exist (de Finetti 1970) – they are not objective properties of a system but assigned values. There are no "true" probabilities and therefore no "true" risk. Risk is *assigned* – it isn't estimated, predicted or calculated. This point has been made forcefully from a different perspective by Douglas and Wildavsky (1982). As opposed to perceived risk (Slovic 1987), risk is justified by consensus among knowledgeable individuals – "experts". A statement about risk, like a scientific proposition, receives credibility by peer judgement. A strong factor in this approval process is replicability; for this reason it is desirable that good practice be promulgated and unified among those who practise risk analysis.

With the growth of risk analysis as an art or a science, there is a corresponding role for risk analysis as a profession (Lind, 1987). A profession is a self-disciplined association of individuals who are licensed to a certain trade that requires specialized knowledge; they autonomously set the qualifications and administer the membership examinations and, most importantly, they subscribe to specific professional *ethics*. Since the general public does not have the specialist knowledge in the matters of the trade of such a profession, the *caveat emptor* rule cannot be applied. The public must be able to trust the professionals; this is why a specialized ethic is necessary for a true profession.

Worldwide there are currently several thousand persons that practise risk analysis with all kinds of academic backgrounds. The public response to published analyses of risk has varied from unconditional acceptance to complete rejection. A risk analysis must either make sense – common sense – or have *a priori* acceptability because of the analysts' qualifications.

There is thus a need, as well as a basis, for a new profession of risk analysis. In particular, the professional ethics arises from the requirement that analysis be unbiased (non-manipulative) and conducive to rational allocation of public resources for the controlled reduction of risk. Further discussion of this topic is outside the scope of this paper, see Lind (1987), but there is one aspect of interest to hazardous waste management that is considered in the following section: Is there a need and a basis for standardization of risk assessment procedures?

Standards for Risk Assessment

Risk is an enormously broad subject, and analyses of risk vary substantially with respect to topic, available knowledge and other resources, credibility and so on. Large sums are allocated to the control of risk; it is estimated that around 10% of our economic output is allocated to mitigate risk or its consequences. The public has therefore a right to know that these sums are allocated rationally. As a first step in this direction, it is desirable that risk analysis be carried out in a controlled way; several agencies have established their own standards in this regard. But this is hardly enough: there should at least be some attempt to have assessments comparable inbetween technologies, so that it may be possible to compare the dollar value received in return for risk control in various disciplines. Are we spending too much on fire safety? Even without wider comparison with, say, hazardous materials control, structural safety of dams or military safety, we would still have to compare the fire provisions in the building code with fire brigade expenses and fire insurance premiums, all as they affect the risk. To begin to answer these questions of comparison of risk, it is desirable that risk analysis be uniform with respect to definitions, fundamental assumptions, and methods. As far as practicable methods and documentation should be standardized.

The need for a unified or generic standard has been expressed from several sides in Canada, and has prompted the Canadian Standards Association to establish a working group to develop a set of Risk Assessment Requirements and Guidelines. The Association is dedicated to the idea that industrial standards should reflect consensus, and the working group is broadly composed with members from various industries (chemical, nuclear, power), federal and provincial government ministries (agriculture, health and welfare), regulatory bodies (Atomic Energy Control Board), and universities.

A large amount of time in the work of this committee has been expended in the tedious but necessary task of establishing a common language and common concepts. Definitions, classifications and descriptions of the overall risk management process and its components were necessary but time-consuming.

It is a fundamental premise for this work that risk assessment in diverse technologies has a common basis – and that this basis consists of more than commonplace generalities. Moreover, it is a premise that all risk assessment procedures must contain a certain minimum of necessary elements in order to be acceptable to the profession such as: clear definition of the system, listing of assumptions. On the other hand, there appears to be a limit to the extent of standardization possible; various disciplines have their own needs, and projects of different magnitude will require different minimum levels of effort. Accordingly, the requirements are supplemented with a set of recommended practices or guidelines, and they are provided with a separate commentary that also is not mandatory. The drafting of this standard is not yet far advanced. It cannot be assumed that it will necessarily be adopted, nor that it will be used if it is adopted. However, it is hoped that it will be published soon enough that it may serve as a paradigm for other (perhaps international) generic standards.

Conclusion

Uncertainty associated with industrial projects can be classified as randomness, indeterminacy, vagueness or model uncertainty. There is no "true" value of a risk; risk is derived from assigned probabilities and is credible only if it reflects expert consensus. Risk assessment is an emerging profession that can gain credibility by establishing standard procedures and consensus guidelines. In the future there is also a need for self-governing and self-disciplining organizations that define curricula, control access to the profession and establish a professional code of ethics. An important component of such a code of ethics is the commitment to efficient use of societal resources to prolong life and improve health.

It is important to search for a unified way to express the different kinds of uncertainty, because all four kinds are present in any project to some extent. Subjective probability, processed in Bayesian fashion, is able to represent randomness and indeterminacy. It is suggested that vagueness be accounted for in the same way, using frameworks such as those established for fuzzy sets but compatible with the mass conservation of probability.

References

Blockley, D.I. and Baldwin, J.F. 1987. Uncertain Influence in Knowledge Based Systems. *Journal of Engineering Mechanics*, 113, 4, April: 467–81.

Cook, I. and Unwin, S.D. 1986. Controlling Principles for Prior Probability Assignments in Nuclear Risk Assessment. *Nuclear Science and Engineering*, 94, 2: 107–19.

de Finetti, B. 1970. *Theory of Probability*. Chichester: John Wiley.

Douglas, M. and Wildavsky, A. 1982. *Risk and Culture*. Berkeley, CA.: University of California Press.

Dubois, D. and Prade, H. 1980. *Fuzzy Sets and Systems*. New York, NY: Academic Press.

French, S. 1986. Comment [on (Genest and Zidek 1986)]. *Statistical Science*, 1, 1: 138.

Genest, C. 1984. A Conflict Between Two Axioms for Combining Subjective Distributions. *Journal of the Royal Statistical Society*, B, 46, 3: 403–5.

Genest, C. and Zidek, J.V. 1986. Combining Probability Distributions: A Critique and an Annotated Bibliography. *Statistical Science*, 1,1: 114–48.

Jaynes, E.T. 1976. Confidence Intervals vs. Bayesian Intervals. *Foundations of Probability Theory, Statistical Inference, and Statistical Theories of Science*, Harper, W.L. and Hooker, C.A., eds. Dordrecht, Holland: D. Reidel. Reprinted in Jaynes, E.T. *Papers on Probability, Statistics and Statistical Physics*, Rosenkrantz, R.D. ed. Dordrecht, Holland: D. Reidel, 204–5.

Kapur, J.N. and Kesavan, H.K. 1987. *The Generalized Maximum Entropy Principle (with Applications)*. Waterloo, Ontario: Sandford Educational Press.

Lind, N.C. 1987. Is Risk Analysis an Emerging Profession? *Risk Abstracts*, 4, 4, October: 167–9.

Lind, N.C. and Nowak, A.S. 1988. Pooling Expert Opinions on Probability Distributions. *Journal of Engineering Mechanics* (ASCE), 114, 2, February: 328–41.

Madsen, H.O., Krenk, S. and Lind, N.C. 1986. *Methods of Structural Safety*. Englewood Cliffs, NJ: Prentice Hall.

Paté-Cornell, M.E. 1987. Risk Uncertainties in Safety Decisions: Dealing with Soft Numbers. *Reliability and Risk Analysis in Civil Engineering*, Lind, N.C., ed. 1: 538–45.

Rosenblueth, E. *et al.* 1987. *Processing Doubtful Information*. Waterloo, Ontario: Institute for Risk Research, University of Waterloo.

Slovic, P. 1987. Perceptions of Risk. *Science,* 17 April, 236: 280–5.

von Neumann, J. and Morgenstern, O. 1947. *Theory of Games and Economic Behaviour*. Princeton, NJ: Princeton University Press.

20

Fault Tree Analysis and its Application to an Exothermal Reaction

U. Hauptmanns

General Description of the Fault Tree Methodology

A fault tree analysis reveals the logical connections existing between an undesired event in a technical system and component and operating failures which lead to it.[1,2] In the case of safety analyses for process plants the undesired event usually is a fire, a release of toxic substances or an explosion. In other cases it is simply an outage of the system. The method of analysis is deductive and is normally used for calculating the probability of occurrence of the undesired event. A qualitative investigation of the system is a prerequisite for it. This requires knowledge about the dynamic behaviour of the system as a consequence of deviations from its nominal operating conditions. Such knowledge may often be derived from experiments or model calculations. Should this not be possible engineering judgement has to be exercised. After determining the undesired event, questions of the "how can it happen" type serve to establish those functions and subsystems (e.g. cooling or electricity supply) whose failure produces the undesired event. These failures are in turn related to the failure of components and wrong human interactions with the system. The result of this process of analysis is represented by a fault tree.

The procedure is best illustrated by an example. It is shown in Figure 20.1. The system under analysis consists of two trains for pumping a fluid. Each of the trains is made up of a pump (P_1 and P_2, respectively), which is driven by an electric motor (M_1 and M_2, respectively). The valves on the pressure side (V_1 and V_2), permit a stoppage in the fluid transport. The mission of the system is fulfilled if at least one of the trains operates successfully. Obviously the undesired event in this case is the failure of the system to transport the fluid. This may occur in several ways. For example, the electricity supply may fail (event: x_g). Then both pumps would be out of operation. As an alternative there

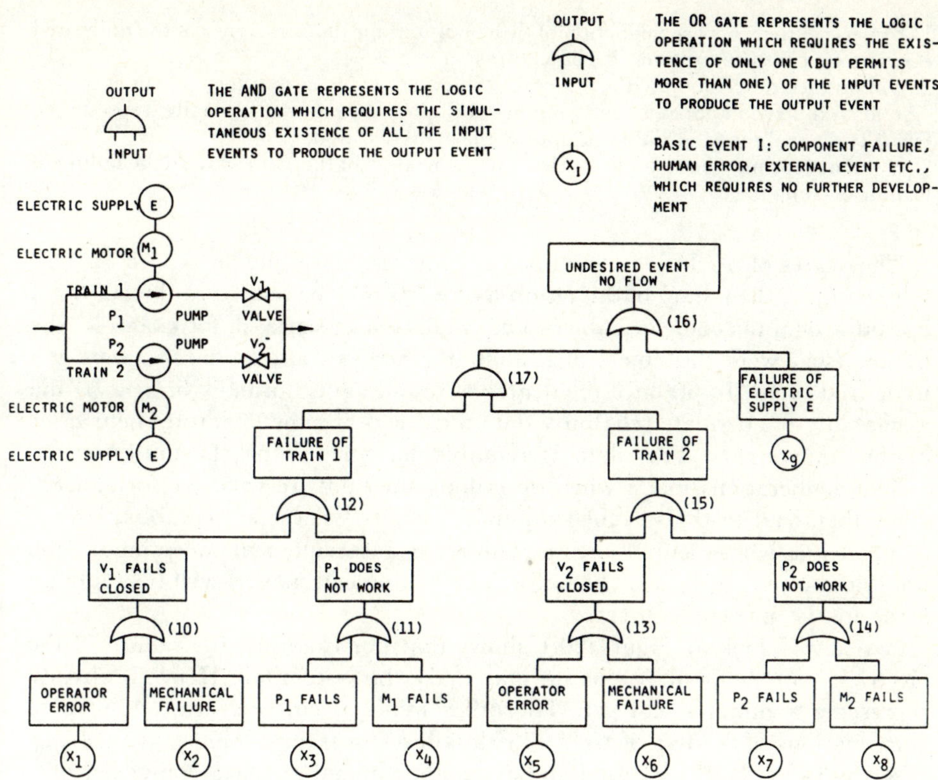

Figure 20.1 Fault Tree for a System Transporting a Fluid

may be other failures which cause an interruption of flow in both trains. For this reason gate 16 is an "OR" gate. Its exit is fulfilled if one or both of its input conditions (failure of electricity supply, or simultaneous failure of both trains for other reasons) are satisfied. Since both trains are identical it is sufficient to explain how train 1 can fail. Analogous reasoning then applies to train 2. The flow of the fluid may be interrupted, if valve V_1 fails closed or pump P_1 does not work, or if both events occur simultaneously. In the fault tree the two events are thus related by an "OR" gate. If we ask ourselves how valve V_1 can fail the answer may be: "It has failed mechanically in closed position or the operator has closed it by mistake". Both events in turn are connected by an "OR" gate, gate 10. The pump P_1 may fail because its electrical motor M_1 stops or because the pump proper fails. The "OR" gate (11) connects both events. As mentioned before, both trains have to fail simultaneously in order to make the system fail. Therefore the corresponding partial fault trees are connected by the "AND" gate (17). The output event of an "AND" gate only occurs if all its input events are fulfilled, i.e. in the present case the failure of both trains to transport the fluid. Several conclusions may be drawn from this simple example, which nevertheless apply in general:

1. Because of the stepwise deduction of the conditions for the occurrence of the undesired event even complex systems become transparent
2. A simplified description of the system is achieved by only permitting two states for the system and its components: functioning or failure (these are normally represented formally by a binary variable adopting the values 0 and 1)
3. Human error may be treated – at least in principle – in the same way as the failure of technical components

The degree of resolution of an investigation may naturally be carried further. For example, the failure of the pump motor M_1 may be related to the failure of its stator or rotor coils, or cables. This would make sense, if, for example, the motor itself, were the object of a fault tree analysis and its failure were the undesired event. In practice the degree of resolution is normally limited by the availability of adequate reliability data for describing the operating behaviour of the components. This data is required for quantifying the fault tree.

The deliberations made when describing the fault tree may be formalized. They then lead to the so-called minimal cut sets.[3] These are combinations of components whose simultaneous failure is just sufficient to produce the undesired event. Naturally the failure of a component may as well be a human error in the present context.

A renewed look at Figure 20.1 shows that, for example, the failure of the electricity supply alone (event x_9) may make the system fail. Hence it already represents a minimal cut set. The other minimal cut sets are made up of combinations of failures of train 1 with failures in train 2. Hence $x_1.x_5$, $x_1.x_6$, $x_1.x_7$, $x_1.x_8$, $x_2.x_5$ . . . In total there are 16 minimal cut sets composed of a combination of two failed components so that the fault tree is represented by 17 minimal cut sets on the whole. Every minimal cut set represents one way in which the undesired event may come about. The corresponding failures are compatible with each other. Their sum is formed according to the rules for the sum of probabilities of compatible events and leads to the so-called structure function of the system.[3]

Without proof it is communicated here that the probability for the occurrence of the undesired event is obtained if the binary variables x_1 in the structure function are replaced by their corresponding probabilities.[3] For example x_1 by the probability of operator error or x_2 by the probability for the valve failing mechanically in closed position.

For completeness' sake it should be mentioned that all products of a binary variable x_1 with itself must be replaced by this variable as such in order to obtain correct results. This follows from the fact that powers of the numbers 0 and 1 in turn result in 0 and 1. If the technical system is large its minimal cut sets and structure function will necessarily have to be found with the aid of a computer programme. In the present context this topic is not discussed in detail and reference is made to the literature.[1,2] Instead a few remarks on the calculation of probabilities for the failure of technical components and human error are in place.

The operating behaviour of technical components is generally described by an exponential distribution. Its characteristic parameter is the constant failure rate λ. This parameter is evaluated by observing a great number of components of the type in question and dividing the number of failures by the accumulated time of operation, (i.e. the product of the number of components and the time of observation). The exponential distribution just mentioned describes the time dependent probability of failure of a component without maintenance, a parameter usually called unreliability. It is given by equation (1). The mean unavailability of a component subject to maintenance in time intervals of θ hours is described by equation (2). The parameter reflects the mean probability of the component not functioning when required. The result (2) is obtained forming the mean value of relation (1) over the time interval θ. Equation (2) is valid only under the assumption that the time required for maintenance is small compared with the expected mean

$$(1) \qquad q(t) = 1 - e^{-\lambda t} \qquad (t > 0)$$

$$(2) \qquad \bar{u} = 1 + \frac{1}{\lambda\theta}(e^{-\lambda\theta} - 1)$$

θ = MAINTENANCE INTERVAL

lifetime of the component and that the component is "as good as new" after maintenance. Equation (2) is frequently used in fault tree analyses, since values for the duration of maintenance are rarely available.

Probabilities for human acts not being carried out or being performed in the wrong way are derived from psychological experiments and the evaluation of operating experience. They are given in the handbook by Swain and Guttman and represent the basic data for the assessment of human error in plant operation.[4] In a concrete fault tree analysis this basic data is modified taking into account the specific circumstances of the human intervention in question. Among these special circumstances figure, for example, the time available for the intervention, stress, degree of hazard, preparation of the operator. Naturally the assessment of human error suffers from larger uncertainties than the evaluation of the probability of failure of technical components.

Following this sketchy presentation of the method, its application to the process step nitration of a plant for the production of the explosive hexogen is presented. The details are taken from a study performed by the Gesellschaft für Reaktorsicherheit which was sponsored by the Federal Environmental Agency in Berlin West (UBA).[5]

Fault Tree Analysis for the Process Step Nitration of the Production of Hexogen

Generalities

The first step consists of a familiarization with the processs. Flow sheets and the process description formed the basis of this task for the two analysts involved. At a later stage this was supplemented by discussions with experts from the plant operator which lasted for one week. These experts represented the fields of design, operation and measurements, and control. During the discussions the conditions of functioning of the system and the expected behaviour of the reaction in case of a deviation of process parameters from nominal were established. In parallel gross structures of the fault tree were developed. These helped to increase the efficiency of the conversations. In addition, they already led to a number of proposals for the improvement of the plant. During the further process of analysis only occasional clarifications from the plant operator and a visit to the plant were necessary. The analysis described below omits those parts of the investigation which refer to failures not contributing substantially to the undesired event.

Short Description of the Nitrator

The Process

Hexogen is produced according to the SH-process by nitrating hexamine with an excess of 8 to 10 of concentrated nitric acid (98.5%). The reaction temperature should not rise above 20°C because otherwise a runaway might occur. In addition to the reacting substances and the product ammonia, formaldehyde and other materials are present in the reactor. The reaction is exothermal and the mixture is chemically unstable. Therefore the excess of nitric acid and the reaction temperature have to be maintained within the permissible range in order to avoid an explosion. The reaction takes place in the nitrator whose process flow sheet is shown in Figure 20.2.

Nitric acid is supplied to the reactor at a temperature of 5°C. Hexamine is introduced into the process via a dosification screw driven by the electric motor M1. The necessary ratio of the two substances is fixed at the beginning of the operation. Since the reaction is exothermal, the reactor has to be cooled. This is achieved by a mixture of water and methanol with an inlet temperature of 5°C circulating through a cooling coil inside the reactor. The mass flow of the coolant is controlled by the pneumatic valve TV1 in such a way that the reaction temperature is maintained close to 10°C. At this temperature the reaction yield is sufficient. At the same time it is far enough away from the critical temperature. The position of the control valve TV1 is fixed by a temperature control circuit. This circuit consists of the resistance thermometer TE1, the transducer TY1, which converts the electric signal from the thermometer into a pneumatic one, and the temperature indicator and regulator TC1, which provides the necessary signal for the pneumatic control valve TV1. In order to obtain a mixture as homogeneous as possible and to avoid local heating the reactor is equipped with a stirrer.

Safety Devices

From the process description it is clear that there are three conditions that have

present design
--- electric transmission
—— pneumatic transmission
-·-·- proposed modifications

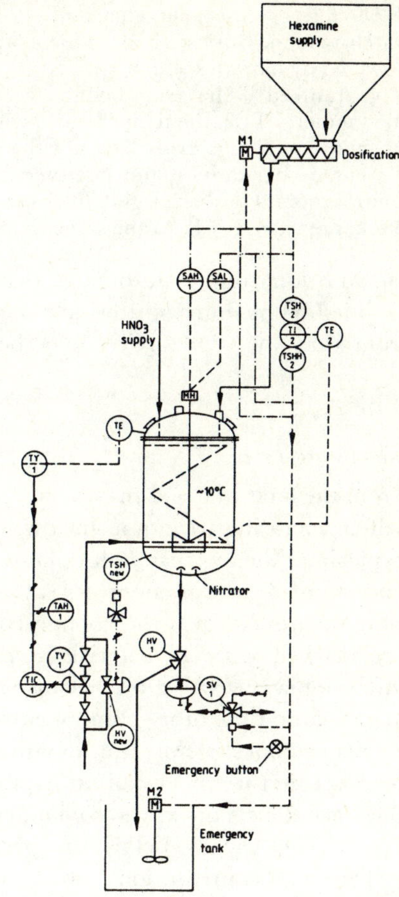

Figure 20.2 Flow Sheet of the Process Step Nitration in the Production of Hexogen

to be satisfied to keep the hazard potential of the substances in the reactor under control. First, the excess of nitric acid has to be assured. The temperature inside the nitrator has to be kept close to 10°C. Local heating has to be avoided by stirring. The important safety features for fulfilling these conditions are:

1. Reaction temperature – the temperature control circuit is equipped with the temperature alarm TA1. Its signal should prompt the operator to switch off the hexamine supply and to discharge the contents of the reactor into the emergency tank. The latter is filled with water. On discharging the stirrer of the emergency tank, which is driven by the electric motor M2, has to be switched on by pushing a button
2. Stirrer revolutions – if the number of revolutions of the stirrer falls below a lower limit the instrument SAL1 gives an alarm in the control room. Should the number of revolutions of the stirrer rise above an upper limit, because for example its shaft has broken, then the alarm is activated by the instrument SAH1

3. Discharge to the emergency tank – the discharge to the emergency tank is activated by an electric measuring chain. This chain is independent from operational devices and powered by batteries which are charged continuously from the grid. Hence power supply is guaranteed if the grid should fail. The measuring chain consists of the resistance thermometer TE2, the temperature switch TSH2, which cuts off the hexamine supply, if its limiting value is exceeded, and the temperature switch TSHH2 which activates the discharge. In addition the instrument TI2 indicates the reactor temperature in the control room. For discharging the solenoid valve SV1 is opened which in turn opens the discharge valve HV1. At the same time the stirrer of the emergency tank is switched on

In addition to the automatic discharge just described the reactor may be emptied by pushing buttons installed in both the control room and *in situ*. Apart from that the valve HV1 may be opened by turning a handwheel.

The Analysis

The Fault Tree

From the facts above it may be deduced that the undesired event in this case is a self-accelerating reaction in the nitrator, which will henceforth be called explosion. The corresponding fault tree is shown in Figure 20.3. Since normal operation of the system is safe, an explosion may only occur if active components fail or if the cooling coil loses its integrity. These initiating events are marked with an asterisk in Figure 20.3. As can be seen not all of the initiating events will directly cause an explosion. Some of them will only do so, if at the same time other components fail. For ease of presentation two initiating events are not developed in detail, since they make but a small contribution to the expected frequency of an explosion, as the analysis showed. In particular these are the events x_7 (no coolant supply or supply in adequate conditions) and x_{12} (HNO_3 supply below the permissible lower limit).

The small contribution made by failures in the nitric acid supply comes about because of the highly redundant control system. In the case of a failure of the coolant supply the contribution is low because of buffer deposits which allow an ample margin of time for possible countermeasures.

In developing the fault tree a number of conservative assumptions were made. For example it is supposed that a failure of the stirrer will invariably lead to an explosion.

Reliability Data

The quantification of the fault tree of Figure 20.3 requires failure rates for technical components and probabilities for human error. Failure rates for technical components should ideally proceed from the plant under analysis or at least have been obtained in a plant where similar components work under comparable conditions. This ideal can – at present – not be fulfilled for analyses of process plants. For this reason recourse had to be taken to data which have been obtained in a systematic and comprehensive observation in nuclear power plants. These involve, for example, details on technical characteristics of

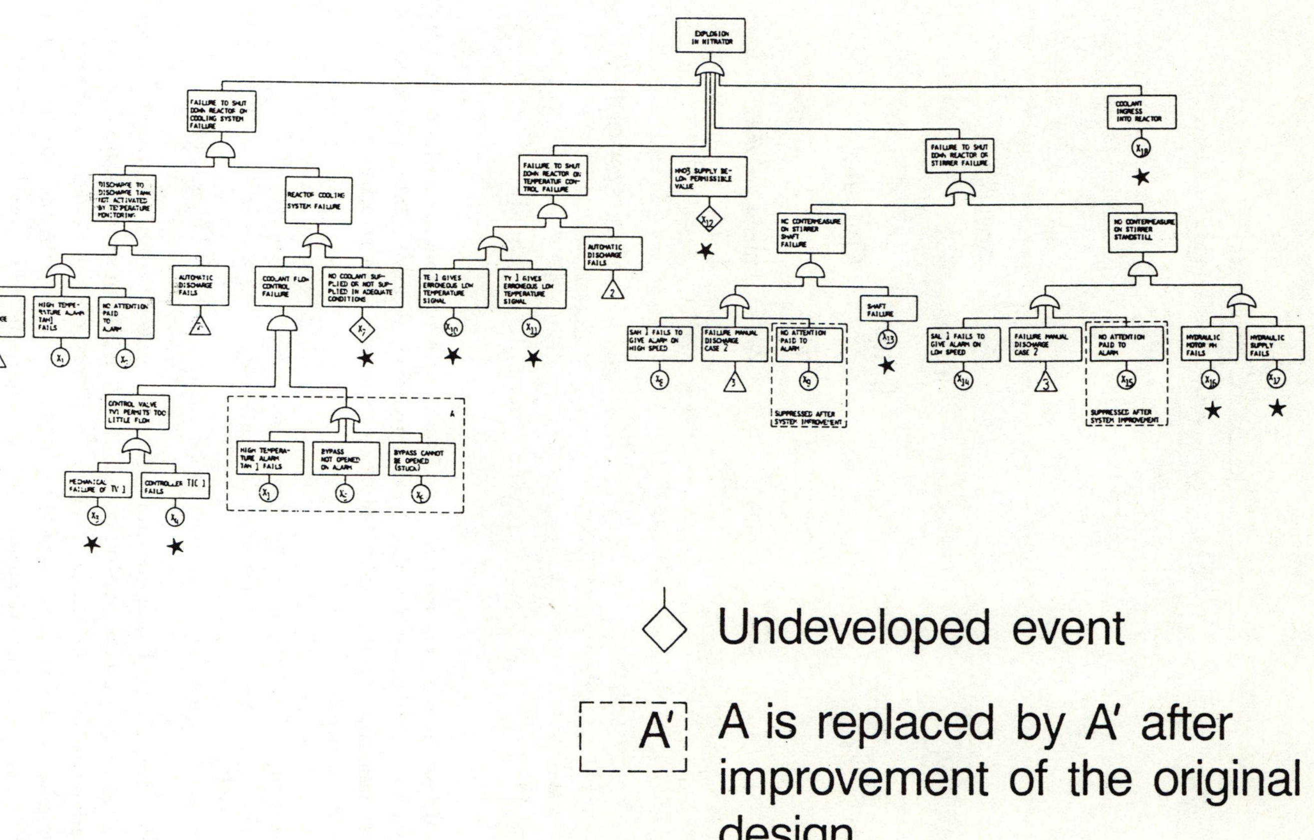

Figure 20.3 Fault Tree for an Explosion in the Nitrator

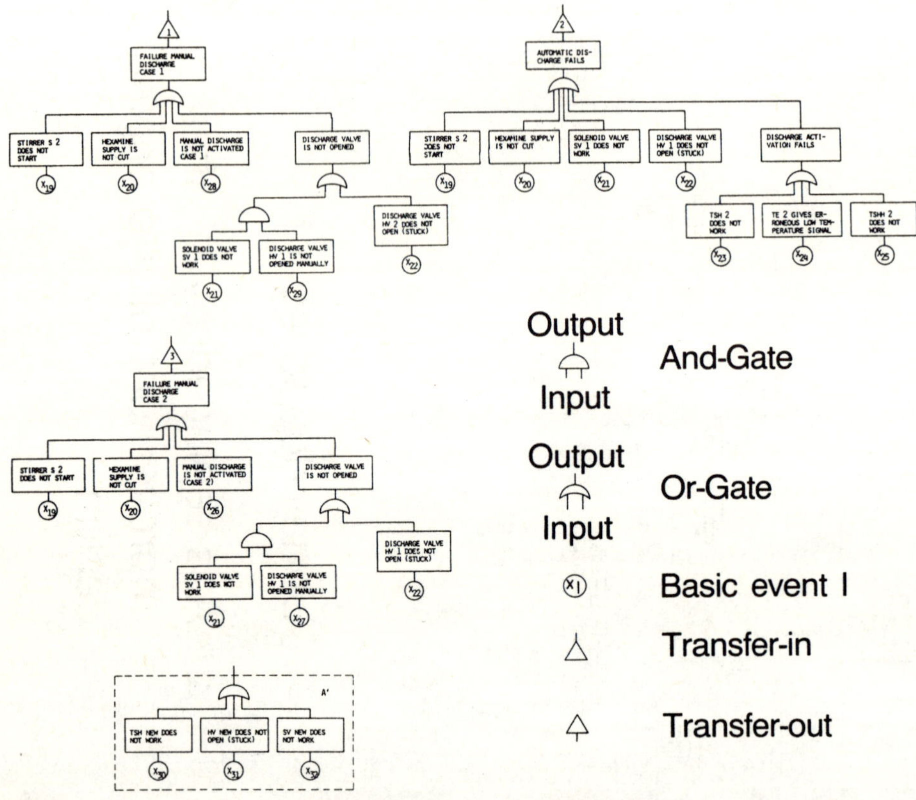

Figure 20.3 Fault Tree for an Explosion in the Nitrator (contd)

components, the working pressures and temperatures, media to which they are exposed. In addition, information from conventional power plants was used. In order to make these data applicable, the components of the plant were divided into three classes:

1. Components without contact with the medium,
2. Components exposed to process media which may be encountered as well as in conventional or nuclear power station,
3. Components exposed to loads only to be encountered in chemical plants.

Data for components belonging to the first two classes were taken over directly from the aforementioned investigations. Occasionally this information had to be supplemented by data from the scarce literature on component behaviour in process plants. These references supplied information on the environmental factors used for adapting reliability data for components exposed to water to the specific conditions of the plant characterized, for example, by the presence of acids. Even so some of the data had to be estimated.

Dispersion factors for failure rates were calculated if several values for a component were available. In doing this it was assumed that these values were samples from a log-normal population. In case there was only one value available the dispersion factor had to be estimated. The procedure sketched above is not considered satisfactory. However it is the best possible method, given that the results of a current research project on the evaluation of reliability data for components in a process plant environment are not yet available.

The reliability of human actions was estimated analyzing the interventions which would be required. The description of the process formed the basis for this task. Specific circumstances caused for example by the ergonomic conditions and the time available for the act in question were taken into account. The procedure by Swain and Guttman was used.[4] It led to failure probabilities between 10^{-4} for very simple acts and 1. The latter value was used for tasks whose execution was considered most unlikely. Among these are figures, for example, on the manual opening of the discharge valve of the nitrator after a rupture of the stirrer shaft. Little time is available for this action and the danger is of course very high.

In order to provide an idea of the order of magnitude of the reliability data, some values for technical components and human failure are listed in Table 20.1.

Table 20.1 Some Values for Failure Rates of Technical Components and Probabilities for Human Failure

Event no.	*Description*	*Median in* $10^{-6}h^{-1}$	*Mean in* $10^{-6}h^{-1}$	*Factor of dispersion*
	Failure Rates			
1	Temperature alarm TAH1 fails low*	6.5	9.5	4.2
3 4	Mechanical failure to control valve TV1 producing too little flow†	21	29	3.6
	Failure of controller TIC1 producing too little flow+	43	44	1.5
6	Bypass cannot be opened (stuck)	0.2	0.25	3.2
	Failure Probabilities			
Event no.	*Description*	*Median*	*Mean*	*Factor of dispersion*
2	No attention paid to alarm TAH1	0.0005	0.0013	10
5	Bypass B1 is not opened on alarm	0.5	0.8	5.0

* Calibration error with a median of 0.01 and a dispersion factor of K=5 is added in the calculations

† Initiating events

Fault Tree Evaluation and Results

The fault tree was evaluated with a computer programme which generates minimal cut sets by simulation. The dispersion of the reliability data was propagated through the fault tree by a Monte Carlo calculation.[2] It was assumed that the components emerge from maintenance "as good as new". Maintenance acts were carried out in some cases in time intervals of 672 hours and of 168 hours in others.

The contributions to the total expected annual frequency of explosion made by the different initiating events are listed in Table 20.2. They sum to a total of

$$h = 4\times 10^{-2}a^{-1}$$

Table 20.2 Expected Annual Frequencies of an Explosion in the Nitrator for Different Initiating Events

Initiating event no.	*Description*	*Expected Frequency of the Initiating Events s_j in a^{-1}*	*Unavailability u_j*	*Expected Frequency of the Undesired Event h_j in a^{-1}*
3	Mechanical failure of control valve TV1	2.5·E−1	4.7·E−3	1.2·E−3
4	Failure of controller TIC1	3.8·E−1	4.8·E−3	1.8·E−2
7	No coolant supply or inadequate supply	6.3·E−2	5.7·E−3	3.6·E−4
10	Temperature measurement TE1 fails	3.1·E−2	4.6·E−2	1.4·E−3
11	Transducer TY1 fails	5.5·E−1	4.6·E−2	2.5·E−2
12	HNO_3 supply below permissible value	5.0·E−	1.0	5.0·E−7
13	Stirrer shaft rupture	1.8·E−3	0.11	2.0·E−4
16	Failure of hydraulic stirrer motor	8.8·E−3	0.11	9.7·E−4
17	Failure of hydraulic supply	7.1·E−2	0.11	7.8·E−3
18	Coolant ingress into reactor	8.5·E−4	1.0	8.5·E−4
	Total			4.0·E−2

The important contributions to this value stem from the initiating events listed in Figure 20.4.

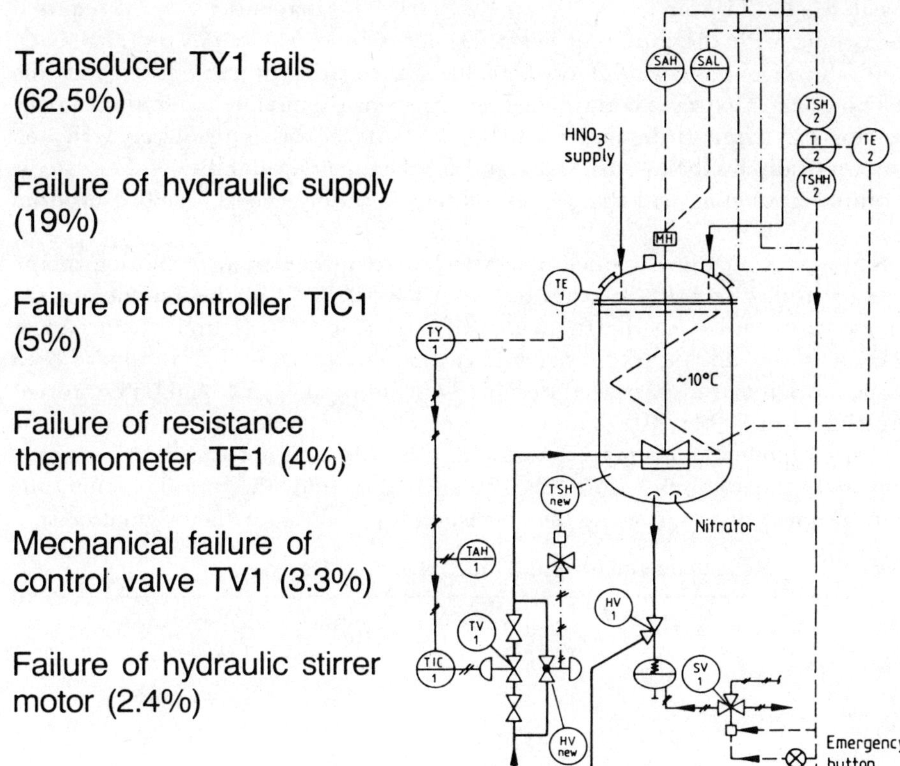

Figure 20.4 The Contribution of Important Initiating Events to the Total Expected Annual Frequency of an Explosion

An examination of the corresponding minimal cut sets shows that in the case of the initiating events "TE1 fails" (10) and "TY1 fails" (11), substantial contributions are made by the failure of the temperature switches TSH2 and TSHH2. In the case of the initiating event "rupture of stirrer shaft" (13), important contributions stem from the "failure of the hydraulic stirrer motor" (16) and "failure of the hydraulic supply" (17) owing to an operator error after the alarm caused by stirrer revolutions outside the permissible range.

To remedy the first case the installation of an independent temperature switch for opening the bypass of the cooling system is proposed. This should become effective in case of an excessive temperature in the reactor. In this way a system which is redundant to the operational system is provided. It will frequently make it unnecessary to discharge the reactor contents into the emergency tank, if disturbances in the cooling control occur. The product would then be saved. This modification of the original design reduces the unavailability of the systems which are necessary to cope with the initiating events 10 and 11 from 4.6×10^{-2} to 1.2×10^{-2} and that of the systems required for coping with the initiating events 3 and 4 from 5.0×10^{-3} to 1.6×10^{-4}.

In the second case, that of disturbances related with the stirring of the reactor, the important contribution stems from the manual activation of the reactor discharge (primary event No. 26). If the discharge were activated automatically by the alarms SAL1 and SAH1 the unavailability of the systems required for coping with the corresponding initiating events would drop from 0.11 to 2.5×10^{-2}.

These modifications reduce the expected frequency of an explosion in the reactor from $h = 4.0 \times 10^{-2}\ a^{-1}$ to $h = 4.1 \times 10^{-3}\ a^{-1}$. In the new design the major contribution to the frequency of explosion derives from disturbances related to the stirring of the reactor. It amounts to 51%. The results of both cases – original and improved designs – including their 5% and 95% centiles are shown in Table 20.3. The improvement of the system resulting from the proposed modifications may be considered as real and not within the range of statistical uncertainty, since the mean value and the centiles after the improvement are well below the corresponding values for the original design.

Table 20.3 Results for the Original and Improved Designs

	Expected Frequencies of Explosion *Mean Value*	*Confidence Interval*
Original Design	$h=4.0\cdot10^{-2}\ a^{-1}$	$h_{05}=8.7\cdot10^{-3}\ a^{-3}$ $h_{95}=0.1\ a^{-1}$
Improved Design	$h'=4.1\cdot10^{-3}\ a^{-1}$	$h'_{05}=3.6\cdot10^{-4}\ a^{-1}$ $h'_{95}=1.3\cdot10^{-2}\ a^{-1}$

Conclusions

The investigation has shown how the safety of the system can be improved and how its availability, hence its economy, can be increased, although the latter was not the objective of the study. Some of the results were already obtained in the qualitative part of the analysis. The quantification of the fault trees brought further insights and revealed areas in which the safety measures were unbalanced. These areas are characterized by strongly differing contributions of the different initiating events to the expected frequency of explosion. The proposals for the improvement of the plant lead to a reduction of this frequency by a factor of 10. They may be put into practice without much effort. However, owing to the qualitative shortcomings still existing for the reliability data of process plants, the results should not be viewed in absolute terms. Reliability analyses should preferably be used for comparing design alternatives.

References

1. Vesely, W.E., *et al.* 1981. *Fault Tree Handbook.* NUREG-0492. Washington.
2. Hauptmanns, U. 1986. *Análisis de Arboles de Fallos.* Barcelona: Editorial Bellaterra.
3. Barlow, R.W. and Proschan, F. 1975. *Statistical Theory of Reliability and Life Testing–Probability Models.* New York.
4. Swain, A.D. and Guttman, H.E. 1983. *Handbook on Human Reliability Analysis with Emphasis on Nuclear Power Plant Applications.* Final Report NUREG/CR-1278m. Washington.
5. Hauptmanns, U., *et al.* 1985. Ermittlung der Kriterien für die Anwendung systemanalytischer Methoden zur Durchführung von Sicherheitsanalysen für Chemieanlagen GRS-59. Köln.

21
Computer-based Information and Decision Support Systems for Management of Hazardous Substances and Industrial Risk

Kurt Fedra

The Problem

Risk from the production and use of hazardous substances is ubiquitous. In addition to hazardous wastes, there are a large number of commercial products that are also hazardous. Their production, transportation, and use – before they enter any waste stream – is clearly also of concern. Industrial production processes that involve hazardous raw materials, feedstocks, or interim products, which may reach the environment after an accident, causing direct health risks to man, are a major component of overall industrial activities.

Further, the transportation of hazardous substances such as chlorine poses considerable risks to public health and the environment. The analysis of alternative policies for transportation, storage, and the location of production and consumption facilities should be based on a detailed scientific assessment of numerous alternatives, involving technological, environmental, socio-economic and political elements in a comprehensive and directly usable form of risk analysis.

The problems of managing hazardous substances then, are neither well defined nor reducible to a small set of relatively simple subproblems. They always involve complex trade-offs under uncertainty, feedback structures and synergistic effects, non-linear and potentially catastrophic systems behaviour – in short, the full repertoire of a real-world situation. The classical methods of operations research and control engineering, which require a complete and quantitative definition of the problem from the outset, are certainly insufficient.

While only the combination of a larger set of methods and approaches holds promise of effectively tackling such problems, the subjective and discretionary human element must also be given due weight. This calls for the direct and interactive involvement of users, allowing them to exert discretion and judgement wherever formal methods are insufficient.

Project Background and Objectives

The background information required for strategic planning and policy making is characterized by a broad range of disciplines and is subject to a variable degree of resolution and uncertainty. The management and decision-making process therefore requires a strong element of human expertise and judgement in addition to the more formal, scientifically-based, analytical techniques. Methods of applied systems analysis and risk assessment, implemented using modern information processing technology, can now support such a comprehensive, interdisciplinary approach to the management of industrial risk. This approach can provide a powerful interactive tool for planners, regulators and policy makers, because it makes access to a large number of relevant databases, problem simulation modules, and decision support tools easy.

The research and development carried out by the Advanced Computer Applications group (Fedra, 1985 1986a,b; Fedra and Otway, 1986) concentrates on integrated systems of software tools to make the scientific basis for planning and management directly available to planners, policy and decision makers.

The objective of the project is to design and develop an integrated set of software tools, building on existing models and computer-assisted procedures. This set of tools is intended for non-technical users and should provide them with easy access to methods of analysis and information management which have previously been restricted to a small group of experts. To facilitate access to complex computer models by the non-expert user it is necessary to build much of the accumulated knowledge of the subject areas into the user interface. The interface therefore incorporates elements of knowledge-based expert systems which assist the user to select, set up, run and interpret the specialized software relevant to his needs.

By providing a coherent user interface, the interactions between different models, their databases and auxiliary software become more transparent to the user and more experimental, educational style of computer use can be obtained. Extensive use of high-resolution graphics and menu-driven operations aids this transparency and makes the system user friendly, greatly facilitating the assessment of alternative policies and strategies for the management of industrial risk.

A Decision Support Approach

The approach to decision support proposed here is based on information management and model-based decision support. It envisions experts as its users, as well as decision and policy makers. The computer is seen as a mediator and translator between expert and decision maker, between science and policy. It is thus not only a vehicle for analysis, but even more importantly, a vehicle for communication, learning, and experimentation.

The three basic functions of such decision support systems are:

1. To supply factual information, based on existing data, statistics, and scientific evidence, or to educate and inform the user
2. To assist in designing alternatives and to assess the likely consequences of such new plans or policy options, that is, to generate and analyze a set of possible solutions, often referred to as scenario analysis, and
3. To help in a systematic multi-criteria evaluation and comparison of the alternatives generated and studied, that is, to structure and assist the actual decision process

The framework foresees the selection of criteria for assessment by the user, and the assessment of scenarios or alternative plans in terms of these criteria. Value judgements are made by the user – he must make the decisions; the computer supports this task.

Hybrid Systems: Embedded Artificial Intelligence (AI) Technology

The application- and problem-oriented rather than methodology-oriented system is designed as a hybrid system, where elements of AI technology are combined with the more classical techniques of information processing and approaches of operations research and systems analysis. Traditional numerical data processing is supplemented by symbolic elements, rules, and heuristics in the various forms of knowledge representation.

The basic approach employed is rapid prototyping together with and around well established operations research techniques. The ACA group designs and develops an integrated set of software tools, building on existing models and computer-assisted procedures.

AI techniques are embedded in the overall system at various levels. The object-oriented overall design and problem representation employs several concepts of expert systems for systems integration. The user interface includes various elements of expert systems technology, for example, natural language parsing, rule-based input checking and error correction. Context-dependent help and explain functions are implemented throughout the system. Selected system components are based on symbolic simulation techniques, inference procedures and rule-based heuristic procedures.

There are numerous systems components where the addition of a small amount of "knowledge" as defined, for example, to an existing simulation model, may considerably extend its power and usefulness and at the same time make it much easier to use. Expert systems relying on huge knowledge bases of thousands of rules are not necessarily purely knowledge driven. Applications containing only small knowledge bases of a few dozen to a hundred rules can dramatically extend the scope of standard computer applications in terms of application domains as well as in terms of an enlarged non-technical user community.

Model Integration and the User Interface

From a user perspective, the system must be able to assist in its own use, that is, explain what it can do, how it can be done, and where a result comes from. The basic conceptual elements of this menu-driven system are described below. The interactive user interface handles the dialogue between the user(s) and the machine; this is largely menu-driven, that is, at any point the user is offered several possible options which he can select from a menu provided by the system (Figure 21.1). A task scheduler or control programme interprets the user request – formulating and structuring it – and co-ordinates the necessary tasks (programme executions) to be performed; this programme contains the "knowledge" about the individual component software modules and their interdependencies; the control programme can translate a user request, into, for example, a data/knowledge base query or a request for scenario analysis. The latter request will be transferred to a problem generator, which assists in defining scenarios for simulation and/or optimization. Its main task is to elicit a consistent and complete set of specifications from the user, by iteratively resorting to the database and/or the knowledge base to build up the "information context" or "frame" of the scenario. A scenario is defined by a delimitation in space and time, a set of (possibly recursively linked) processes, a set of control variables, and a set of criteria to describe results. It is represented by a set of process-oriented models, that can be used in either simulation or

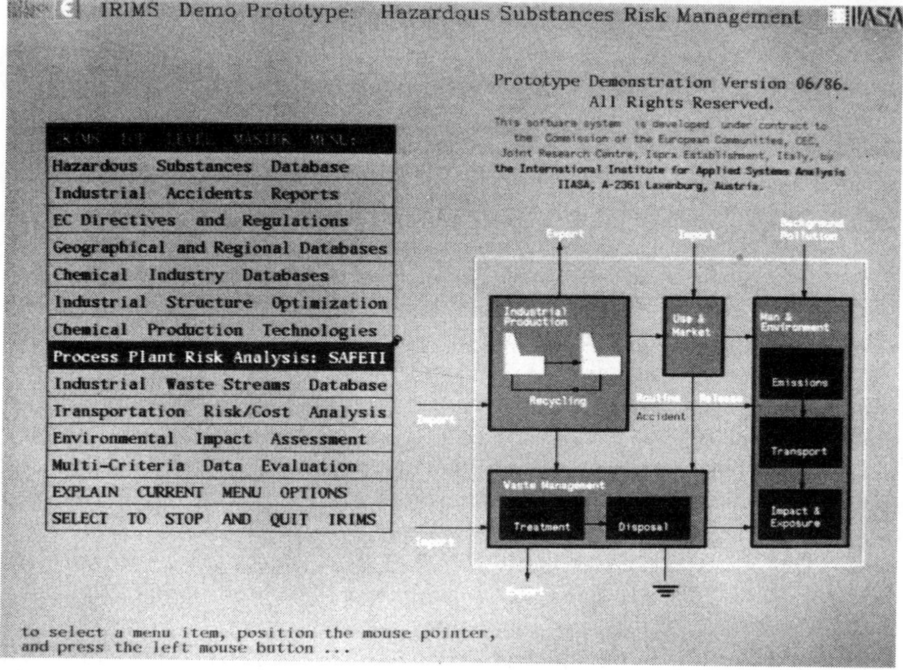

Figure 21.1 Top Level System Master Menu

optimization modes. The results of creating a scenario and either simulating or optimizing it are passed back to the problem generator level through an evaluation and comparison module, that attempts to evaluate a scenario according to the list of criteria specified, and assists in organizing the results from several scenarios. For this comparison and the presentation of results, the system uses a graphical display and report generator, which allows selection from a variety of display styles and formats, and in particular enables the results of the scenario analysis to be viewed in graphical form. Finally, the system employs a system's administration module, which is largely responsible for housekeeping and learning. It attempts to incorporate information gained during a particular session into the permanent data/knowledge bases and thus allows the system to "learn" and improve its information background from one session to the next.

It is important to note that most of these elements are linked recursively. For example, a scenario analysis will usually imply several data/knowledge base queries to provide the frame and necessary default parameters transparently. Within each functional level, several iterations are possible, and at any decision breakpoint that the system cannot resolve from its current goal structure, the user can specify alternative branches to be followed. The system must, however, on request, "explain" where a result comes from and how it was derived, for example, from a database, inferred by a rule-based production system, or as the result of a model application.

AI technology is embedded in this integrated software system at various levels, in several modules. They range from heterarchical, frame-based databases to rule-based pre-processors and input generators for classical numerical simulation models, rule-based heuristic feasibility and consistency checking of interactive input, to symbolic simulation and intelligent parsers for language-oriented input. The emphasis, clearly, is on a broad set of problem- and knowledge-representation techniques, integrated into one coherent framework.

While most operational examples of expert systems work in a relatively small and well-defined problem domain (computer systems configurations, interpretation of chromatographic experiments, diagnosis of a small set of illnesses; for a recent review see Weigkricht and Winkelbauer, 1987) our system spans the very large and not-so-well defined composite problem area of strategic risk management.

The model for our object-centred system's design is therefore based on the composite concept of a team of experts, co-ordinated by a systems analyst, who orchestrates the tasks of the individual domain experts.

Primary interaction is through the systems analyst, represented by the menu-driven and largely symbolic user interface. The user interface translates the user's request and specifications into tasks the system can perform, calls upon the domain experts (models and databases), and communicates their results to the user.

Symbolic Representation and Computer Graphics

To almost everybody but the developer, a model is what one sees of it. And here a picture is certainly worth a thousand words, or numbers. It needs little persuasion to convince most people that a colour-coded overlay of, say, the path of a cloud of toxic or radioactive material over the map of Europe is more immediately understandable than a table with numbers. This, however, is only the most obvious part of the user interface.

The user interface, from a more technical point of view, is characterized by its bandwidth, that is, the amount of information that can pass between man and machine per unit of time. Clearly, serial interfaces such as speech or typing and reading are very slow and cumbersome. Graphical representations, on the other hand, are less precise; they convey patterns or *gestalt*, but the individual element, the pixel, cannot be decoded as a unit.

It is quite obvious that any efficient system will need all forms of encoding: numbers, text, and pictorial or symbolic formats, and, in fact, even sound. In addition to the visible (or audible) parts of the user interface, it also needs some logic structure, or some intelligence. A very promising approach is the object-oriented model of a team of experts, as mentioned above, providing consultation to the user.

While command-driven interfaces can be very effective for the experienced user, menu-driven systems offer the advantage of self-teaching and error-proof features. To return to the model of the expert team, the team initiates the dialogue, and guides the user through the set of possible actions that could be taken to describe the problem and the search for feasible solutions. It will also assist the user to ask meaningful and sufficiently formulated questions only, it will provide background information on request, and it will explain its reasoning and suggestions.

For such a complex system to be easy to use, it has to be self-explanatory. This can partly be achieved with a well-structured system of menus, and additional explain functions at each of the system's functional levels. Responsiveness and speed are another key element: whenever the user does something, there must be an immediate response acknowledging the user's entry, offering further options or instructions, or indicating the system's state resulting from the user's action: The user should not be left in any doubt about his last input.

As a related issue, the control logic must be straightforward and obvious. Consistency in the logical structure of menu options, as well as in the pictorial layout of the interface, are important in order to minimize a novice's frustration as well as to maximize the efficiency of an experienced user.

Uncertainty, Robustness, and Qualitative Representation

Every model will always include a considerable element of uncertainty. Some variation in the assumptions will result in somewhat different answers (Gupta *et al.*, 1985; Goodman and Nguyen, 1985). However, we must offer the assurance that any result the system might communicate to the user is valid over a reasonable range of conditions, or in other words, that solutions are robust.

The graphical techniques of communication make it easy to incorporate uncertainty in the interface: scaling of symbols or colour-coding allow only a relatively low resolution. The basic qualitative aspects of results however, can be communicated very effectively by symbolic formats, and they should certainly be the more reliable aspect of any solution. Detailed numerical information certainly requires numerical formats – and interpretation, at least in the form of some qualification as to its precision.

Graphical representation can also imply the use of a more coarse yardstick, which, in most cases, is also the more appropriate and honest yardstick to use. But the use of symbolic and declarative, rather than numerical and procedural representations, also allows human experience and knowledge, both very important factors, to be tapped for inclusion in computer-based systems.

Application Examples

Under contract to the Commission of the European Communities' Joint Research Centre (JRC), Ispra, the ACA group is working on a project entitled "Decision-oriented Software for the Management of Hazardous Substances and Industrial Risk". The primary intended application of the decision support system is for regulatory purposes, for example, within the framework of the EEC post-Seveso Directive or similar legislation. Based on a demonstration prototype developed in 1985 and installed at the JRC in early 1986, the system has now been extended towards a full-scale pilot system by the inclusion of several additional modules, new linkages between the modules, and a number of improvements as a result of first tests at the JRC. An assessment of the potential benefits of this modern approach was started with the development of a Demonstration Prototype System, called the IRIMS (Ispra Risk Management Support) system.

In a related project sponsored by the Netherlands Ministry for Housing, Physical Planning, and the Environment (VROM), an interactive, graphics-based intelligent interface to a large fault-tree analysis and consequence modelling system is being developed to address problems of production, transportation and use of chlorine in the Netherlands. In parallel, this serves as a test case for the concept and approach of the IRIMS system.

These two research and development projects implement several examples of the decision support approach described above.

Several databases, numerous simulation and optimization models, and specific decision support tools are integrated in an interactive and graphics-based user environment. The current stage of development of these systems is discussed below.

Databases

Structures, test implementations, and integration with either interactive browsing programmes, graphical display options, or operational simulation models have been completed for a number of databases.

Geographic Data

The basic background map of the demonstration prototype is a contour map of Europe; the contents of various databases used in one or several of the simulation models can be viewed as interactively constructed map "overlays". They include:

1. Political boundaries
2. Major settlements (>100,000 inhabitants; the actual settlements database is more than twice as large, including more than 1000 entries, which are used in the transportation network database)
3. European highway and national roads network (the complete European highway network as well as selected national roads, connecting the above settlements and numerous auxiliary towns; this database also provides direct input to the transportation risk/cost analysis module)
4. Major industrial plant locations (concentrating on phenol and chlorine as major feedstocks or products)
5. Chemical storage facilities (concentrating on phenol and chlorine)
6. Major water bodies (rivers, lakes)

Hazardous Chemicals Database

The chemical substances database includes a subset of EC's ECDIN database (Figure 21.2). Its structure and contents are specifically geared towards the data requirements of the simulation models used. The preparation of a useful operational subset requires considerable input from the end user, in particular for compiling and processing the physical properties of selected substances, according to the substance description questionnaire developed. Detailed descriptions for a few individual substances as well as the allocation of substances to a few substance classes (organized largely by chemical taxonomy), defined together with the end user, have been included in the database (Fedra *et al.*, 1987).

Our approach foresees the use of a basic list of about 700 substances (or individual substances, that is, entities that do not have any subcategories), constructed as a superset of EC and US Environmental Protection Agency (USEPA) lists of hazardous substances. In parallel we have constructed a set of substance classes which must have at least one element in them. Every substance has a list of properties or attributes; it also has at least one parent substance class in which it is a member. Every member of a group inherits all the properties of this group. In a similar structure, all the groups are members of various other parent groups (but only the immediate upper level is specified at each level), where finally all subgroups belong to the top group of hazardous substances.

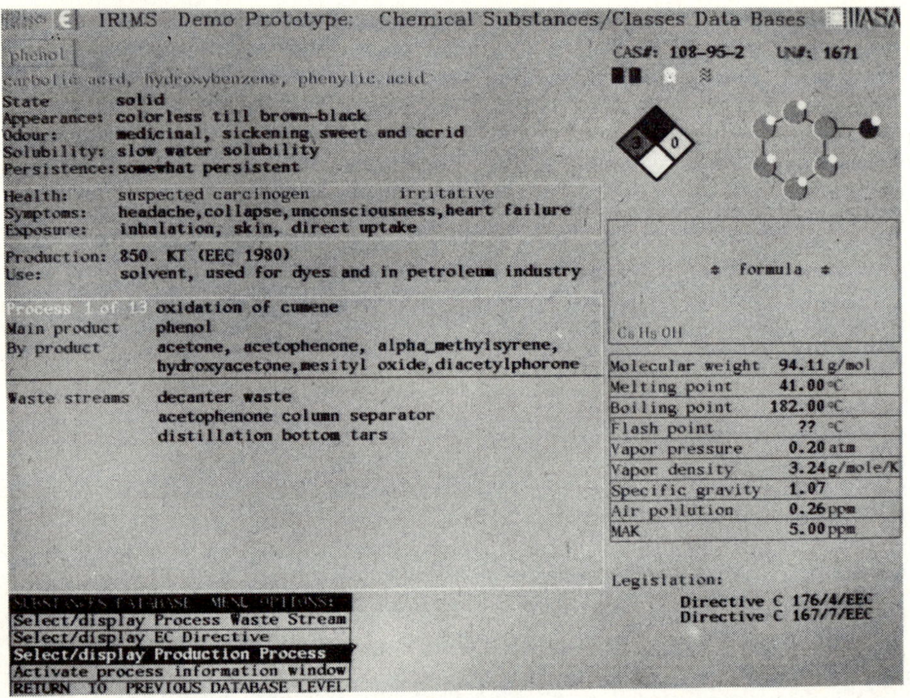

Figure 21.2 Summary Page Description for a Basic Substance

While attributes of individual substances are, by and large, numbers (for example, a flashpoint or an LD_{50}), the corresponding attribute at a class level will be a range (flashpoint: 18–30°C) or a symbolic, linguistic label (for example, toxicity: very high).

The structure also takes care of unkowns at various levels within this classification scheme. Whenever a certain property is not known at any level, the value from the immediate parent–class (or the composition of more than one value from more than one immediate parent–class) will be substituted. The structure is also extremely flexible in describing any degree of partial overlap and missing levels in a hierarchical scheme. An interactive query and display facility for the existing substance and substance classes structure and the subset of individual substances included is implemented and linked to the overall framework. It provides access to individual substance descriptions either from a listing of substances or from substance classes.

Industrial Waste Streams Database

This database uses a set of descriptors similar to the RCRA (US Resource Conservation and Recovery Act) database (ICF, 1984). Access to individual waste stream descriptions is either from a list of waste streams, through industrial origin, waste stream names (interpreted by an intelligent parser), or waste stream properties.

From the individual waste streams, connections are implemented to the above substances database for the constituents of concern, as well as to the industrial sites database for the sources of these waste streams; the latter can be selected by identifying industrial locations from a map of the respective regions.

Chemical Process Technology Database

A chemical processes database, including unit processes, combined process technologies, and unit equipment, has been developed. A test example describing phenol chlorination with plant hardware configuration has been completed. The database is accessible from its own menu-driven interface, providing various access mechanisms, including parsed language input, as well as from the hazardous substances database.

The database, in turn, provides access to the waste streams database and the hazardous substances database. For the example of phenol chlorination, a complete knowledge base to drive the symbolic process simulator is included.

Industrial Production Sites

A small subset of European producers (mainly related to phenol and/or chlorine production) has been compiled and integrated in a structure similar to the industrial waste streams database. Using the list of products and hazardous substances (as defined by the post-Seveso Directive) of a given site as a menu, the corresponding substance descriptions from the hazardous substances database can be displayed.

Chemical Storage Facilities

Chemical storage facilities provide a structure similar to that of the production facilities database, concentrating, however, on major storage rather than production facilities. For the substances in storage, a connection to the hazardous substances database is provided, that allows the appropriate database entries for the respective chemicals to be called.

Major Industrial Accidents

A simple display programme for text files structured according to Appendix VI of the post-Seveso Directive has been developed. A short description of the Seveso accident is included as an example. From the accidents database, direct connections to the industrial site database and the hazardous substances database are provided.

Regulations and Legislation

Along the lines of the above accident reports, the text files accessible through this module cover selected EC directives. As an example, excerpts from the post-Seveso Directive on major-accident hazards of industrial activities and the proposed dangerous substances listing for Directive 76/464/EEC are included.

These text files are also accessible from the hazardous substances database; a list of directives applicable for a specific substance is displayed as part of the information on that chemical. From this list, directives can then be selected for display.

Simulation/Optimization Models

The simulation models of the production system can be configured to describe the comprehensive life-cycle of hazardous substances. The major components of the simulation system are: the industrial production sector; use and market; waste management, including treatment and disposal; the cross-cutting transportation sector; and man and the environment. Each of these major components is represented by several individual models, covering a variety of possible approaches and levels of resolution. Each element of the simulation system can be used in isolation, or it is linked with several others as pre- or post-processors into increasingly larger systems of interconnected models. None of the complexities of the system's integration are obvious to the user; the style of the user interface and interactions with the system are always the same at the user end.

Several simulation and/or optimization models have been integrated into the demonstration prototype and are outlined below.

PDA (Production Distribution Area)

PDA is an interactive optimization code (based on DIDASS, one of a family of multi-criteria decision support tools developed at the International Institute for Applied Systems Analysis) and a linear problem solver, for chemical industry structures, configured for the pesticide industry (12 processes, 13 major products) of a hypothetical region (Dobrowolski *et al.*, 1982, 1984).

The user can select optimization criteria, define allowable ranges or constraints on these criteria, define reference points for the multi-criteria trade-off, and display various levels of model output, including the waste streams generated by the different industrial structure alternatives. These waste streams can then be used to provide input conditions for the environment impact models.

Industrial Process Simulation

The industrial process simulation module provides a rule-driven dynamic simulation of a production process, implemented in CommonLisp. It is based on a set of production rule packages representing knowledge of the production processes and the necessary plant equipment driven by an inference engine which performs the forward-chaining of the rules.

The production process starts as soon as input material is provided to the

Operating Units which are connected to the external input streams. These Operating Units perform their Unit Activities depending on the input materials, the operating conditions of the Unit and the constituents of the Unit, and by this produce some output material, which they send (via the linked input/output streams) to the Operating Units, which are activated on receipt of input material. They too perform their Unit Activities and produce output; this activates other Operating Units and so on. After the production and the release of output material an Operating Unit is deactivated until it gets new input material. This sequence of activation and deactivation of Operating Units by materials terminates when there is no more input material for any of the Operating Units, for example, all external input has been transformed to the desired products, byproducts and waste.

During the simulation of the production process the operating hazards of the Units and the hazards caused by the materials used and produced (e.g. input materials, interim products, end products, waste materials), the material hazards, are recorded and dynamically updated in the form of Hazard Ratings.

The simulation module features an animated display of the basic steps in the chemical manufacturing technology described in the technology database, and includes the dynamic display of risk ratings for the individual production steps and process streams (Winkelbauer, 1987).

LRAT (Long-range Atmospheric Transport)

A Lagrangian trajectory model (Eliassen and Saltbones, 1982) for large instantaneous sources describing, for example, major accidents, using a subset of the EMEP European synoptic wind is implemented on a European scale with completely interactive problem definition, context driven auto-startup feature, and extended (animated) graphical display for the simulation.

The user can select the location for the accident by simply dragging a cross-hair cursor over the map of Europe. The magnitude of the emission, season, time of the day, and weather pattern can be selected by simply pointing at the appropriate description or icons symbolizing, for example, a repertoire of characteristic weather patterns. The appropriate input parameters are then automatically selected, scaled, or interpolated by the model system.

River

A simple river water quality model for toxic substances, extracted from the generic screening level USEPA model system TOXSCREEN (Hetrick and McDowell-Boyer, 1984) simulates pollutant dispersion in an arbitrary river segment (Figure 21.3). The model features extensive interactive input modification based on predefined default values as well as an animated graphical display.

The model is connected to the hazardous substances database, so that the parameters for specific substances can be loaded from this database after identifying a substance by one of the database access mechanisms.

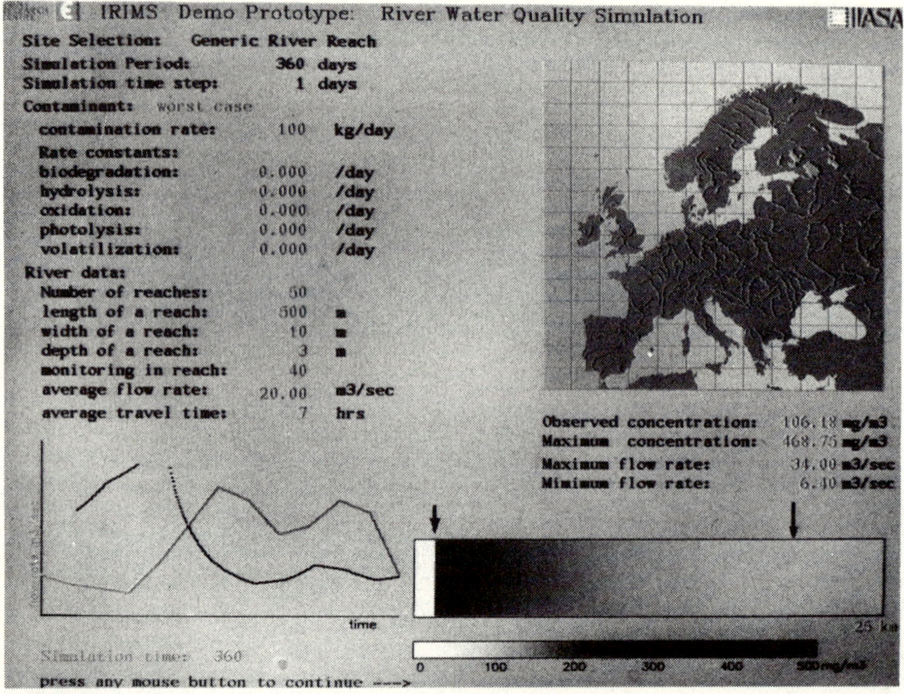

Figure 21.3 River Water Quality Simulation Model

FEFLOW

A 2D finite element groundwater contamination model, configured for a set of generic problem situations. Problem descriptions can be modified interactively, by setting pumping rates, activating or deactivating pumps or well galleries, specifying the concentration or mass flux of the pollutant source, and setting material parameters such as decay and absorption rates. The model generates animated graphical output of flow fields and time-varying concentrations in the observation or pumped wells defined in a given problem (Diersch, 1980).

A more advanced standalone version has been developed, which combines a fast 1D screening level model with the 2D finite element model, economic evaluation, the dynamic display of spatially distributed pollutant concentrations, and finally geographical background data including LANDSAT satellite maps where appropriate.

HASTM (Hazardous Substances Transportation Model)

HASTM is a transportation risk/cost analysis model (basic software elements developed by the Ludwig Boltzmann Institute, Vienna). The graphics-based interface developed for the model allows interactive definition of a transportation problem (Figure 21.4). The model is to be used in conjunction with the discrete optimization system described below.

After selecting beginning and end points for a specific transport by pointing at these sites on the map, the user can select a specific substance either in terms of a symbolic description based on the hazard diamond code system, or by selecting a substance from the hazardous substances database and defining the amount to be transported. The model will then generate a set of route alternatives, and evaluate them in terms of transportation costs and risk estimates for alternative vehicle types. A discrete multi-criteria decision support tool can then be invoked for the comparative evaluation of this set.

Multi-criteria Discrete Optimization (Data post-processor, DIDASS based)

This is available as a data post-processor accessible from the main menu level as well as from selected simulation models. The module, based on the reference point approach (Lewandowski and Grauer, 1982; Wierzbicki, 1983) allows interactive problem definition, that is, selection of relevant decision criteria, the setting of constraints from these criteria, identification of the pareto-optimal set (eliminating dominated alternatives) and provides extended graphical display options.

These include projections of the decision space for interactively selected pairs

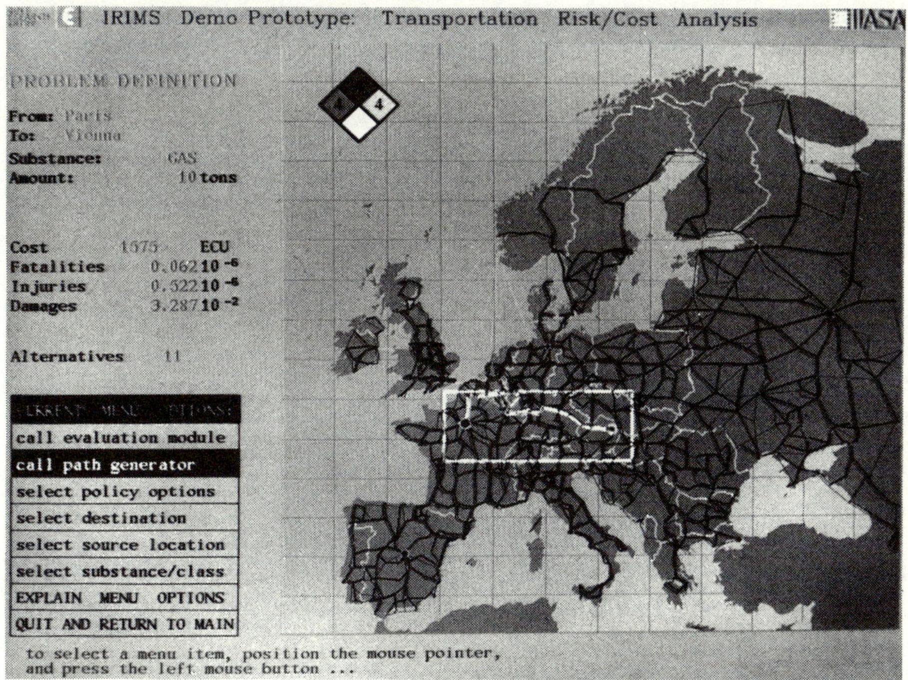

Figure 21.4 Transportation Risk-Cost Analysis Simulation Model

of criteria in scattergrams, histograms for individual criteria, and the possibility to cross-reference alternatives in various projections. Finally, the user can define a reference point or a desired target solution, and then find the efficient solution, that is, the solution closest to this target.

Process Plant Risk Analysis (SAFETI)

The parallel study for VROM develops an interactive and graphics-oriented framework and post-processor for the risk assessment package SAFETI (Technica, 1984) to facilitate the quick generation, display, evaluation and comparison of policy alternatives and individual scenarios.

The SAFETI package is a computer-based system for risk analysis of process plants. The software package was developed by Technica under contract to the Ministerie van Volkshuisvesting, Ruitmeljike Ordening en Milieubeheer, in association with the Dienst Centraal Milieubeheer Rijnmond.

SAFETI begins by generating a plant description. Failure cases are then generated and clustered. Finally, the failure cases are processed by consequence analysis programmes producing: radiation radii for early ignition of flammable gas; dense cloud dispersion profiles and associated flammable mass for late ignition; and toxic effect probabilities as "appropriate" consequence parameters can be combined to produce risk contours and F–N curves. SAFETI is accessible from the IRIMS master menu, and runs under its own interactive, line-oriented menu system.

The graphical interface to SAFETI's databases and consequence modelling results allows for the display of the raw data such as plant locations, weather data, or population distribution as thematic overlays on a map. When risk analysis, using SAFETI's original interface, has been performed for a specific process plant, the results are available for graphical display and interpretation. In addition to the F–N curves, risk contours can be displayed as transparent overlays on a map of the Netherlands. This map allows arbitrary zooming to provide the appropriate level of detail and resolution for a given problem.

References

Diersch, H.J. 1980. Finite-Element-Programmsystem FEFLOW. Programme description. Institut für Mechanik de AdW DDR. Berlin. In German.

Dobrowolski, G., Kopytowski, J., Lewandowski, A. and Zebrowski, M. 1982. *Generating Efficient Alternatives for Development in the Chemical Industry*. CP-82-54. Laxenburg, Austria: International Institute for Applied Systems Analysis.

Dobrowolski, G., Kopytowski, J., Wojtania, J., and Zebrowski, M. 1984. *Alternative Routes from Fossil Resources to Chemical Feedstocks*. RR-84-19. Laxenburg, Austria. International Institute for Applied Systems Analysis.

Eliassen, A. 1978. The OECD Study of Long-range Transport of Air Pollutants: Long Range Transport Modelling. *Atmos. Environ.* 12: 479.

Eliassen, A. and Saltbones, J. 1982. Modelling of Long Range Transport of Sulphur over Europe: a Two-year Model Run and Some Model Experiments. EMEP/MSCW Report 1/82: Blindern, Oslo. Norwegian Met. Inst.,

Fedra, K. 1985. *Advanced Decision-oriented Software for the Management of Hazardous Substances. Part I: Structure and Design.* CP-85-18. Laxenburg, Austria: International Institute for Applied Systems Analysis.

Fedra, K. and Loucks, D.P. 1985. Interactive Computer Technology for Planning and Policy Modelling. *Water Resources Research,* 21, 2 : 114–122.

Fedra, K. 1986a. *Advanced Decision-oriented Software for the Management of Hazardous Substances. Part II: A Prototype Demonstration System.* CP-86-10. Laxenburg, Austria. International Institute for Applied Systems Analysis.

Fedra, K. 1986b. Decision Making in Water Resources Planning: Models and Computer Graphics. Paper presented at the UNESCO/IHP-III Symposium on Decision Making in Water Resources Planning, Oslo, Norway, 5–7 May.

Fedra, K. and Otway, H. 1986. *Advanced Decision-oriented Software for the Management of Hazarous Substances. Part III. Decision Support and Expert Systems: Uses and Users.* CP-86-14. Laxenburg, Austria: International Institute for Applied Systems Analysis.

Fedra, K., Weigkricht, E. and Winkelbauer, L. 1987. A Hybrid Approach to Information and Decision Support Systems: Hazardous Substances and Industrial Risk Management. *Economics and Artificial Intelligence.* Proceedings of the First IFAC/IFORS/IFIP/ARCET International Conference, 2–4 September, Aix-en-Provence, France. Oxford: Pergamon.

Goodman, I.R. and Nguyen, H.T. 1985. *Uncertainty Models for Knowledge-Based Systems.* Amsterdam: North-Holland.

Grauer, M. and Fedra, K. 1986. Intelligent Decision Support for Technology Assessment: The Chemical Process Industry. Paper presented at the VIIth International Conference on Multiple Criteria Decision Making, August 18–22, Kyoto, Japan.

Gupta, M.M., Kandel, A., Bandler, W. and Kiszka, J.B. 1985. *Approximate Reasoning in Expert Systems.* Amsterdam, North-Holland.

Hetrick, D.M. and McDowell-Boyer, L.M. 1984. *User's Manual for TOXSCREEN: A Multimedia Screening-level Program for Assessing the Potential Fate of Chemicals Released to the Environment.* ORNL-6041, Oak Ridge National Laboratory and EPA-560/5-83-024. Washington, DC: Environmental Protection Agency.

ICF. 1984. The RCRA Risk–Cost Analysis Model. Phase III Report and Appendices. Submitted to the Office of Solid Waste, Economic Analysis Branch, US Environmental Protection Agency, Washington, DC: ICF Incorporated.

Lewandowski, A. and Grauer, M. 1982. The Reference Point Optimization Approach – Methods of Efficient Implementation. WP-82-26. Laxenburg, Austria: International Institute for Applied Systems Analysis.

Technica. 1984. The SAFETI Package. Computer-Based System for Risk Analysis of Process Plants. Vol. I–IV and Appendices. London.

Weigkricht, E. and Winkelbauer, L. 1987. Knowledge-based Systems: Overview and Selected Examples. WP-87-101. Laxenburg, Austria: International Institute for Applied Systems Analysis.

Wierzbicki, A. 1983. A Mathematical Basis for Satisficing Decision Making. *Mathematical Modelling,* 3 : 391–405.

Winkelbauer, L. 1987. Symbolic Simulation of Chemical Production Processes.
WP-87-45. Laxenburg, Austria: International Institute for Applied Systems.

Zhao, C., Winkelbauer, L. and Fedra, K. 1985. *Advanced Decision-oriented Software for the Management of Hazardous Substances. Part VI. The Interactive Decision Support Module.* CP-85-50. Laxenburg, Austria. International Institute for Applied Systems Analysis.

Part III

Experiences from Developing Countries

22

Environmental Aspects of Hazardous Waste Management for Developing Countries: Problems and Prospects

Asit K. Biswas

Although generation of hazardous wastes has been an integral part of many of the activities of modern civilization, the problems associated with their safe disposal captured the world's attention in the late 1970s. It was primarily due to the environmental disaster that occurred in the Love Canal area of Niagara Falls, New York. Love Canal was an uncompleted, abandoned nineteenth-century waterway that was used as an industrial waste disposal site since the 1930s. Between 1942 to 1951, Hooker Chemical and Plastics Company dumped more than 20,000 tons of toxic chemical wastes into Love Canal. Some of these wastes were in drums but the rest was discharged directly into the trench. The site was covered over and sold for development in 1953. In August 1978, some 25 years later, with visible and alarming effects on the health of the residents, President Carter declared Love Canal an emergency and a disaster area. This conferred upon the disaster the dubious distinction of being the first man-made event in the United States which was declared to be an emergency.

As the scientists began to investigate the scope, seriousness and overall impact of the various problems associated with the disaster at Love Canal, their findings raised new issues and questions with the existing hazardous wastes disposal practices. Chief among them was the fact that Love Canal was indeed a "state of the art" waste disposal facility of the time and that toxic chemicals were stored and disposed of there knowingly. Accordingly, as the immediate and long-term adverse implications of the Love Canal incident became apparent, societal concerns required answers to an important set of questions. Among the major questions were how many similar disposal sites existed, where were they located, what threats they posed to human health and the environment, and how these problems could be neutralized? Once the search for hazardous wastes disposal sites was initiated, it was soon found that many such sites existed in different parts of the United States as well as in several other countries. In some areas, the problems surfaced many years later after the burial of the hazardous wastes had contaminated soil and water resources of the localities. It was also found that with inadequate legislative requirements and regulatory processes, toxic wastes were occasionally clandestinely discharged into storm sewers or open lagoons, dumped along roadsides, or spread across fields and forested areas. All these findings increased public concern and heightened political activities, which – not surprisingly – moved hazardous wastes disposal much higher in the priority listing of serious environmental issues of the late 1970s and early 1980s.

General Considerations

While there is no universally accepted definition of hazardous wastes, the definitions that are in use in various countries or international organizations are similar. A few select examples will prove the point.

Canada "Waste is any substance for which the owner/generator has no further use and which he discards."

"Hazardous wastes are those wastes which due to their nature and quantity, are potentially hazardous to human health and/or the environment and which require special disposal techniques to eliminate or reduce the hazard."

Philippines "materials which are inherently dangerous to the human body or to animals, including, but not limited to, materials that are toxic or poisonous; corrosive; irritants, strong sensitizers; flammable; explosive, i.e. generate power through decomposition, heat or other means; infectious, i.e. represent a potential source for the transmission of diseases to humans, domestic animals or wildlife; radioactive, i.e. containing sufficient radioactivity to require labels; and pesticides.

UNEP "wastes other than radioactive wastes which, by reason of their chemical reactivity or toxic, explosive, corrosive or other characteristics causing danger or likely to cause danger to health or the environment, whether alone or coming into contact with other wastes, are legally defined hazardous in the State in which they are generated or in which they are disposed of or through which they are transported."

USA "one that may cause or significantly contribute to serious illness or death or that poses a substantial threat to human health or the environment when improperly managed."

Using the definition that is considered to be appropriate, it is possible to compile a list of hazardous wastes. Some countries, for example the United States or Canada, have compiled such lists, which are updated as and when necessary. The Environmental Protection Agency (EPA) of the United States defines as hazardous any waste which meets with one of the following four criteria:

1. *Ignitability* Wastes that pose a fire hazard during routine management. Fires not only present immediate dangers of heat and smoke but also can spread harmful particles over wide areas.
2. *Corrosivity* Wastes requiring special containers or segregation from other wastes because of their ability to dissolve toxic contaminants
3. *Reactivity* Wastes that tend to react spontaneously, to react vigorously with air or water, to be unstable to shock or heat, to generate gases or to explode
4. *Toxicity* Wastes that, when improperly managed, may release toxicants in sufficient quantities to pose a substantial hazard to human health or the environment

Although hazardous wastes are generated in many different segments of society, industry is by far the largest source. While the amount may vary from country to country, generally the chemical and allied products industry account for nearly 50 to 70% of all hazardous wastes produced. For the United States, current estimates indicate that some 60% of all industrial hazardous wastes is generated by the chemical and allied products sector.

Reliable estimates of hazardous wastes currently being generated are not available for most developing countries. However, in the absence of reliable estimates, it can be assumed that as a general rule of thumb some 10 to 15% of wastes produced by industry are likely to be hazardous. On this basis, it is likely that the generation of hazardous wastes is increasing at the rate of 2 to 5% per year.

Estimates of the quantities of hazardous wastes generated by industry indicate that the total production in the OECD countries was some 300 million tonnes annually in 1985, out of which the United States accounted for some 268 million tonnes, European OECD countries 24 million tonnes, and the balance of 8 million tonnes was contributed by the Pacific OECD countries. Very little information is available at present on the extent of hazardous wastes produced in most developing countries and their waste disposal and management practices. A UNEP–WHO *ad-hoc* Working Group of Experts on Environmentally-sound Management of Hazardous Wastes concluded in 1984 that "it is safe to say that virtually all developing countries have yet to develop a comprehensive hazardous waste management scheme, as compared to those already established in most industrialized countries."

From an environmental and health viewpoint it is not only important to determine the source, extent and type of hazardous wastes being produced but also it is essential to have similar information on the amount of such wastes generated in the past, how and where they were disposed, how they are being managed at present, and in case of improper disposals, what the existing and potential problems could be, and how they can be resolved.

The present state of poor information base in developing countries is not surprising. Even for an advanced country like the United States, as late as mid-1979, its Council on Environmental Quality reported that "existing estimates on the extent of the problem vary so widely they are only marginally helpful". In addition, the number of existing disposal sites was "one of the major unknowns". The same year, USEPA estimated that only about 10% of hazardous wastes were being disposed of in a manner that was likely to comply with the regulations the Agency was planning to adopt. Similarly, estimates of the cost of cleaning up abandoned hazardous waste dumps in the United States varied widely, ranging from a low $28.4 billion to a high of $55 billion.

Hazardous Wastes Management

Sound management of hazardous wastes must consider more than only safe storage and disposal practices. It would require a holistic approach that will consider comprehensively all alternatives available to institute a cradle-to-grave management system. It should consider not only the characteristics, volume and location of the wastes, but also how and why wastes are being produced and what effective steps can be taken to reduce substantially the quantity of wastes eventually to be disposed of.

A systems approach to hazardous wastes management will include consideration of the following six steps.

1. *Minimization of hazardous wastes generation* By modifying production processes, it may be possible to reduce the quantity of hazardous wastes finally generated. If the quantity of wastes generated is less, the quantity to be disposed of in an environmentally sound manner will also be less. For example, introduction of a new ferrosilicon furnace in Norway has contributed to higher yields, less raw materials and energy consumption and reduced waste generation. In Bulgaria, the introduction of low-waste technology has reduced the production of industrial wastes by about 5.5 million tonnes annually. Often a move to such low-waste technology is more economical than the processes they replace. It makes good sense in terms of both good economic and environmental housekeeping. Introduction of such technology also enables industry to comply with environmental regulations more effectively.
2. *Reprocessing and re-use of wastes* Many wastes may contain useful materials, which can be reclaimed and re-used. In many cases extraction of resources from concentrated wastes requires less energy and contributes to less air and water pollution and solid wastes generation than the mining and processing of virgin raw materials. Recovery of energy and raw materials from wastes has increased significantly in recent years. For example, in the German Democratic Republic, approximately 30 million tonnes of wastes are recycled annually. In Hungary, of about 22.5 million tonnes of industrial wastes generated every year, some 6.5 million tonnes are recycled. The re-use of wastes from the chemical, pharmaceutical, food processing and mining industries in Bulgaria has increased significantly in recent years. The recycling of aluminium cans in the United States increased more than twenty-one-fold, from 24,000 tonnes to 510,000 tonnes,within a ten-year period from 1972 to 1982. However, since only a small fraction

of wastes is economically suitable for recovery at present, new low-waste technologies need to be developed before recovery can become a more widely viable alternative for many different industrial processes.

3. *Transfer of wastes to another industry* In certain cases, hazardous wastes from one industry can be transferred to another industry which can use them as raw materials. An information exchange system can be set up to act as a clearing-house, which can assist in matching producers of wastes with potential purchasers and users. The first such information exchange was set up in the Netherlands in 1972. Similar information exchanges and materials exchanges now operate in Canada, United States and several other countries. It should, however, be noted that probably only a small fraction of hazardous wastes may be suitable for exchange.

4. *Separation of hazardous and non-hazardous wastes* If hazardous and non-hazardous wastes can be separated at source, only a limited quantity of hazardous wastes need to be handled. This will naturally reduce the final handling, transportation, storage and disposal efforts and costs.

5. *Transform hazardous wastes to non-hazardous* Hazardous wastes can be made non-hazardous by incineration and physical and biological processing. Some toxic organic wastes can be destroyed by incineration. If carried out properly, it will not contribute to environmental degradation. Incineration, however, is an energy-intensive and expensive process. Chemical and biological processes can be used to transform hazardous to less hazardous or non-hazardous wastes. Among the physical processes used are carbon or resin adsorption, distillation, centrifugation, flocculation, sedimentation, reverse osmosis and ultrafiltration. Chemical processes include neutralization, oxidation, precipitation and material exchanges to remove heavy metals. Some of the biological processes that can be used are activated sludge treatment, trickling filter and controlled land application.

6. *Disposal of wastes in a controlled landfill* After consideration of the above five options, whatever hazardous waste remains has to be disposed of in secure landfills. Siting, planning and operation of secure landfills must be properly carried out. Regular monitoring of disposal sites is essential, even after they are closed, to ensure potential health hazards and environmental contamination do not occur. Sound disposal of hazardous wastes is not cheap, but in the long run it will invariably turn out to be significantly less expensive than inadequate disposal. For example, the estimated cost of sound disposal at Love Canal would have been less than $2 million, but by May 1980, more than $36 million had been already spent for the clean-up and associated expenses. The final bill was much higher.

Education and Training

One of the essential and immediate requirements for sound hazardous wastes management in developing countries has to be education and training. As mentioned earlier, interest in hazardous wastes management in developing countries is of comparatively recent origin. With increasing industrial activities and serious accidents like at Bhopal, management of hazardous wastes and toxic chemicals has rapidly become a priority concern. While interest and priority in this area has increased, one major factor that is hindering institution of proper and reliable hazardous wastes management in all developing countries is the shortage of required number of trained personnel. Without adequate expertise, hazardous wastes management will always have significant risks associated with it.

Training is required for people at three levels:

1. *Professional* generally holding a university degree in an appropriate subject and may also have postgraduate qualification
2. *Technician* having a diploma from a technical school, usually on the basis of 2–4 years of attendance and
3. *Vocational* having completed a primary school education, followed by up to two years of certificate course or training by a government-sponsored course

While training programmes at all three levels are necessary, only training at professional level will be discussed in this paper.

There are seven different types of professional training that can be identified. These, however, are not necessarily mutually exclusive but in most cases they complement each other. They are discussed below.

Institutional Training

Institutional training is an essential prerequisite to developing professional skills in the area of hazardous waste management. Accordingly, on a long-term basis, each nation should ensure that its colleges and universities are capable of imparting the type of training that is consistent with national needs. In most developing countries, however, appropriate universities and/or colleges where people could be properly trained in hazardous wastes management do not exist.

Not surprisingly educational institutions share many of the same constraints of other public bodies. Such constraints include limited operating budgets, inadequate possibilities for career development of teaching staff, under-staffing, inadequate library, computer, research and other support facilities, inexperienced staff and poor staff–student ratios. These factors generally contribute to theoretical and conventional orientations to teaching which often are not appropriate for the solution of practical problems. Furthermore, in an area such as hazardous wastes management where the knowledge-base is improving rapidly, the lack of training facilities available to teachers during their working lives is contributing to the obsolescence of the material being taught.

Even assuming adequate funds are available, producing good teaching and professional staff is a long-term process. A place has to be made available now in a suitable university in order that an experienced teacher or a project manager can emerge in 12 to 15 years' time. Allowance should be made for the continuous manpower losses during the various phases of manpower development, therefore more than one place per project manager will be necessary.

On a long-term basis, countries generally have no other alternative but to ensure that an adequate number of educational institutions are available to produce professionals in the various disciplines necessary for hazardous wastes management. Once national educational capacities are developed for professional training, specialized training abroad in specific subject areas, if considered necessary and desirable, can be considered. General professionals, however, must be produced within each country.

Since institution-building takes time, in the short and medium term, many developing countries may find it necessary to use external educational institutions in order to produce the requisite number of qualified professional people. Training abroad, developed or developing, for undergraduate and postgraduate qualifications has certain advantages as well as disadvantages. On the positive side, training can be initiated immediately in any country where appropriate training facilities exist, with well-qualified teachers as well as laboratory, computer and library facilities. On the negative side, training abroad may not be very relevant to the problems in countries at different stages of development and having different institutional, legal and socio-cultural requirements. In addition, especially for longer term training in developed countries, there are always a significant number of professionals who after training will decide to stay abroad because of better career and financial prospects.

Cost effectiveness of external training is also an important consideration. Generally training in developed countries tends to be more expensive. Some estimated comparative unit costs are given in Table 22.1. It should be noted that unit costs vary widely from one country to another, and even from one institution to another in the same country. Accordingly the costs given in Table 22.1 should be considered indicative rather than definitive. Irrespective of any special consideration, it should be noted that only a limited number of people from any developing country can be trained abroad. Hazardous wastes management will require a significant number of professionals in a country, the bulk of whom have to be trained nationally.

Table 22.1 Estimated Education and Training Costs for Hazardous Wastes Management

Type of Course	*Cost in US $*
Long course at professional level in UK (BSc, MSc, PhD), fees only	11,200/man–year
Short course (2 weeks) in Europe, fees plus air travel	950/man–week
Short course (2 weeks) in Africa, fees plus local travel	575/man–week
Individual professional development programme in Europe, fees only, 1 to 52 weeks	310/man–week

In-service Training

All work or projects have a training component, but this is mostly implicit rather than explicit. Many developing countries like China or India have some training activities in the hazardous waste management sector. However, these activities, whenever they exist, tend to emphasize theoretical aspects; those aspects, whose importance has been acknowledged only comparatively recently, such as the environmental, legal and institutional, are mostly absent or dealt with in a general way.

What is required is for each nation, following completion of an employee's formal education, to implement a training programme on a planned and structural basis throughout employment. Thus, training should be considered by all nations as a continuing activity in various departments related to hazardous wastes management. Only in this way can new ideas, concepts, instruments and equipment be brought to the attention of the staff enabling them to perform the tasks assigned to them more efficiently.

Short Courses

Well designed short courses (1–12 weeks) can be of immense benefit to all staff, improving their level of knowledge and complementing the formal education received during their earlier training. Such courses can be used to familiarize staff in new areas, such as environmental aspects of wastes management, instrumentation or use of computers and systems analysis. They can also serve periodically to update technical knowledge and skills and to broaden interdisciplinary understanding. These courses can be held either within a country or abroad.

Counterpart Training

In most multilateral or bilateral projects for hazardous wastes management, where expatriate professionals are used, counterpart staff are appointed with the objective of knowledge and experience transfer. Properly done, counterpart training can play an important role in developing countries.

Generally, however, the training of counterpart personnel who are supposed to take over from expatriate staff is unsatisfactory. Expatriate staff are frequently used as substitutes for local manpower and spend much of their time performing routine and mundane duties. The weakness of the national institutions often renders the foreign experts, even when they are suitably chosen, ineffective. Frequently there are not enough adequately qualified counterpart staff, and even when they exist, bureaucratic delays in their recruitment may not leave sufficient time for training.

Individually Tailored Programmes

Many multilaterally or bilaterally supported projects now contain specifically tailored programmes for training of a limited number of select individuals. These training programmes are invariably abroad. They may range from 2–8 weeks of study tours in various institutions in one or more countries in specific areas of specialization to more formal training at educational and research institutions leading to diplomas or degrees.

Conferences, Symposia and Workshops

Under certain conditions, participation at conferences, symposia and workshops to present papers, discuss experiences and share knowledge can have a training role. However, considering the fact that most conferences or symposia have durations of one week or less, they are not planned to play an explicitly educative role. If not suitably organized, their role in training may not be cost-effective. One could argue that the perusal of documentation from a conference may in certain cases be a more cost–effective training alternative.

Information and Resource Materials

Pursuit of knowledge is seriously handicapped in all developing countries by lack of access to up-to-date books, manuals, journals and data bases. Inadequate funding means that many of the most recently published books and periodicals reporting new findings and methodologies are not available, and the few publications that may have been purchased are not readily accessible because of poor documentation and library facilities. Similarly, the lack of micro-computers and audiovisual facilities means that staff cannot remain informed about the new developments in a rapidly changing technical area such as hazardous wastes management.

Selection of Trainees

Training should be designed to equip national personnel to meet the various challenges posed by the hazardous wastes management sector. An essential first step in designing a training project has to be a clear and explicit definition of the personnel to be trained. Their needs should be reviewed in detail, along with the constraints they face, in order to obtain a clear and unambiguous picture of what is required. While this may be stressing the obvious, the point needs emphasis because most of the training programmes sponsored by different national and international agencies are primarily based on assumptions rather than on actual investigations.

Trainees should be selected on the basis of vocational motivation, level of formal and informal preparation including their ability to absorb training, proficiency in the language being used for training, potential for the trainees to use the knowledge in solving real problems. The tendency to favour civil servants in central ministries over other personnel should be avoided. Thus, specific criteria should be stipulated for selecting trainees.

Since the selection of trainees is usually made by governments subject to the pressures and conflicts that exist within any civil service structure, it may often be difficult for an organization such as UNIDO to ensure that the appropriate selection criteria are adhered to. However, by stipulating selection criteria explicitly in advance, UNIDO can encourage countries to adhere to these criteria. Occasional rejection and/or questioning of the nomination of apparently unsuitable trainees could help. In any case, significant departures from the stated criteria in the actual selection should be clearly stipulated in subsequent evaluation reports.

In many international and national institutions, it appears that training is viewed by all concerned as something intrinsically good and that the trainees should be grateful without questioning the underlying reasons for its provision. Trainees should be clearly notified in advance of the type of training they will receive, its objectives, and what will be expected from them on completion. In the interest of fairness and equity, trainees should be advised of the specific selection criteria, so that their motivation does not suffer. This is especially important for short-term training courses.

Potential Role of UNIDO

UNIDO has already made a useful contribution in the area of hazardous wastes management and industrial safety and undoubtedly will have a major role in the future. However, as the extent, magnitude and complexities of the management issues and risks associated with hazardous wastes management become known, the role and potential contribution of UNIDO become even more crucial and significant.

Although there are many areas where UNIDO can make a contribution to developing countries, the discussion below focuses on four areas of particular importance where either existing programmes can be intensified or new ones created to meet emerging needs. It should be noted that the suggested action areas are not listed in any order of priority since priorities may differ from country to country.

1. *Environmental impact assessment of hazardous wastes management projects and programmes* There is an urgent need to develop operational guidelines to carry out environmental impact assessment which would meet the special requirements of developing countries. It is essential that environmental impact assessment processes be developed that can be used with limited expertise and resources available in developing countries and which can be carried out within a reasonably short period of time. Currently this type of operational guideline does not exist. UNIDO can play an important role, either directly or through co-operation with organizations like UNEP, to develop such guidelines and then promote their use.

2. *Capacity to react immediately to catastrophic accidents* Serious accidents such as those at Bhopal, Chernobyl, the Rhine and Seveso have clearly demonstrated the need to react immediately to contain adverse health and environmental

impacts which may arise. When serious accidents occur in developing countries, urgent assistance may be necessary. UNIDO could play a major role by being able to respond immediately to any request for assistance from member countries as and when such accidents take place. While UNIDO is currently capable of advising and assisting countries on the prevention and mitigation of routine industrial accidents relating to the release of chemicals to the environment, the responses to catastrophic accidents such as at Bhopal or Seveso have to be different. Since the health and environmental implications will obviously vary according to the nature of the accident, it would be imperative to identify the best expertise available globally, to assemble a team of eminent experts and transport them to the site of the accident within a very short period of time after its occurrence. Thus, all the relevant institutional arrangements need to be in readiness to ensure a speedy response after any request for assistance is received.

3. *Training* Training is an important component of any successful hazardous wastes management programme, and at present is generally much neglected. Training should be considered simultaneously at different levels. UNIDO should consider further intensifying its efforts in this area in collaboration with national counterparts, international organizations and bilateral aid agencies. Existing and future UNIDO-sponsored institutes such as the Pollution Control Research Institute in Hardwar, India, should be encouraged to provide training, both nationally and regionally, on all aspects of hazardous wastes management, including environmental impact assessment. Furthermore, UNIDO should assist in the preparation of appropriate training manuals, guidelines and books which would be relevant to the needs of developing countries.

4. *Network* The establishment of a network of national counterparts, international organizations, bilateral aid agencies and non-governmental organizations for technology transfer, information determination and other activities should be considered. If such a network is established, a small secretariat will be necessary, either within or outside UNIDO, to service it effectively.

23

Hazardous Waste Management in ASEAN: With Emphasis on Small and Medium Industries

Filemon A. Uriarte, Jr.

ASEAN co-operation in the environment field was initiated in 1977 when the draft ASEAN Sub-Regional Environment Programme (ASEP I 1978–1982) was prepared with the assistance of the United Nations Environment Programme (UNEP). The ASEAN Committee on Science and Technology (COST) recommended the convening of a meeting of ASEAN experts on the environment to consider the proposed ASEP I. Consequently, the First Meeting of the ASEAN Experts Group on the Environment (AEGE) was held in Jakarta on 18–20 December 1978. The AEGE has met annually since then and it has become a permanent body under the purview of ASEAN COST.

Among the major challenges confronting ASEAN is the problem of providing its people with their basic needs in the face of increasing population pressures. In less than two decades, the population of ASEAN is projected to increase by 35% from 280 million in 1984 to 378 million in year 2000.

The response of ASEAN to the problem is to accelerate its development activities. This means rapid industrialization, more intensive and extensive agriculture, and increased exploitation of natural resources. While these efforts have made ASEAN among the fastest developing regions in the world, they have also resulted in industrial and agricultural pollution, depletion of natural resources, and conflicts in the use of the environment. To resolve the apparent conflict between economic development and environmental protection, ASEAN has adopted the principle of environmentally-sound and sustainable development. Consequently, the harmonization of economic development with environmental protection, the integration of economic and environmental planning at all levels, and the implementation of environmentally-sound and sustainable development strategies constitute the central theme of the ASEAN objective and policy on the environment.

For the next five years, ASEAN has set four goals for industry and environment, namely:

1. Implementation of demonstration projects in industrial pollution control with emphasis on clean technologies in specific industries, resource recovery, waste recycling and utilization, and proper waste disposal
2. Establishment of proper management programmes for the control of hazardous substances and wastes
3. Development and implementation of a regional programme on the prevention of industrial accidents which may have adverse environmental consequences through proper risk assessment, adequate handling procedures, safety measures, emergency plan, and the like
4. Development of methods for the assessment of highly pollutive industries and potentially harmful chemicals entering ASEAN

The importance that ASEAN places on hazardous waste management and industrial safety is reflected in the fact that three out of four goals under the priority area of industry and environment have relevance to the subject.

To achieve goals, 2, 3, 4, ASEAN member countries plan to undertake the following regional projects:

Project 1 Training Workshops for the Transportation, Collection, Use, Treatment, and Disposal of Hazardous Substances
Project 2 ASEAN Workshop on the Control and Use of Chlorofluorocarbons
Project 3 ASEAN Workshop on Contingency Planning for Chemical Accidents or Spills
Project 4 Preparation of Guidelines for the Assessment of Highly Pollutive Industries and Potentially Harmful Chemicals Entering ASEAN
Project 5 Development of a Notification and Assessment Scheme for Industrial Chemicals

These five projects are aimed at establishing the proper management and control of hazardous substances, preventing industrial accidents, and avoiding the uncontrolled entry of potentially harmful chemicals into ASEAN.

Hazardous Wastes: Generation, Disposal and Management

In most ASEAN countries, reliable information on the generation, storage, transport, use, and disposal of hazardous substances and wastes is often not available. The following sections present some of the available information. The gaps in the available information point to the need for greater regional and international co-operation in this field.

Generation and Disposal of Hazardous Wastes

Singapore

In April 1985, the Ministry of the Environment assumed overall responsibility for the control of hazardous substances. A section was established within the Anti-Pollution Unit (APU) of the ministry to review and formulate policies, procedures, and standards in consultation with other ministries. Among the first activities of the section was a study of the control of hazardous substances at the various stages of transport, storage, use, and disposal.

In 1985, a total of 28,180 tonnes of hazardous wastes were generated by 2188 industrial firms, as shown in Table 23.1. The largest amount consisted of spent solvents, both halogenated and non-halogenated, coming from the manufacturing of electronic components, metal finishing, servicing of aircraft, printing, and pharmaceutical industries.

The metal finishing industry is a major source of toxic and hazardous wastes in Singapore. There are about 26 manufacturing firms which have in-house electroplating and metal finishing facilities and more than 50 small and medium independent metal finishing plants which mostly service the electronics and appliance industry.

Table 23.1 Sources and Amounts of Hazardous Wastes in Singapore (1985)

Type of Waste	*Sources*	*Number of firms*	*Amount tonnes/year*
Industrial wastewater with toxic substances (e.g. heavy metals, cyanides, phenols)	Electroplating, metal finishing, manufacture of printed circuit board, batteries, etc.	184	980
Solvents	Manufacture of electronic components, metal finishing, printing, paints, varnishes, inks, etc.	122	10,100
Oil	Ship repairs, motor vehicle workshops, petrol service stations, manufacture of pipes, tubes, etc.	1512	6720
Etchant	Manufacture of printed circuit board	16	4320
Acids and Alkalis	Manufacture of batteries, micro-chips, industrial chemicals	11	660
Photographic wastes	Printing, publishing, photo studios	341	5400
Highly toxic wastes (e.g. arsenic, Hg compounds)	Manufacture of electronic components	2	<5
	Total	2188	28,180

All medium to large metal-finishing plants have wastewater treatment facilities varying from simple, conventional chemical treatment to highly sophisticated, automatically controlled facilities with in-plant waste segregation, water recovery, and recycling. A number of small plants discharge untreated rinse water into the central sewer system while the others have simple, batch-type wastewater treatment plants.

The majority of the other large companies which generate hazardous wastes have in-house treatment and/or recovery facilities. The rest, including the small and medium companies, are served by nine waste disposal contractors with facilities to recover, process, and/or treat hazardous wastes. The facilities include distillation units, incinerators, and equipment for the fixation and electrolysis to recover and/or detoxify wastes.

In 1985, the nine contractors handled 14,215 tonnes of hazardous wastes, approximately half the total amount generated in Singapore. These wastes include oils, etchants, solvents, acids, and alkalis as shown in Table 23.2.

In-plant segregation of wastes for recycling and recovery is practised widely in Singapore. Industries using large amounts of solvents normally segregate the spent solvents for recovery in-house or by external contractors. Printed circuit board manufacturers also segregate the spent etchants for recovery by outside contractors. Waste lubricating oils from vehicle repairing and servicing are also segregated, collected, and reprocessed for reuse.

Several photographic processing centres have electrolysis units to recover silver from spent fixer solution and one contractor collects spent fixer solution from the smaller photographic processing centres for the same purpose. Another waste processing contractor recovers copper from spent etchants.

Table 23.2 Types and Amounts of Hazardous Wastes Handled by Private Contractors, Singapore (1985)

Type	*Tonnes*	%
Waste oils	7789	54.8
Spent etchants	2548	17.9
Trade effluents	1069	7.5
Halogenated solvents	1068	7.5
Non-halogenated solvents	947	6.7
Acids and alkalis	794	5.6
	14,215	100.0

When recycling, recovery, or reuse are not viable, the wastes undergo physical–chemical treatment or incineration. At present, eight companies operate their own incinerators to dispose of spent solvents, solvent residues, oily sludge, and waste oil. These incinerators operate between 750°C and 1300°C.

Since ocean disposal is not allowed in Singapore, dewatered sludge from treatment facilities and other solid residues which cannot be treated economically by other means are disposed of at the government-operated dumping ground after fixation, usually by mixing with cement in steel drums. Certain hazardous substances, such as PCB-contaminated transfer oil, cannot be treated properly in Singapore and are, therefore, exported overseas for incineration.

Malaysia

The Department of Environment (DOE), Ministry of Science, Technology, and Environment, has the overall responsibility for the control and management of hazardous substances. Since 1981, DOE has been conducting surveys and studies to establish the generation and formulate guidelines and regulations on the handling, storage, treatment, and disposal of toxic and hazardous wastes in Malaysia.

A 1977 survey showed that nearly 8000 m^3/yr of hazardous sludge are generated in Kelang Valley, the most industrialized area in Malaysia. It consisted of 4520 m^3/yr from acid pickling operations, 3000 m^3/yr from anodizing plants, 330 m^3/yr from more than 100 electroplating plants, and 130 m^3/yr from galvanizing plants.

A 1983–4 survey indicated that over 52% of the toxic and hazardous wastes are generated by the electronics industry, 14% by the metals and electroplating industries and the rest by the chemical, rubber, plastic, printing, packaging, tannery, and pharmaceutical industries.

The most recent survey was carried out by the DOE in 1985 and covered more than 700 industries located in the west coast of Peninsular Malaysia where most of the industrial activities are concentrated. The results of this survey are summarized in Table 23.3.

Acid and alkali wastes from small and medium industries are discharged without prior treatment and these constitute 25% of the total volume of hazardous wastes generated. Nearly 18% of the total volume is composed of sludge containing heavy metals. These are mostly residues from wastewater treatment plants of electronic manufacturing and metal finishing plants.

At present, Malaysia has no facilities for the final disposal of toxic and hazardous wastes, aside from in-house wastewater treatment plants constructed by medium to large industrial firms. However, a national plan has now been prepared to deal with the problem. This plan includes the following two important elements:

1. Disposal facilities shall be constructed consisting of:
   a. landfill sites to receive toxic sludges produced by wastewater treatment plants and ashes produced by incinerator plants
   b. incineration plants for wastes that require thermal destruction such as halogenated hydro-carbon solvents and pesticides
   c. physical–chemical treatment plants to serve small and medium metal finishing plants
   d. special storage facilities for highly toxic wastes such as mercury, arsenic, and polychlorinated biphenyl (PCB)
2. Collection and transport facilities from waste generating areas to disposal facilities, including transfer stations and pre-treatment facilities.

Table 23.3 Hazardous and Toxic Waste Production from Industries in Malaysia (1985)

Waste Category	*Volume* m^3/yr
Organic Wastes	
Mineral oil and hydrocarbons	13,030
Paint/dye/ink/pigments	11,300
Photographic wastes	9000
Glue/resin/latex/urea-formaldehyde	3250
Spent solvents	3130
Pesticides/pharmaceutical wastes	290
	40,000
Inorganic Wastes	
Acids containing heavy metals	80,300
Alkalis containing heavy metals	24,200
Other chemical wastes	4700
	109,200
Semi-Solid/Solid Wastes	
Mineral sludge	100,500
Sludge containing heavy metals	75,000
Dust/slag/clinker with heavy metals	42,900
Asbestos dust/sludge	35,300
Paint/pigment/ink/dye sludges	9800
Contaminated paper, plastics, rags, etc	4300
Spent catalyst/batteries/acid sludges	300
	268,100
Total	417,300

This plan is now being implemented and constitutes a major area of concern for Malaysia's Department of Environment.

Thailand

The National Environment Board (NEB) of the Ministry of Science, Technology, and Energy and the Industrial Works Department (IWD) of the Ministry of Industry share the overall responsibility for the control and management of hazardous wastes in Thailand.

There are currently 80,000 factories registered with the IWD and about 20,000 are located in Bangkok and immediately surrounding areas. A 1984 survey shows that a total of 682 factories in Bangkok and nearby areas generate hazardous wastes. These include electro-plating plants, textile mills, tanneries, and paint factories as shown in Table 23.4. These factories generate 22,100 tonnes of sludge and slurry and 84,000 m^3 of effluents annually.

A major contributor to the hazardous wastes generated in the greater Bangkok area are the 200 electroplating plants, the majority of which are small with no treatment plants or with inadequate and ineffective treatment facilities.

The typical electroplating firm in Thailand renders a chome–nickel–chrome service. It has six or more plating tanks lined with either PVC or rubber and the largest tank has a volume ranging from 1 to 3 m^3. The electrical current per tank averages from 200 to 900 amperes. The firm performs pre-treatment of materials and extracts oil using alkaline solutions. Table 23.5 presents data on these electroplating plants.

It is evident that more than 72% of the electroplating plants can be considered small and more than 66% have no or inadequate wastewater treatment facilities. Hexavalent chromium is the predominant toxic component of the wastewater since more than 72% of the firms do chrome plating.

At present, plans are underway for the construction of a central Industrial Hazardous Wastes Treatment Centre on a 40-acre site in the western suburb of Bangkok. This is the first of three hazardous wastes treatment centres planned by IWD. The other two are to be located in the northern and eastern suburbs of Bangkok.

The first treatment centre in Bangkhuntien will have a 200 m^3/day chemical treatment plant for electroplating wastewaters, an 800 m^3/day chemical treatment plant and polishing lagoons for textile dyeing wastewaters, and chemical detoxification and cement mixing facilities for handling hazardous sludges. Since the land surrounding the area is not suitable for a landfill, the cement-mixed wastes will be transported to a final disposal site about 100 km from the centre. The centre is expected to start operations in 1988.

Table 23.4 Factories Generating Hazardous Wastes in the Greater Bangkok Area, Thailand (1984)

Industry	*Number*
Textile	210
Electroplating	200
Tanning	122
Paint	72
Car batteries	16
Dry cell batteries	16
Steel galvanizing	15
Car assembly	15
Fluorescent lamps	5
Integrated circuits	5
Caustic soda	4
Lead smelting	2
	682

Table 23.5 Electroplating Plants in Thailand

Category	% *of Total*	*Category*	% *of Total*
Type of Plating		Pre-Treatment	
Zn	13.9	None	0
Cr/Ni-Cr	72.2	Rarely	0
Pb	0	Sometimes	0
More than 2	5.6	Often	2.8
More than 3	5.6	Very Often	16.7
More than 4	0	Always	80.6
Volume of Largest Tank (m^3)		Oil Extracting Method	
Less than 1	5.6	By solution	2.8
1 to 3	72.2	Emulsion	0
4 to 7	13.9	Alkali	88.9
8 to 10	2.8	Electrolyte	0
11 to 20	5.6	Supersonic	2.8
More than 20	0	Others	5.6
Electrical Capacity Per Tank (amps)		Material of Tank	
Less than 200	8.3	Wood	0
201 to 500	47.2	Lead	0
501 to 1200	27.8	PVC/Rubber	88.9
1201 to 3000	13.9	Fibreglass	11.1
More than 3000	2.8		
Number of Tanks		Wastewater Treatment	
One	0	None	22.2
Two	8.3	Rarely	5.6
Three	11.3	Sometimes	22.2
Four	16.7	Often	16.7
Five	19.4	Very Often	0
Six or more	44.4	Always	33.3

While the government, through IWD, has provided over 30 million Baht for the land, treatment facilities, building, and utilities, the operation of the centre including collection, transport, treatment, and hauling of treated sludge to the final landfill site will be given to a private contractor. A similar arrangement is being considered for the two other treatment centres.

Indonesia

The Ministry of Population and Environment has the overall responsibility for the control and management of hazardous substances and coordinates the efforts of various ministries on this matter. The Ministry has considered the problem since 1983 and is now in the process of conducting surveys and studies to determine the sources, types, and amounts of hazardous wastes generated in Indonesia and to select suitable treatment systems and final disposal sites.

The survey of industrial firms generating hazardous wastes is currently being conducted in nine industrial centres throughout Indonesia. In East Java, one of the nine industrial centres, a total of 202 industrial firms have been identified as generators of hazardous wastes. These include 87 metal industries, 34 basic chemical plants, 14 pharmaceutical companies, and 67 miscellaneous industries.

One company which produces ammonium sulphate generates 170 kg/day of solid wastes containing arsenic and sulphur compounds and 3 kg/day of effluents containing hexavalent chromium. The arsenic- and sulphur-bearing solid wastes are placed in steel drums, sealed with cement, and dumped into the sea. The chromium-laden wastewater is diluted to a final hexavalent chromium concentration of 0.02 ppm and discharged into the coastal water. Another company that produces pesticides has a biological treatment plant for its liquid effluents. The solid wastes are also mixed with cement in steel drums and dumped into the ocean.

At present, plans are underway for the construction of a temporary disposal site for hazardous and toxic wastes in Pulo Gebang, East Jakarta. The construction of a larger and permanent final disposal site in Cibinong, West Java is planned. These disposal sites are aimed at serving the many small and medium firms in the greater Jakarta area and as final disposal sites for the sludge generated by treatment plants serving the large industrial plants.

Philippines

The National Environmental Protection Council (NEPC) and the National Pollution Control Commission (NPCC), two agencies which have been placed under the newly created Department of Environment and Natural Resources, are jointly responsible for the overall control and management of hazardous substances. The NEPC coordinates the activities of an Inter-Agency Task Force on Hazardous Substances while the NPCC is responsible for the implementation of government standards on industrial wastes. Accurate data on the sources, amounts, and types of hazardous wastes are not available. In the Greater Manila area, however, it is apparent that a major portion of toxic wastes comes from the metal finishing plants and microelectronics industry.

The most common metal finishing activities are electroplating, anodizing, enamelling, galvanizing, and phosphating. In the case of the electroplating industry, the largest demand comes from the wire sector with almost 90% of this demand accounted for by office supplies. Next in size are the electroplating needs of the household goods industry followed by the motor vehicles manufacturers. Ninety per cent of the needs of the motor vehicle sector and 80% of the household appliances sector are supplied by in-house facilities. The electroplating firms in the Philippines utilize simple metal, plastic, fibreglass reinforced plastic or concrete tanks with capacities of 7 to 11 cubic metres. A typical plant has 5 tanks but a few firms have up to 20 tanks. In terms of production volume, anodizing is the largest metal finishing activity. Over 90%

of the market is supplied by three firms with modern production facilities. These three firms produce anodized and coloured aluminium mainly for the building industry. Small enterprises anodize components for household appliances (6.1%), electronic components (2.4%) and household goods (0.6%). There are seven galvanizing plants with a total capacity of 420,000 tonnes/yr. These plants produce both plain and corrugated galvanized iron sheets. All use the conventional hot-dip process.

In 1984, there were 109 electronics firms in the Philippines, 11 owned by multinational companies, 17 by Filipinos, and the rest are joint ventures with foreign companies. Out of the 109 firms, 36 are semi-conductor or micro-circuit components manufacturers, 36 are in consumer electronics, and 34 are engaged in industrial electronics.

Case Studies: Small and Medium Industries

Small Electroplating Plant

Stancor Manufacturing Corporation in Malabon, Metro Manila is a typical electroplating plant in the Philippines. Established in 1969 as a stock corporation, the firm manufactures various kinds of household utensils like pot holders, soap racks, glass holders, record racks, dish drainers, and the like. It employs 45 people, operates six days per week and 16 hour daily in two shifts, and produces an average of 10 to 12 tonnes of products monthly or approximately 45,000 to 55,000 pieces. Its electroplating section, which started operating in 1977, provides copper–nickel–chrome plating for its own products and does not normally accept outside jobs.

As in most small electroplating plants in ASEAN countries, the Stancor plant is located in a residential area. It is enclosed by a high concrete fence with a high metal gate. The electroplating section occupies an area of about 85 m^2 and has one degreasing tank, one phosphoric acid tank, three copper plating tanks, one nickel plating tank, one chrome plating tank, six drag-out tanks and six rinse tanks. It consumes monthly 20 kg each of copper pyrophosphate and chromic acid, 25 kg each of potassium pyrophosphate and potassium hydroxide, 10 kg of nickel sulphate, and undetermined amounts of nickel fluoride, boric acid and other chemicals.

Water for both domestic and process use comes from its own deep well. The plant produces 3 to 5 m^3/day of wastewater coming mainly from its electroplating section. Table 23.6 presents the characteristics of the effluent. The concentration of cyanide and hexavalent chromium are not high which is typical of effluents from small electroplating plants in the Philippines. These plants normally use excessive amounts of water for rinsing thus producing relatively dilute effluents.

The wastewater treatment plant of Stancor Manufacturing Corporation was constructed at a cost of P65,000 in 1978 (approximately US$9000). It was designed to handle about 6 m^3/day of wastewater on a batch basis.

As in most wastewater treatment plants serving small electroplating firms in the Philippines, the Stancor wastewater treatment system does not provide for the segregation of alkaline and acid wastes, for the separate treatment of cyanide-bearing effluents, nor for the treatment of concentrated, spent plating liquor. Invariably, the owners of these small plants claim that the plating solutions are never replaced completely but are merely replenished with new chemicals. In some instances, the owners claim that the spent, concentrated solutions are sold to buyers but are not willing to provide additional information.

Medium-scale Tinplate Plant

The National Steel Corporation operates a medium-sized tinplate factory in Pasig, Metro Manila with two electrolytic tinning lines. The first line started operating in 1963 with a capacity of 75,000 tonnes/yr and uses the Halogen process developed by E.I. du Pont de Nemours and Co. The second line, with a capacity of 50,000 tonnes/yr, started operating in 1969 using the Ferrostan process developed by Carnegie-Illinois Corporation.

In the Halogen line, the electrolyte is composed of chlorides and fluorides with 35 g/l tin content. It contains 1.5 g/l of an organic agent which promotes the deposition of a compact coating with good flow-brightening characteristics. The optimum pH is 2.7 and the operating temperature is 65°C. The sources, flow rates, and characteristics of the wastewaters produced by the Halogen line are given in Table 23.7. The pH, fluoride, cyanide, and chromium levels do not meet government standards.

The Ferrostan line uses stannous sulphate electrolytes called Ferrostan solution which is essentially a solution of stannous tin in phenol sulphonic acid plus additive, dihydroxydiphenylsulphone and monobutylphenylphenol sodium monosulphanate. The metallic tin content is about 35 g/l in an operating temperature of 35–55°C. The sources, flowrates, and characteristics of the effluents produced by the Ferrostan line are shown in Table 23.8. The pH, suspended solids, chromium, and iron levels do not meet government standards.

Table 23.6 Typical Analysis of Untreated and Treated Effluents (Source: Inspection Reports, National Pollution Control Commission)

Parameter	*Untreated*	*Treated*
Colour	170	20
Turbidity, SJU	190	25
pH	2.8	8.0
Suspended solids, mg/l	100	5
Dissolved solids, mg/l	2400	3800
Cyanide, mg/l	0.08	0.01
Chromium (VI), mg/l	0.6	0.02
Copper, mg/l	10.0	0.07
Nickel, mg/l	18.0	2.34
Iron, mg/l	18.9	0.59
Zinc. mg/l	4.8	0.31

Table 23.7 Sources, Flow Rates, and Characteristics of Wastewaters From ETL 1 (Halogen Line)

Parameter	*Alkali Cleaning Section*	*Acid Pickling Section*	*Reclaim-Overflow Rinse*	*Chemical Treatment Rinse*
Flow rate, m^3/hr	14.4	3.6	5.8	30.4
pH	9.5	4.2	6.8	4.6
Fluoride, mg/l	–	–	1410	–
Chloride, mg/l	–	–	2500	–
Cyanide, mg/l	–	–	4.7	–
Cr^{+6}, mg/l	–	–	–	0.25

Table 23.8 Sources, Flow Rates, and Characteristics of Wastewaters From ETL 2 (Halogen Line)

Parameter	*Alkali Cleaning Section*	*Acid Pickling Section*	*Chemical Treatment Rinse*
Flow rate, m^3/hr	13.1	46.9	11.0
pH	10.0	4.9	5.7
Suspended solids, mg/l	80	–	–
Cr^{+6}, mg/l	–	–	5.2
Fe^{+2}, mg/l	–	17	–

From 1963 to 1976, the tinning lines were operating without treatment facilities for the effluents. In 1977, treatment facilities were constructed but these were improperly designed and were grossly inadequate to handle the toxic and hazardous components of the wastewater. In 1986, the company retained the services of a Filipino consulting company to design an effective wastewater treatment plant, using, as far as possible, the existing facilities. The treatment plant has been constructed and is now operational.

A conventional chemical treatment process has been selected for the plant consisting of a two-stage alkaline chlorine oxidation of cyanide; reduction of hexavalent chromium ion using sodium metabisulphite at low pH; neutralization; chemical coagulation-flocculation using alum and an organic polyelectrolyte; clarification; and sludge filtration.

The entire project, including design, supervision, construction, training of operators, preparation of operation manual, and plant start-up costs about P6 million (approximately US$300,000).

Summary and Recommendations

The preceding discussion indicates that ASEAN countries have made hazardous waste management a high priority in both regional and national environmental protection activities. While data are often inadequate, it is nevertheless quite evident that as a result of rapid industrialization in ASEAN during the last decade, substantial amounts of hazardous and toxic wastes are generated and discharged into the environment. The rapid growth of the microelectronics industry in ASEAN, the presence of a large number of small electroplating plants in the urban areas of ASEAN, and the upsurge in per capita energy consumption all indicate that ASEAN countries have good reason to afford high priority to the management of hazardous materials and wastes.

Because the majority of industrial plants in ASEAN are small- or medium-scale, waste management assumes an added dimension not normally found in most developing countries. Small electroplating plants, small battery manufacturing firms, and small paint factories, to name a few, are scattered throughout the major urban centres of ASEAN. Invariably, these factories have neither the financial means nor the technical expertise to treat and handle the hazardous effluents generated by their operations properly. The amount of waste generated by one plant is often too small to justify the recovery of valuable components. Very often there is also the problem of space for the location of the treatment plant and/or recovery facilities. These factors must be taken into consideration in the preparation of guidelines and/or of action, as well as in identifying the training needs.

To manage hazardous wastes effectively, it is imperative that adequate and reliable data are available on the sources, amounts, types, and characteristics of the wastes. While it is apparent that a significant amount of information is available for the metal finishing industry, data on the other types of industries, for example, those generating halogenated and non-halogenated solvents, pesticides, etchants, spent oils, and the like, appear to be very limited.

Under the Third ASEAN Environment Programme (ASEP III), five regional projects on the management of hazardous substances, industrial safety and emergency planning have been proposed. These projects need funding support. In particular, the training workshops for the transportation, collection, use, treatment, and disposal of hazardous substances should be given priority. Emphasis should be placed on the *technology* and *economics* of resource recovery, recycling, and re-use of hazardous wastes. The recovery of solvents, the re-processing of used oils, and the reclamation of valuable metals from various types of hazardous industrial wastes are some of the priority topics that should be taken up during these training workshops. Case studies should be presented to indicate the economic viability of the various technologies and alternative means for adapting these technologies to suit the conditions in ASEAN should be examined. These workshops should be aimed at training the professionals and in creating awareness among the decision makers in both the government and the private sector.

Finally, there is a need to develop a mechanism for the continuing exchange of technical information, transfer of technology, and upgrading of technical skills. The ASEAN Secretariat and the ASEAN Experts Group on the Environment are two possible conduits for these activities. UNIDO with its existing information network and access to expertise in the developed countries, could serve as the counterpart.

References

1. Proceedings of ASEAN–Australia Regional Training Course for Water Quality Management in Tropical Regions, Chiangmai, Thailand, 25 June–10 July 1986.
2. Proceedings of ASEAN–UNEP–CDG Workshops on Developing Policies and Strategic Guidelines for Managing Hazardous Wastes, Singapore, 7–9 May 1986.
3. National Pollution Control Commission. 1986. Engineer's Report: Wastewater Treatment Plant for National Stul Corporation, submitted by TEST Consultants, Inc., Manila, Philippines.
4. ASEAN Environment Programme III (1988–1992). 1987. Prepared by the ASEAN Experts Group on the Environment, with assistance from UNEP–ROAP and ASEAN Secretariat, Jakarta, Indonesia.
5. Uriarte, F.A., Jr. 1985. Metal Finishing and Coating in ASEAN: Current Practice and Future Prospects for Clean Technologies, International Symposium on Clean Technologies, Karlsruhe, Federal Republic of Germany, October 7–18.
6. Uriarte, F.A., Jr. 1985. UNEP Design Manual: Pollution Control Facilities for Small Battery Manufacturing Plants. Bangkok: UNEP–ROAP.
7. Uriarte, F.A., Jr. 1984. UNEP Design Manual: Pollution Control Facilities for Small Electroplating Plants. Bangkok: UNEP–ROAP.
8. Uriarte, F.A., Jr. 1982. Environmental Quality Management for Selected Small and Medium Scale Industries in Urban Areas of ASEAN: Assessment, Problems, Recommendations. Bangkok: UNEP–ROAP.

24

Hazardous Waste Management in India: Policy Issues and Problems

S.P. Mahajan

The Bhopal tragedy will be remembered for many years. Often in cases of industrial accidents such as occurred at Bhopal, the significance of warning signs are not fully appreciated until after the disaster has struck. In the past few years such localized disasters have taken place in India but the basic reasons for their occurrence has not been scientifically analysed nor any corrective measures outlined for action by public, government agencies, social and political organizations. Chlorine leakage in the suburbs of Bombay, sulphuric acid fumes from a Delhi industry, fish kills in the Tungabhandra river and the Gomti are some of the examples of minidisasters which in the absence of emergent action by all concerned would have led to larger-scale disasters. One can learn from these occurrences; and the results, if disseminated widely, would be of immense importance to those planning and co-ordinating disaster management activities elsewhere as well as to those who become the first victims of these disasters.

Some policy issues together with the problems encountered in hazardous waste generation and disposal/management are addressed in this paper.

Background

India has primarily an agriculture base with more than 70% of the population living in small villages. There is still illiteracy in many parts of the country, but people are becoming more knowledgeable and conscious of their rights with the help of information acquired through radio and television media. Small-scale industries do not give much thought to pollution they may generate and more often than not, pollution protection equipment is more costly than that of the production unit itself.

Even sewage collection and treatment facilities are inadequate and together with industrial pollution place considerable stress on the natural environment. Sludges and solid wastes from industrial units and households are eyesores and a cause for concern.

Uncontrolled sources of pollution like household grates, cattle sheds, slum areas and automotive exhausts are also matters requiring immediate attention. As most of the larger towns in India are situated on the river banks, their run-offs and also those from agricultural operations find their way to the rivers, making them unfit for drinking purposes. Drinking water quality is of prime importance and every effort is being made to maintain it for all the fresh water resources such as rivers, lakes, underground water.

Poverty among people is the biggest source of pollution in India. Where existence itself is a struggle and illiteracy widespread, environmental protection is not a priority. This is where the United Nations with its poverty alleviation programmes have substantially contributed to environmental conservation.

Perspectives

India has missed the two historical industrial revolutions, it cannot afford to miss the third. In fact, some making up is in order. Technologies have to be acquired, developed, absorbed and digested. Hazardous materials will have to be handled, wastes produced and treated. There are already innumerable small- and medium-scale industrial units consuming and producing hazardous materials. The methods of their safe handling and disposal have developed by experience and innovation. The one sad experience has been with a turnkey project in which the problems were kept secret along with the operations. With a full flow of scientific and technological information and with sincere efforts, many of these mishaps can be avoided.

Problems

For hazardous waste management in India and other developing countries in the region, the following problems exist:

1. Lack of information on the hazardous waste
2. Lack of co-ordination of different agencies
3. Lack of adequate resources
4. Lack of specific legislation and standards for hazardous wastes
5. Lack of incentives and the will to enforce

Information

In India data are being collected on the type, nature and sources of hazardous wastes from different process industries, the quantities produced, inventories, methods and sites of disposal. These data will be invaluable for the formulation of rational policies and pragmatic disposal/management procedures. In addition, data on growth in the industrial sectors envisaged in the planned documents will serve as indicators on the specificity of hazardous wastes and overall implications as regards air and water pollution.

Co-ordination

The data outlined above are being collected by a number of organizations and establishments including government and semi-government bodies. Co-ordination among these organizations is of utmost necessity in order to avoid duplicity of effort and expense.[1,2,3,4] Hazardous waste management operates with the same controlling procedures as those hitherto applied by municipal and local bodies to sewage treatment and solid disposal methods. There is scope for improved co-ordination among different organs of the government, such as the Ministries of Health, Chemicals, Industry, Energy, Environment and so on. The Ministry of the Environment has initiated the overall policy guidance to improve co-ordination among different ministries and to improve follow-up measures.

Legislation

The legislation for the containment of hazardous waste is still embodied in a variety of other laws relating to, for example, health protection, factory safety. There are no specific laws to deal with hazardous wastes. The criteria used to define hazardous wastes are still under discussion. An exclusive list (and/or inclusive) for classifying hazardous wastes has been visualized and work started thereon. One specific area which required exclusive attention in the recent past relates to the transportation of hazardous waste/products through rail, road and waterways. The responsibilities are being clearly delegated and accountability ensured.

Penalties/Incentives

More often the penalties imposed for noncompliance with measures under the existing laws are too small to warrant any additional expenditure on treatment of wastes. It is more cost–effective and expedient to continue to pay fines than incur the cost of corrective measures. Incentives such as tax rebates, low interest rates and selective subsidies have been more effective in bringing about a change in attitudes and providing at the same time the necessary finance.

Waste Management

Reduction at Source

Experience gained on pollution abatement in process industries has shown that preventive policies turn out to be most cost-effective in the long run, even though initially there is reluctance for changing over from conventional processes and practices. In the existing installations, however, the change over to improved processes has not been possible in view of high costs and potential risks involved.

Treatment and Disposal

Reducing dependence on land disposal, though a technical possibility, has always presented problems. Wastes containing organic compounds have in some cases, been completely destroyed for hazardous constituents; the inorganic wastes, however, can at best be solidified to produce environmentally acceptable treatment residues. Wastes containing mixtures of organic and inorganic constituents often present technical difficulties. Mackie and Niesen[4] identify many treatment options commercially available (see Table 24.1). Among the physical treatment methods, one can consider magnetic separation, liquid–solid separation, and membrane techniques such as reverse osmosis and dialysis.

Table 24.1 Some Commercially Available Treatment

Treatment	*Typical Applications*
Physical	
Centrifugation	separates liquids and solids
Filter Presses	removes moisture from solids, sludges
Distillation	solvent purification
Carbon Absorption	removes organics
Reverse Osmosis	removes metals and organics
Chemical	
Precipitation	removes metals
Oxidation	destroys organics
Reduction-dechlorination	reduces chlorine content of hydrocarbons
Photolysis	destroys dioxin and cyanide
Biological	
Aerobic/Anaerobic	removes metals and organics
Land Treatment	degradation of organic sludges
Thermal	
Liquid Injection	destroys organics in liquid wastes
Rotary Kiln	destroys organics in sludges and solids
Stabilization/Solidification	
Sorption	uses variety of material to solidify inorganic liquids (e.g., fly ash, lime, clays and carbon)
Pozzolanic Reactions	uses lime-fly ash or portland cement to solidify inorganic wastes.

Chemical treatment options have been available for a long time and attention has mainly focused on new applications and operations and the efficient use of the reactions taking place. Some commercial applications include oxidation–reduction reactions and degradation of trace organic compounds, photolysis to destroy cyanides and dioxins, and precipitation of metals for recovery and/or disposal.

Biological processes are sensitive to the presence of toxic elements and non-biodegradables but some degree of success for hazardous wastes, essentially for refinery sludges, has been achieved by this method.

Traditional incineration methods are useful primarily for organics with fairly high calorific value. They are efficient but expensive as compared with physical, chemical and biological methods.

Solidification/stabilization methods are best used for wastes containing inorganic constituents. There is no change in toxicity of the waste but the potential mobility of the constituents is subtantially reduced.

New Technologies

New technologies are being developed for handling hazardous wastes. A recent review by the Hazardous Waste Consultant identifies these innovative technologies, some beyond the laboratory stage of development and considered ready for pilot and commercial development.[2] They are outlined in Table 24.2.

Conclusion

It may appear that the current level of industrialization in India does not warrant the handling of large quantities of hazardous waste. The problems lie with the small- and medium-scale industries who generate small but significant quantities of waste that cannot be subjected to uniform treatment. In the future dependence on land disposal must be reduced and therefore it is essential that innovative approaches and new technologies are developed to deal with the increasing quantities of waste being generated by industry.

Table 24.2 New Technological Developments

Process	*Process description*
Enzyme Destruction	Biological destruction of organics; does not involve living organisms; can be maintained in immobilized systems or applied directly to wastes or contaminated material.
UV Photolysis	Used to detoxify liquids containing dioxin, being developed for application on contaminated solids; dioxin mobilized by surfactants and subjected to UV photolysis; can reduce concentrations by 90 to 99%.
Pyroplasma Processes	Breakdown of waste fluids to element constituents; being developed as a mobile unit; tested for destruction of chlorinated organics; low power consumption and rapid start–stop mode.
Plasmadust Process	Recovery of metals from iron and steel mill baghouse dust; reduces metal oxide to elemental forms; iron removed with molten slag; zinc and lead removed as gas; tests resulted in yields of 96% for iron, zinc, and lead.

Table 24.2 (cont.)

Process	*Process description*
Plasma Arc	Destruction of PCBs and PCB-contaminated equipment; destruction and removal efficiency of 99.9999%; possibility for metal recovery from molten slag.
Circulating Bed Incineration	High heat-transfer and turbulence allow operation at temperatures lower than traditional incinerators; accommodates solid and liquid wastes; complete destruction of organics at relatively low temperatures; no need for scrubber system to remove acid gases; particularly cost-efficient for homogeneous wastes from oil and petrochemical processes.
High-temperature Fluid Wall Reactor	Most suitable for contaminated soil; liquid wastes require a carrier; pyrolyse organics to carbon, carbon monoxide and hydrogen; equipment not attacked by inorganic components; mobile units possible; reaches destruction efficiencies of 99.9999%.
Penberthy Pyro-Converter	Glass-melting furnace technology adapted for destruction of organics; suitable for liquids, vapours, solids and sludges; solid residues (inorganics) incorporated into glass matrix; current use for production of HCI and destruction of chlorinated organics.
Pyrolysing Rotary	Operates in oxygen-free environment and at lower temperatures than conventional kiln; produces gas suitable for energy recovery or further treated to recover condensed hydrocarbons; recovery of metals possible without volatilization; reduced need for air pollution control; need to verify destruction efficiencies of hazardous constituents.
Rollins Rotary Reactor	Suitable for viscous and high-solids content wastes; no need for supplemental fuel; reduced gas scrubbing requirements; high-transfer efficiencies may increase destruction efficiencies at lower temperatures.
Supercritical Water Oxidation	Oxidizes organics to carbon dioxide and water; high pressure steam or electricity produced, inorganic salts precipitated; especially efficient with highly concentrated organic wastes; for water containing 10% organics, destruction efficiency greater than 99.99%; suitable for chlorinated solvents and PCBs.
Wet Oxidation	Suitable for dilute aqueous waste that cannot be incinerated or biologically treated; destruction efficiencies expected in range of 99% or 99.99%; oxidizes organics and inorganics; not appropriate for halogenated aromatics.
Vertical-tube Reactor	Adaptation of wet oxidation into 1-mile deep well system; operates at lower pressure than conventional process; currently applied to municipal wastewater.

References

1. Sundaresan, B.B., Subrahmanyam and Bhinde, A.D. 1983. An Overview of Toxic and Hazard Waste in India. *Industrial Hazardous Waste*, Htum Nay, and Huismans, J.W., eds. Industry and Environment Special Issue, 4, Paris: Industry and Environment Office, UNEP.
2. The Hazard Waste Consultant: A Guide to Innovative Hazardous Waste Treatment Processes, January/Feb, 1985. Cited in Pirages, S.W. 1987. Restriction for Land Disposed Wastes: Can the Industry Readily Comply? *Encology*, 1, 8: 22–7.
3. Mahajan, S.P. 1985. *Pollution Control in Process Industries*. New Delhi: Tata-McGraw Hill.
4. Mackie, J.A. and Niesen, K. 1984. Hazard Waste Management: The Alternatives. *Chemical Engineering*, 19, 16: 50–64.

25 Management of Hazardous Materials and Wastes in China

Shi Qing

Management of Hazardous Materials

Management of hazardous materials and wastes in China has been separately implemented by the individual departments of government. In 1961 the State Council promulgated the following regulations and rules; "Regulations on the Management of Safe Production for the Medium and Small-size Chemical Plants", "Interim Rules on the Management for Storage of Hazardous Chemicals", "Interim Rules on the Permit for Managing and Purchasing Hazardous Chemicals", "Regulations on Transportation of Hazardous Material by Railway", "Regulations on the Management for Preventing Fire of Hazardous Chemicals", and "Interim Rules on Punishment for Violation of Regulations on the Management of Explosive and Combustible Materials". These regulations and rules were put into effect respectively by the Ministry of Chemical Industry, the Ministry of Railways, the Ministry of Commerce and the Ministry of Public Security. The municipalities of Shanghai, Tientsin, Beijing and the other large cities have also issued their own Regulations and Rules on the management of hazardous materials.

Due to a rapid increase in the production and usage of chemicals in China, the State Council further promulgated "Regulations on the Safe Management of Hazardous Chemicals" on February 17, 1987, which referred to the developed countries experience in managing chemicals. The regulations define as hazardous: explosives; compressed gases and liquefied gases; combustible gases, liquids and solids; spontaneously combustible materials and combustible materials in wet condition; oxidants and organic peroxidants; toxicants; corrosives. They are classified as seven categories according to the national classification standard "Classification of Hazardous Goods and Number of Items". The regulations deal with the production, usage, storage, management, transportation and package of the hazardous chemicals as well as punishment for violation of the regulations.

The production of hazardous materials is controlled by the national uniform production programme. The production of toxic and hazardous materials by small town plants is prohibited because of their lack of effective pollution control. All plants for the construction of new plants or the extension of existing plants producing hazardous chemicals, must be considered and approved by the higher government authority reporting to the Ministry of Chemical Industry.

The plants which produce, use, store, transport and manage hazardous chemicals must take adequate safety measures, have facilities for treatment of wastewater as well as liquid and solid wastes, implement the regulations under the National Environmental Protection Law. Emergency and fire prevention procedures must be established in the plants and operators must be equipped with preventive appliances.

The registration and permit system for the production, usage, storage, transportation and management of hazardous chemicals has now been implemented. The permit must be applied for through the government authority above the county level. It is issued by the relevant provincial or municipality departments which function directly under central government. Inspections will be carried out every 2 to 3 years.

Special regulations on the transportation of hazardous chemicals have been issued by the relevant ministries of the State Council. Specific conditions for different kinds of hazardous materials have been established. The hazardous materials cannot be transported if they do not meet the requirements of the regulations.

Violation of the regulations and rules mentioned above will be punished by the administration and blame will be attributed by the judicial office in accordance with the seriousness of the case.

Management of Hazardous Waste

China produces around 40 million tons of hazardous waste annually. The National Environmental Protection Agency (NEPA) is now formulating the legislation for the regulation of wastes. It is also implementing policies for resource recovery and for the treatment of waste to eliminate harmful contaminants. The Pollution Control Standards of Wastes containing Cd, Pb, Hg, As and Cr were already issued by the same Agency.

The Standard and Regulations of Hazardous Wastes Pollution Control

The State Council promulgated the control standards of pollutants containing heavy metal and the standard methods for testing the toxicity of liquor on 1 October, 1985. (See Table 25.1). The development of standardized chemical analysis followed shortly after. At present the State is in the process of establishing national standards to control hazardous wastes containing cyanide, organic chlorine and organic phosphorus and to impose economic penalties for discharging toxic wastes.

Table 25.1 Identification Standard of Leaching Toxicity

Toxic Substance	*Highest Permissable Concentration in Leaching Liquor in mg/l*
1. Mercury and its compounds	0.05 (calculated as Hg)
2. Arsenic and its compounds	1.5 (calculated as As)
3. Lead and its compounds	3.0 (calculated as Pb)
4. Cadmium and its compounds	0.3 (calculated as Cd)
5. Hexavalent Chromium compounds	1.5 (calculated as Cr^{6+})
6. Copper and its compounds	50 (calculated as Cu)
7. Zinc and its compounds	50 (calculated as Zn)
8. Nickel and its compounds	25 (calculated as Ni)
9. Beryllium and its compounds	0.1 (calculated as Be)
10. Fluoride and its compounds	50 (calculated as F)

The Treatment, Disposal and Recovery of Hazardous Wastes

According to the principle of "Whoever causes pollution shall be responsible for its elimination" as specified by national environmental protection law, hazardous wastes should be treated and disposed of by large- or medium-size factories and enterprises. Some chemical and petrochemical plants have been equipped with incinerators for treatment purposes. Waste liquids such as waste acids and alkalis are treated or recovered by chemical, biological and physical processes in certain factories. The sludge generated from the treatment of wastewater is used as forage for earthworms at the vinylon plant and bleaching and dyeing mill in Sichuan. Although earthworms concentrate the heavy metal contained in the sludge in their body tissue, the level of heavy metal concentration to be found in their execreta falls within the permissible limits of pollution control standards. Earthworm excreta can therefore be used as fertilizer for flowers and trees and the income generated can be balanced against the expense of feeding the earthworms. Protein, amino acids and other substances can also be separated from the toxic substances in earthworm tissues and used as materials for producing medicines and cosmetics. Comprehensive utilization of sludge can be obtained by biological engineering.

In recent years industry has developed rapidly in the small towns and villages of China but it is impossible for small-scale industrial plants to treat their hazardous wastes. NEPA is therefore organizing the construction of treatment centres to process various hazardous wastes for whole cities or a specified region. In Shenyang a prototype engineering project investigating the feasibility of hazardous waste treatment was established in 1986 by TDP of the United States.

The Treatment, Disposal and Utilization of Wastes Containing Chromium.

There are eighteen chromate manufacturers in China. Chromium-plating plants exist in most large cities and small towns and the pollution from hexavalent chromium-bearing waste is a relatively serious problem. Many institutes have successfully developed more than ten technologies for the treatment, disposal, recovery and utilization of chromium-bearing wastes. The principal treatment disposal technologies are:

1. Dry calcinating for removal of toxicity
2. Wet sodium sulphite for removal of toxicity
3. Detoxification process with ferroferric sulphate
4. Detoxification process with complex compound
5. Detoxification process with aqueous vapour
6. Cement solidification
7. Isolating process with anti-leaching concrete partition

The principal technologies for utilization of the chromium-bearing waste are:

1. Use as colouring agent of glass
2. Producing chemical fertilizer – calcium superphosphate
3. Producing cast stone
4. Producing mineral wool
5. Replacing magnesium limestone iron-smelting flux
6. Producing slag brick
7. Producing aggregate of ceramsite concrete
8. Use as material in cement manufacture.

All technologies mentioned above have been applied in different scale plants in China.

Recovery of Chemical Products from Chromium-containing Depleted Electrolytes and Residues

There are many kinds of ageing chromium-containing waste electrolytes in the electroplating industry. The content of CrO_3 varies from 50g/l to 350g/l. In addition, in the wastes there are various metal ions such as Cr^{3+}, Fe^{3+}, Zn^{2+}, Cu^{2+}, Al^{3+} and an ion such as $SO4^{2-}$, $PO_4{}^{3-}$. There are also large quantities of sludge, and residues, such as trivalent chromium-containing residues, from chemical reduction and active carbon processes, and residues from barium salt and electrolysis reduction processes. The mix of components in the residues is complex, making them difficult to treat. However, using the technology illustrated in Figure 25.1 it has been possible to convert these types of waste and recover various chemical products. They include lead chlorate, lemon yellow, iron oxide, red bottom mud, sodium nitrate, aluminium sulphate.

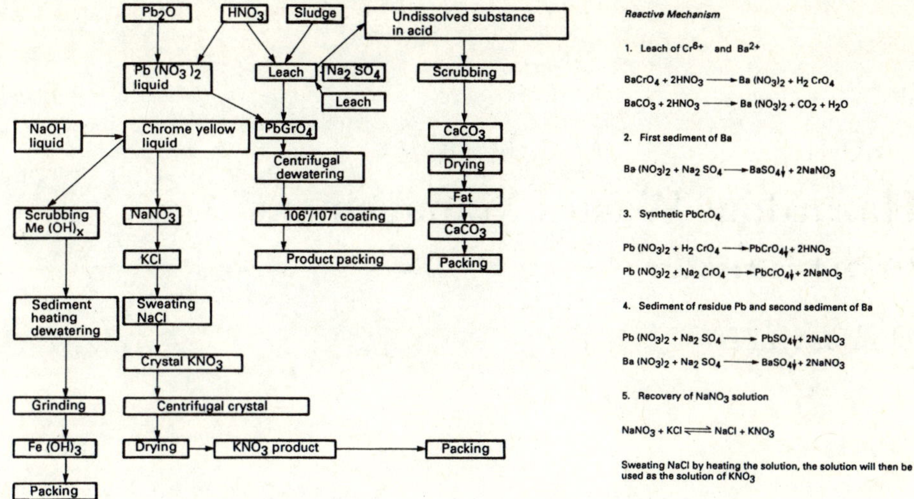

Figure 25.1 Converting and Recovering Chemical Products from Residues and Sludge

26

Hazardous Wastes Management in Sri Lanka

D.P. de Alwis

Rapid industrialization during the last few decades has adversely affected the natural environment of Sri Lanka because of the discharge of different types of raw industrial wastes into the environment. Even domestic solid wastes, which are relatively inert, have become a public health hazard as a result of the careless way in which they have been handled.

Hazardous untreated wastes generated from a variety of large and small industries in Sri Lanka include paper manufacturing wastes, textile wastes, electroplating wastes, fertilizer manufacturing wastes, and spillages in pesticide manufacturing plants.

Overview of Problems Encountered with Hazardous Wastes

Surface Water Pollution

It has been observed that pollution of water resources has steadily increased in recent years. The main river Kelani, which flows through the capital city, receives large quantities of raw industrial liquid wastes from a number of factories situated on its banks. Wastes other than those with high oxygen demand also enter the river without treatment. They include chromium derived from the leather tanning and electroplating industries, lead from petroleum processing, dyes from textile processing and leather finishing and liquid waste from paper factories. Fish in the river are frequently affected by these discharges.

Ground Water Pollution

A number of urban areas have been identified with the problems of ground water pollution. The industrial areas of Ratmalana and Ekala and the vicinity of the Hunupitiya Fertilizer Corporation are particularly significant. Liquid wastes deposited in the low-lying areas, and solid wastes stored on factory premises could, on leaching, be a threat to the ground water quality. Land fill sites containing municipal solid wastes are often potential sources of surface and ground water pollution. This method of disposal is also unsightly and unhygienic.

Pesticides

The use of pesticides has risen dramatically in Sri Lanka during the past two decades as has the incidence of pesticide poisoning. The records of the medical statistical unit of the Department of Health reveal that approximately 13,000 patients are admitted annually to government hospitals for treatment for acute pesticide poisoning with a mortality rate of about 8%.

Current Legislation on Hazardous Waste Management

In Sri Lanka there are a number of principal acts which legislate against health hazards and environmental pollution. Among them three acts in particular play an important role in establishing the basis for the preparation of standards, norms and guidelines. They are the Greater Colombo Economic Commission Law of 1978, National Environmental Act of 1980, and the Control of Pesticides Act of 1980. A legal framework has been also coupled to the above legislation.

The Pesticide Act regulates the import, packaging, labelling, storage, formulation, transport and sale of pesticides whereas the remaining two acts are concerned with industrial wastes. A registrar of pesticides and a formulatory committee comprising of members drawn from government institutions dealing mainly with agriculture and public health has been appointed to ensure that the Pesticides Act is implemented.

The preparation of guidelines detailing the parameters governing the discharge of liquid wastes to inland surface waters, coastal waters, public sewers and irrigated lands was a major step towards improving the quality of life while endeavouring to decrease and control environmental degradation.

There is, however, a wide gap between pollution control legislation and its actual implementation.

Problems Associated with Hazardous Waste Management in Sri Lanka

Three major problems were identified as being associated with the management of hazardous industrial wastes on the island: a lack of trained personnel, a lack of financial resources to fund the waste treatment facilities, and a shortage of space within existing factories. Collective treatment of effluent appeared to be the answer. Industrial zoning was therefore introduced to prevent the haphazard siting of industry in residential areas.

Steps Taken to Control Environmental Degradation

The industrial zoning concentrates on one site the treatment of wastes from a number of factories, thereby reducing the treatment costs. In addition, it is the most cost-effective means of utilizing the meagre financial resources available to fund pollution abatement schemes since the need for extra space on factory premises for industrial treatment facilities is eliminated.

A planned industrial zone was established outside Colombo taking into account environmentally sound safeguards. This industrial estate which covers an area of 500 acres (202 hectares) is endowed with infrastructural facilities such as power, roads suitable for motor vehicles, treated water supplies, treatment facilities for combined industrial and domestic sewage and disposal sites for solid wastes.

Investors in the industrial zone were required to meet certain enforceable discharge standards for liquid wastes. To achieve these environmental quality standards a programme for monitoring industrial effluent discharges was implemented on the estate in 1986.

As well as monitoring at source the industrial wastes released into the sewerage system, the quality of ground water used for drinking and industrial purposes was also monitored. The effluent discharged by the treatment plant into the nearby river was also monitored to check that it conformed to national standards.

Fifteen physico-chemical parameters were selected for monitoring the industrial effluent. World Health Organization chemical guidelines for drinking water were used when monitoring ground water quality on this industrial estate.

The majority of the factories were able to meet the standard specifications set for discharged effluent. Short-term unsatisfactory discharges were occasionally seen but did not appear to have any disruptive effect on treatment operations at the common treatment plant. The biological process used in the treatment plant in combination with sufficient retention time resulted in a significant reduction in the BOD of the incoming (domestic and industrial combined) liquid wastes. Most of the time, the effluent from the treatment plant conformed to tolerance limits set for wastewater discharges into inland surface waters and had a minimal effect on the aquatic life and self-purification capacity of the river. No threat to ground water quality resulting from industrial activities on the estate was observed.

Conclusion

Appropriate on-site pre-treatment followed by combined biological treatment seems currently to be a practicable and environmentally sound option for disposing of industrial liquid wastes in Sri Lanka.

References

CEA. 1985. Industrial Pollution in the Kelani River. Preliminary Survey and Interim Report. Volumes 1 and 2. Colombo, Sri Lanka.

Colombo Landfill Resource Recovery Study. Interim Report. Prepared for United Nations Development Programme, Swiss Co-operation and The World Bank by Environmental Resources Ltd., April 1987.

Environmental Management in Sri Lanka. 1973. Report of the sub-committee appointed by the Ministry of Planning and Employment to study Environmental Management in Sri Lanka.

Environmental Protection and Management: Final Report. 1978. Colombo, Sri Lanka: Foundation Institute.

Ministry of Local Government Housing and Construction. 1984. Report of the Committee on Pollution of Drinking Water Sources. Colombo, Sri Lanka.

Jeyarnthnam, J., de Alwis Seneviratne, R.S. and Copplestone, J.E. 1982. Survey of Pesticide Poisoning in Sri Lanka. W.H.O. *Bulletin*, 60, 4.

Ellepola, K. 1984. National Paper on the National Register of Potentially Toxic Chemicals.

27

Methodology for Risk Assessment in Handling Petrochemical Products in Developing Countries

Sergio A.S. Almeida and Ricardo L. Cunha

The accident rate for transporting petrochemical products has been increasing in developing countries, especially in Brazil where recent accidents have caused many fatalities.

In 1983, at the city of Pojuca, near Salvador, Bahia, a train transporting gasoline derailed and caused the death of approximately 100 persons. Also in 1983, an 18-ton rock, displaced by the explosion of dynamite during the construction of a highway in the State of São Paulo, damaged an oil pipeline owned by Petrobrás, the State Oil Company. A large quantity of oil spilled from the pipeline polluting about 20 sq km of estuarine mangroves along the coastal area and causing ecological damage to several beaches in the Bertioga region.

In February 1984, at Cubatão, São Paulo, 100 persons were killed by a fire caused by extensive leakage from a 14 ft pipeline transporting gasoline and other light petroleum products. The leakage occurred in the slum (favela) area of Vila Socó and the resulting fire destroyed the houses of over 1000 persons. The reports prepared by Petrobrás and the environmental agencies concluded that the disaster was the result of a lack of adequate urban planning. Such planning would have ensured that haphazard housing developments did not spring up in areas with high industrial risk. In addition, the reports demonstrated operational errors during fluid transportation. Pumping had been initiated with valves closed, causing a large pressure rise in the pipeline and consequent pipe failure. The old pipeline was progressively replaced, as it reached a phase where leaks occurred at a logarithmic rate. This paper analyses the replacement of the pipeline causing the Vila Socó accident.

The methodology developed for environmental risk analysis of the new pipeline included measures to ensure a greater degree of safety than that afforded by the old pipeline.

However, sophisticated models were not used in the analysis of possible pipe failure and fluid mass outflow. Instead a methodology compatible with the resources of a developing country was utilized.

The pipeline under study connects the Utinga Petrochemical Terminal, located in the city of São Paulo, to the Cubatão Pumping Station, and from there to the Alemoa Terminal, located at the Port of Santos. The methodology described in this paper was employed by Multiservice Engenharia Ltda. under contract to Petrobrás. It does not include specific calculations.

System Description

The Santos–São Paulo Oil Pipeline (OSSP) begins at the Alemoa Terminal, at Santos, S.P. The multipipe, 15 metre-wide pipeway runs through the Santos Mountains, passes the Capuava Refinery and ends at the Utinga Terminal. The OSSP pipelines were installed in 1951 by the former Santos–Jundiaí Railroad to transport light petroleum products imported through the Port of Santos to be consumed in São Paulo. This was the first long-distance pipeline constructed in Brazil. In 1976, the pipeline system was transferred to Petrobrás and in 1978, the initial phase of a protection system was implemented to reduce leakages.

The system begins at Alemoa Terminal, at the Port of Santos, crosses the RFFSA Railroad and the Casqueiro River and proceeds to a multipipeline 9.8 km distant. From the multipipeline the system extends to the Cubatão Pumping Station (EBC) and from there continues as a single 14 inch pipeline (Pipeline "A"), which carries light products and liquid petroleum gas . The distance from the EBC to the Capuava is 44.89 km. From EBC the pipeline continues to the Utinga Terminal located at São Caetano do Sul, São Paulo. The new system is equipped with a cathodic protection system, scrapper-traps and instrumented pigs for the control of internal quality. At Vila Socó, where the accident occurred, warning signs giving information on contingency plans in case of an emergency have been placed along the pipeway for the people living in the area. In addition, valves have been installed as a safety measure upstream and downstream of the zone.

Methodology Objectives

The main objective of the risk analysis is to fulfil the legislative requirements of Environmental Impact Assessment, as prescribed by the Brazilian National Environmental Council. The analysis includes consideration of population welfare, safety evaluation, as well as the protection of the natural ecosystem, in case of environmental accidents in the industrial area. Secondly, the risk analysis provides data for environmental impact assessment studies including information on the causes and effects of point-source pollution loads in water, air and soil. The study also seeks to minimize the potential risks in industrial facilities, for the pumping and transportation of toxic and flammable fluids and to reduce industrial costs for system maintenance by detecting critical areas, especially the equipment and parts of the system with the greatest accident potential.

This methodology evaluates the potential risk of accidental discharges in the internal and external areas of industrial plants and oil terminals as well as areas designated for fluid transportation. The evaluation includes analyses of toxicity, flammability, and explositivity, in relation to potential danger to which the population and the environment are exposed. The studies include analyses within the facilities as well as in the surrounding areas.

According to Kletz and Lawley (1982), some risks to the environment are obvious and are immediately identifiable, others require a critical period of observation before they can be recognized. In other cases codes of professional practice play an important role. Checklists are widely used as instruments for easy risk identification, but one disadvantage of this tool is that important items may be omitted.

In the case of the pipeline under study, the vulnerable areas were first identified by the "obvious" criterion based on the analyst's experience. Information was also obtained from the environmental protection agencies (FEEMA-RJ and CETESB-SP) about similar problems.

Consideration was given to faults in the pipeline system that could cause significant environmental damage or damage to life inside or outside of the facilities, resulting from a single event. The magnitude of the damage potential is reduced by using a combination of the evaluation methodologies described above to identify and analyse potential faults.

The classical methodologies of fault and event tree analysis, failure modes and effect analysis, and common mode failure were used according to the need.

Three non-classical methodologies were employed. Fluid spillage was estimated for extreme and medium situations (maximum and medium discharge). Checklists were used for environmental assessment analysis. Risk management was based on the experience of environmental protection agencies–FEEMA/RJ and CETESB/SP (Dept. de Accidental Pollution).

The study was organized around the establishment of a systematic and sequential logic, formed by a model in which the study was divided into three phases: environmental assessment, risk identification, and risk management, constituting the risk model.

Environmental Assessment

Human actions and the most significant natural ecosystems in hierarchical order are considered in this phase. The ecosystem is classified into physical, biotic and socio-cultural environments, where living beings, flora and fauna and man exchange energy within the system supported by the physical environment.

Risk Identification

The risk analysis using fault causing events is based on studies of the cause–effect models (Event Trees) and effect–cause models (Fault Trees), which are summarized below.

The event trees are used to evaluate causes in a cause–consequence model, in relation to the type of fluid discharge and its toxicity and flammability, in order to investigate the initial events and those provoked by others, in series or in parallel. This method has been successfully used for small to medium sized subsystems. For complex systems, such as petrochemical industries, application of this method is difficult. At the Santos–Utinga pipeline project this method was only used at critical areas, from an initial causing event.

Three basic options are considered for the event tree in relation to the hazard, that is, for flammable fluids, toxic fluids and both. In addition, many other aspects may be considered.

The system fault analysis uses consequence interpretation from the consequence–causes model. The systems are subdivided on the computer. The subsystems are classified by their known components (parts, connections, pumps), and after grouping by similarity the faults in the systems and subsystems are analysed by their effects. The risk criticality is determined by means of probability calculations for the risk occurring alone or combined in the fault tree. The fault trees are essentially consequence–cause models utilized to develop the many qualitative and quantitative aspects of fault occurrence.

Calculations for the fault occurrence frequency should take into consideration similar accidents already observed in existing facilities. However, often such data do not exist. In such cases, critical probability criteria are established, with classifications of high, average and lesser probability. Such probabilities are estimated based on the frequency of accident occurrence for each component or system based on obtainable information or on the analyst's experience. The interactions of the many fault possibilities are made in the fault tree from left to right, following many alternatives using Boolean algebra.

For subsystems where fault causes may be derived from the intersection between other adjacent subsystems (for example, the case of an oil pipeline next to an alcohol pipeline), by the degeree or frequency of one fault causing another, or by a number of faults resulting from a single simple cause, the classical analysis of Common Mode Failure is employed to evaluate the synergetic effects of the combined systems.

The criteria for environmental assessment used to evaluate the degree of risk severity for flammability, explosivity and toxicity, in relation to adjacent populations and to the environment in general have been established by the

National Fire Protection Association (United States), and have been adopted by many national environmental protection agencies.

Estimates for fluid discharge from the many systems and sub-systems used the Bernoulli Theory and simplified equations, with coefficients recommended by The World Bank.

Environmental Risk Management

In this phase the mitigative measures and/or recommendations for risk minimization are established following an evaluation of the consequences of various courses of action and computation of mass discharges.

Selection of Risk Areas

The risk (or vulnerable) areas are classified in accordance with grading criteria ranging from zero to four, depending on the potential to cause environmental damage. Such damage is identified according to the social or ecological risk, wherever the pipeline crosses rivers, mangrove areas, highways, railroads, other pipelines, energy transmission lines and by its proximity to housing. The criteria used are classified as follows: low criticality – no interference, 0; low interference, 1; medium interference, 2; high criticality – high interference, 3; excessive interference, 4.

The environment is analysed according to its physical, biological and social components both within the zone directly concerned and its area of influence. The information is stored in data banks for future reference.

In the subsystems, identification of the main risk areas is made in accordance with potential damage to the environment, using methodology based on the analyst's experience (obvious methodology), also taking into consideration the simplicity of the facilities and road access. Preliminary checklists are made and registered in data banks. After classification, risk areas are separated and only those presenting high criticality are studied.

Evaluation of Social and Ecological Risk

Social Diagnosis

Social risk evaluation in the directly and indirectly influenced areas is made using concepts of social and individual risks for the exposed population. Of the many types of risk, those utilized are defined as follows:

1. Voluntary risk is that assumed by any individual, voluntarily, such as the risk associated with smoking, being overweight
2. Non-voluntary risk is that which society does not accept but to which it is exposed due to human activities

3. Social risk is interpreted as being the product of accident occurrence frequency for a determined event by its occurrence magnitude (consequence per event). Individual risk is the occurrence probability for any event, and is determined by dividing the social risk by the potentially affected population.

It has been suggested that a criterion of probability of 10^{-5} year for voluntary risk (i.e., one death per 100,000 persons per year) and 10^{-7} for non-voluntary risk.

The British Standard of the Advisory Committee on Major Hazards defines the Fatal Accident Frequency Rates (FAFR) as the number of fatal accidents for a group of 1000 workers at an industrial plant during a lifespan. The FAFR for England (U.K. Factory Act) is equal to four for an ordinary industrial plant. For Petrochemical plants a reduction to 0.4 is suggested.

In terms of risk the Alemoa Terminal was considered compatible with a petrochemical plant and a FAFR of 0.4 was admitted, which is equivalent to a death risk probability of 3×10^{-5}/man–year.

The risks to the public are more restrictive than for workers (voluntary risk). The World Bank suggests that the maximum acceptable death risk for the population should be between 10^{-5}/ and 10^{-6}/person–year. These factors are lower than those proposed by the official publication on population risks in relation to oil refineries and similar industries located in the islands of the Thames estuary in the United Kingdom (Canvey Island Report). For the case under study, the estimated probability for accidental spillages from the pipeline not necessarily involving fatalities was 0.011 per year or 1.1% per year. For the calculations of consequences per event, data from the serious accident at Vila Socó were used.

Ecological Diagnosis

The ecological risk from an environmental accident is equal to the probability determined for accidental discharge from the pipeline, that is, 1.1% per year, or one accident each 92.6 years. The accident magnitude is considered as equal to the maximum discharge of fluid, corresponding to about 8.600 kg/sec (diesel oil), 3.500 kg/sec (alcohol) and 6.400 kg/sec (PGL).

The probability of a spillage whereby oil would flow to the rivers Casqueiro and Cubatão through the natural drainage system exists, even for small oil discharges. Although the probability is small for the new pipeline such an accident could happen carrying oil along the rivers to the sea. Aquatic flora and fauna along the littoral would be affected and the oil slick on the sea would be carried south by the currents and wind patterns in the Bay of Santos.

References

Awazu, L.A., de Mello, *et al.* 1985. *Análise Histórica da Ocorrência de Acidentes Ambientais no Estado de São Paulo.* CETESB/ABES.

Cunha, R., Lisboa, *et al.* 1986. *Poluicão Acidental, uma nova área de atuação.* Rio de Janeiro: Publicações FEEMA.
Grollier Baron, R., Lombre, R. and Rochinas, V. Hazard Analysis for a Liquid Propane Pipeline.
Kletz, T.A. and Lawley, H.G. 1982. *High Risk Safety Technology.* Green, A.E., ed. Chichester: John Wiley.
World Bank. 1985. *Manual of Industrial Hazard Assessment Techniques.* Washington, DC.

28
Management of Hazardous Materials in Chemical Industries in India

J.M. Dave

Chemical industries have suffered many mishaps in India where large quantities of hazardous substances have been accidently released causing minor injuries to large-scale death and destruction. Three such incidents of recent occurrence – the release of methylisocyanate from the Union Carbide plant at Bhopal, MP, the oleum release from the plant of MS Shriram Food and Fertilizer Industries, Delhi and the chlorine leakage at Bhadrachalam Paper Mills at Bhadrachalam, AP, represent differing magnitudes of damage. In this paper the various factors contributing to these accidents are identified and preventive and curative measures that need to be implemented by industry, government authorities and agencies and by the population of industrial areas are suggested.

Case Studies

Union Carbide India (UCC), Bhopal, MP

Methylisocyanate (MIC) was used as intermediate product in manufacturing "Sevin" (carboryl). On the night of the accident a sufficient quantity of water was accidentally introduced into one of the three storage tanks (No. 610) of 45 tons capacity each, to start exothermic polymerization. This event caused the rapid rise in temperature and explosive discharge of about thirty tons of MIC in approximately half an hour.[1] The MIC cloud moved towards the inhabited area with a slow blowing wind. The death toll was over 2000 persons and about 150,000 to 200,000 persons were affected. Most of the dead were the slum dwellers sleeping in shacks with no doors or windows, or people caught in the open, in the path of the cloud.

Shriram Food and Fertilizers Industries (SFFI), Delhi

Shriram Food and Fertilizers Industries manufactures sulphuric acid single superphosphate. Oleum from the H_2SO_4 plant to be sold to other customers was stored in MS tanks. As a result of corrosion, oleum tank supports failed resulting in the spillage of 30 tons of oleum on the ground and drainage systems. The acid mist travelled as a cloud almost touching the ground up to a distance of 20 km in a heavily populated area. Over 200 persons were affected and there was possibly one death from the exposure.[2]

Bhadrachalam Paper Industries (BPI), Bhadrachalam, AP

Bhadrachalam Paper Industries makes a variety of papers by processing wood and imported pulp. Chlorine is used to bleach pulp to produce white paper. The chlorine leakage occurred at the base of the bleaching tower. The chlorine injection line to the tower has a shutoff valve and gaskets on the valve joint failed, thus releasing 100 kg of chlorine in about thirty to forty minutes.[3]

Based on information obtained from each organization and other agencies on the events, an outline of the state of awareness, precautions and public information system is given in Table 28.1.

It can be seen that all three accidents share common features which can be broadly summarized as:

1. The absence of hazardous substances classification in industry and of zoning to separate industrial activities from areas of population.
2. Unawareness on the part of industrial managers, supervisors and other staff of the dangers of handling hazardous substances and the wider consequences of their release on the population and environment.
3. The absense of warning or alarm systems to inform neighbours of the accident.
4. The ignorance of the local population about the health hazard of an accidental release and ignorance of how to protect themselves and their property.
5. Lack of information on antidotes and treatment of the exposed and unavailability of basic medications.
6. No emergency plan of action by industry or administrative authorities and agencies.
7. Inadequate monitoring of the safety of the production operations by the factory inspectorate system.

Many of the fatalities and much of the damage could have been avoided if the necessary precautions had been taken.

The major problems faced by government agencies and the industries in all three events were:

1. The difficulty of obtaining complete information on the nature of the hazardous substance being used possibly because of trade secrecy, ignorance on the part of the management, or unavailability of data/literature on, for example, toxicity, hazards,
2. The unavailability of staff trained or experienced in the handling of hazardous materials,
3. The absence of suitable training facilities for workers, supervisors and managers in industry and government agencies,

Table 28.1 State of Safety Awareness and Preparedness for an Emergency (Sources: Dave, J.M. 1985. Bhopal Methylisocyanate Incident: An Overview; Oleum Accident at the Shriram Food Fertilizer Plant, unpublished report of enquiry committee; chlorine leak at Bhadrachalam Paper Mill, committee report)

Safety Measures	*U C C* *Bhopal*	*S F F I* *Delhi*	*B P I* *Bhadrachalam*	*Comments*
Factory Management				
Awareness by staff of hazardous nature of operation				
Executives/Manager	Yes, limited	Yes	Doubtful	
Supervisors	Doubtful	Doubtful	No	
Operators/workers	No	No	No	
Plant safety in design	Not adequate	Nil	Nil	
Protection measures in case of accidental release	Nil	Nil	Nil	
Location/siting factory safety consideration	Doubtful	Yes	Yes	SFFI was located 10 km outside city 20 years ago, but city grew around it
Skill requirements of operators	No	No	No	
Public information programme	No	No	No	No alarm, no data on toxicity, effects, etc.
Disaster action plan	None	None	None	No medication or evacuation plans
Government Authorities/Agencies				
Classification of hazardous materials/Industries and categorization	No	No	No	
Statutory powers for				
Risk analysis	No	No	No	Not required by law
Landuse/zoning	Not applied	Not applied	In remote area	Laws exist
Safeguards/protective measures	Not applied	Not applied	Not applied	No laws for it
Skill/certificate for operators	Not required	Not required	Not required	No law
Inspectorate system 1948 Factory Act	Failed	Failed	Failed	
Emergency action plan for accidental release	None	None	None	No co-ordinating agency or authority
Any specific legal controls	Yes	Yes	Yes	Licence to operate factory and capacity
Public Awareness				
Risk and hazards/danger due to chemical	Nil	Nil	Nil	
Alarm-understanding and protection action	Nil	Nil*	Nil*	* No alarm exists
Antidote/simple remedial measure knowledge	Nil	Nil	Nil	
Causes of Accident	-			
Equipment failure				
Human error	-	-		
Any other unknown factor				

4. The rapid turnover of staff, and
5. The absence of a suitable legal and statutory framework for controlling the operation of industries using and producing hazardous materials and wastes.

Since chemical production is vital to the manufacture of fertilizers, pesticides, pharmaceuticals, textiles, inevitably the number of chemical plants in all developed and developing countries will rapidly increase. Those industries using hazardous substances will also increase in number. Therefore, proper actions to manage them confidently and to minimize or avoid the possibility of industrial accidents is becoming increasingly urgent and government agencies and industry must organize themselves to meet the challenge. UNIDO can play an important role in this by providing guidelines on the management of hazardous materials/wastes in industry and by setting standards for training methods for the administrators, factory managers, supervisors, workers and other staff. The guidelines may broadly cover the responsibilities of and actions to be taken by industry, government authorities and the local population.

Industry Guidelines – Actions and Responsibilities

1. *Awareness*. There should be understanding of the degree of hazard/toxicity from the material/waste and its possible impact on the population and environment in the case of release.
2. *Risk Analysis*. When analysing the risk involved all potential damage to the population, property and the environment, together with an evaluation of the processes and substances used and produced should be accounted for. The long term effects of operational releases and minor upsets as well as acute effects from large-scale accidental discharges storage should be fully evaluated. The adverse effects of topographical and metereological factors should also be taken into account.
3. *Preventive Actions*.
a. *Zoning*. The location of industry and hazardous material should be a safe distance, as determined by the risk analysis, from populated and or special interest areas, such as green belts. Vegetation and other types and sizes of physical barrier should be considered along with specific local topographical and climatic conditions.
b. *Safeguards*. In processes likely to release hazardous materials during routine operations or by accident adequate safeguards should be included. They may be in the form of codes for providing backup systems or two or three pieces of equipment in series, for example, a scrubber followed by an incinerator or such chemical reactions as will remove or destroy the material (e.g., for Cyanides, Cyanates, Dioxins, PCBs) or shutdown interlocks must be provided. Inbuilt protective measures in the plant such as curtains of water, showers, neutralization chemical release can be built in at the design stages, so that in case of accident they would operate automatically. They should be located at all critical and vulnerable points in the plant (e.g. valves, pressure reliefs, rupture diaphragms).
c. There should be training of managers/supervisors and operators in safety consciousness and precautionary measures to minimize the release of such materials. Only trained and certified persons should be employed as operators/workers in critical operations and units.
4. *Public Information Arrangements*. Detailed information on the nature of the material, its effects on health, toxicity should be produced in pamphlet forms in cases of exposure. They should include the antidotes or remedial measure and protective action to be taken

when the alarm is raised. It should prepare the public for any group action required in the case of an emergency or accident.

5. *Medical knowledge and facilities.* Industry must obtain information on all toxicological and harmful effects of the materials and should be medically prepared to meet any eventuality in terms of maintaining an adequate stock of medicines, equipment and medical staff. Full information should also be supplied to nearby local hospitals and healthcare centres, including all data on toxic effects, and the treatment required by exposed persons, type of medication and facilities needed.

6. *Emergency Plan of Action.* The plan of action must include a special type of alarm, evacuation programmes, routes and traffic controls, actions to be taken by district and local authorities, fire brigade, and methods of rapid communication with all concerned. Mock exercises may be necessary to evaluate the plan and also to educate the public. The plan should define the responsibilities of factory staff for each of the necessary actions. Arrangements should be made for the use of radio and television as a means of informing the population.

7. *Negotiation with all administrative and local bodies for assistance and co-operation in the case of an accident.* The officers to be contacted and their facilities should be identified.

Central, State and Local Government – Actions and Responsibilities

The action of government administrative agencies' may include legislation or statutes requiring:

1. Classification of hazardous materials in different categories based on degree of risk and extent of damage potential, for the purpose of control and zoning, safety or green belts width of separation for inhabited areas.
2. Risk analysis requirements for such material, such as the methods and items to be included.
3. Enforced code of practice for handling, storage, transportation and transfer of hazardous materials such as types of containers, their labelling, colour coding pipes, safety precautions while transporting by road or transferring, including antidote, remedial and protective actions required in case of breakage or spillage.
4. Plant designs to incorporate safeguards; failsafe control devices with backups (e.g. scrubbers in series followed by flare or incinerator) to prevent accidental release with the capacity to handle anticipated emergency discharges.
5. Protective measures in case of a release, such as water sprays, neutralizing chemicals applications personal safety gear.
6. Setting of standards on minimum skills or qualifications for different categories of workers (e.g., plant managers, supervisors and plant operators), and organizations to train and issue with certificates to such persons.
7. The establishment of an inspectorate system to monitor the installation, operation and maintenance of control measures.
8. Prescription of the form, substance and dissemination of the information to be provided to the public around the plants. It should include details of toxicity of the substances in use, the long- and short-term effects on health and the environment of acute exposure, and the antidote and remedial measures needed in the event of exposure.
9. A variety of types of *warnings and alarms* for different categories of releases and protective measures to be taken by the population (stay indoors, evacuate, cover faces with wet towels, etc.). Use of radio, television and other media to inform the public.
10. *Emergency Action Plans* for large-scale accidental release, identification of the role and

functions of individuals in the factory and other external government, administrative and safety agencies (e.g. police, fire brigades); other requirements are methods of contact and co-operation, and the establishment of machinery to implement a plan of action (evacuation, protective measures, etc), use of the media to disseminate information to the public on protective measures for themselves and their property.
11. Preparedness of medical help: knowledge of the nature of the effects and type of damage, availability of medicine stocks, equipment and other facilities, identification and assignment of hospitals and medical centres for emergency action.

General Public

The general public must take on the responsibility of:

1. Becoming aware of the nature of the hazard and education of the population in their responsibilities for self protection when the alarm is raised,
2. Becoming aware of the significance of each type of alarm and the different actions required,
3. Acting to form non-governmental voluntary organization or groups to help execute the emergency plan of action in the case of a major release of the hazardous substance.

UNIDO has a unique role to play in classifying the processing of hazardous materials and offering guidelines on the zoning and siting of plants. UNIDO is also in a position to develop a manual of hazardous materials management and to provide training in the development of skills for operating and managing industrial installations.

References

1. Dave, J.M. 1985. Bhopal Methylisocyanate Incident : An Overview. *Proceedings of International Symposium on Highly Toxic Chemicals Detection and Protection,* Sept. 25–7, 1985, Saskatoon, Canada. H. Bruno Schiefer, ed. : 1–37.
2. Shriram Food Fertilizer Limited. Oleum accident. Unpublished report of the enquiry committee.
3. Chlorine leak at Bhadrachalam Paper Mill, AP. Committee report (Chairman, J.M. Dave). Unpublished.
4. Scientific Studies on Factors Related to Bhopal Toxic Gas Leakage. "Varadarajan Committee". Unpublished.

29

Strategies for Major Accident Prevention in Chemical Industry: The Case of Thailand

D. Phantumvanit and N.C. Thanh

Many toxic chemicals are regularly in use in our daily lives. Because of the rapid increase in both the number and the quantity of chemicals that are manufactured, transported, stored and used, it is inevitable that the number of accidents involving the production and use of chemicals will rise. Several international accidents in the chemical industry clearly point out that accidents involving toxic chemicals are not just a few isolated incidents, they happen regularly, and each time cause the loss of life or serious property damage.

Given the rapid increase in the use of chemicals in Thailand, serious accidents are bound to occur if there are no adequate plans to prevent them. It is essential that a national strategy for preventing accidents involving the use of chemicals be formulated now.

A study has been made to examine these issues in perspective. A technical survey was conducted on 27 selected chemical plants and the toxicity and potential hazards of the chemicals used and processes employed was assessed. The result of already completed studies in this area were analyzed. The opinions of local experts were also sought at a seminar to discuss the issues involved in formulating a national strategy for major accident prevention in chemical industries. This paper outlines the present status and profile of accident prevention in Thailand, and the strategies developed for major accident prevention in the chemical industry resulting from this study.

Present Status of Accident Prevention in Thailand

Existing Regulations and Legislation

Existing regulations and legislation in Thailand that will be pertinent to any new strategy concerning accident prevention in the chemical industries are detailed below.

1. Poisonous Substances Act (PSA) in 1967 amended in 1973 PSA was enacted for the control of import, export, manufacture, sales, storage, transport, and use of poisonous substances.
2. National Environmental Quality Act in 1975 amended in 1978 This Act created the National Environmental Board (NEB) and the Office of the National Environmental Board (ONEB). It authorizes NEB to perform functions which mostly concern policy development and coordination with other government agencies in matters relating to environmental quality.
3. Announcement No. 103 of the National Executive Council in 1972 This was promulgated for the general protection of workers.
4. Factories Act in 1968 This Act empowers the Ministry of Industry to control the establishment and operation of factories. Under this Act, the Ministry of Industry can issue regulations limiting waste discharges from factories, air emissions, occupational safety and the working environment inside the factories.
5. Regulation of the Office of the Prime Minister on Accident Prevention in 1983 This regulation created the National Safety Council of Thailand (NSCT). The primary duties of NCST are to advise the Council of Ministers in matters related to accident prevention, coordinate the work of various government agencies, propose new laws or regulations to the Council of Ministers, and promote public awareness and public relations.
6. Other Acts Several other related acts have also been passed governing or controlling the use of gasoline, liquid petroleum gas, fire codes and controlling of nuisance.

Existing Standards Relevant to Toxic Substances

Of all the environmental quality standards promulgated to date, only three have direct relevance to toxic substances as listed below.

1. Notification No. 12 of the Ministry of Industry in 1982 This contains the wastewater effluent standards for factories. The standards limit the concentrations of several pollutants in the wastewater effluents including several toxic substances such as chromium, arsenic and pesticides.
2. Announcement of the Ministry of the Interior, May 30, 1977 This sets out standards limiting the concentrations of chemicals, dusts, fibres, vapours, fumes and gases in the workplace.
3. NEB Standards for Surface Water Sources These standards classify surface water sources into five classes according to the intended uses and specify the maximum allowable concentrations of several pesticides, heavy metals and toxic substances.

Activities of Government Regulating Agencies

Authority for regulating accident prevention measures and potentially dangerous chemicals lies with a number of government agencies. Major activities already carried out by these agencies are described below.

1. National Environment Board (NEB) NEB is primarily a policy and advisory organization responsible for formulating policies for conservation and enhancement of the national environment. NEB also coordinates the work of various other governmental agencies involved in the conservation of the environment. It has set up a Toxic Substances Committee which oversees matters related to toxic substances.

2. Department of Industrial Works (DIW) DIW is responsible for the control of industrial pollution and occupational safety in the factories. The Factory Control Division of DIW is responsible for the registration of all factories in Thailand. Newly proposed factories must submit information including drawings and specifications on the plant and the processes as well as pollution control facilities to this Division. Matters related to environmental pollution are reviewed by the Industrial Environmental Division.
3. Department of Agriculture (DOA) DOA of the Ministry of Agriculture and Co-operatives handles toxic substances through its Division of Toxic Substances for Agriculture. This Division is responsible for research in the use of toxic chemicals in agriculture and for the monitoring of residues of toxic substances in the environment.
4. Ministry of Public Health (MOPH) In the area of accident prevention, MOPH is responsible for occupational health of workers in the factories, for controlling the quality of food consumed by the public and for the general environmental health of the people in both urban and rural areas.
5. National Safety Council of Thailand (NSCT) NCST concentrates on raising public awareness of accident prevention issues and measures. It publishes and disseminates information on various kinds of accidents including those from automobiles, LPG and fire. The NSCT handbook on "Accidents" (1984) covers chemical labelling, body injuries caused by chemicals, pesticides poisoning and the associated medical treatments. NSCT has been effective in media communication, particularly in conveying its message on television.
6. Department of Labour (DOL) The Department of Labour of the Ministry of the Interior is responsible for the protection of workers. It enforces various legislation related to labour. Its daily functions include carrying out inspections, improving industrial relations, conducting occupational safety and health, and other technical and laboratory work.

Profile of Accident Prevention in the Chemical Industry

Past Development

Industrialization has been progressing rapidly in Thailand during the past decade. In 1978 the Ministry of Industry counted 60,384 factories in the country; in 1983 the total number had increased to 91,214.

Using the Standard Industry Classification (SIC) system, there are 2023 factories of category 35, which are manufacturers of chemicals, petroleum, coal, rubber and plastic products. Of the total, 1438 are located in Bangkok and the five surrounding provinces, namely Nonthaburi, Samut Prakarn, Nakhon Pathom, Pathum Thani and Samut Sakhorn. These statistics indicate the concentration of chemical factories in and around Bangkok.

It should be noted that the manufacture of chemicals is increasing, and that Thailand also imports a substantial quantity of chemicals from abroad. This is particularly true for pesticides, whose processes are limited to formulation and repackaging. Thailand manufactures only a small quantity of intermediate compounds, most of those needed are imported.

Injuries from the Chemical Industry

Thailand has a population of approximately 50 million. The labour force in

1984 was estimated to be about 24.9 million, out of which 23.5 million were in employment and 1.4 million were unemployed. The majority of those in employment (68.4%) are in the agricultural sector, with only 8% in manufacturing.

Statistics on injuries from industrial activities have been compiled by the Department of Labour in administering the Workmen's Compensation Fund. In 1984, 39,128 injuries were reported, of which 2–3% were considered serious involving the loss of limbs, eyesight or death. It is estimated that if smaller establishments are taken into account the actual number of serious injuries would be 3 to 5 times higher. It should also be noted that the economic implications of occupational injuries can be expensive. The amount of compensation paid quadrupled, from 62.17 million Baht in 1978 to 246.79 million Baht in 1984 (US$1 = 26 Baht). The risk of accidental death in the chemical industry in Thailand is more than 20 times that of the United Kingdom and the Netherlands.

Survey Findings

Although the field survey was limited to 27 plants, an effort was made to ensure that they were representative of each category of industry (basic industrial chemicals except charcoal and fertilizer – 37.3%; pesticides – 37.1%; explosives and ammunition, paints, varnishes and lacquers, matches – 22.2%; synthetic resins, plastic materials and artificial fibres except glass – 3.7%; petroleum refineries – 3.7%). Emphasis was given to small and medium factories based on the Department of Labour's classification.

In trying to identify potential major hazard installations, the World Bank Guidelines for Identifying, Analyzing and Controlling Major Hazard Installations in Developing Countries were used as the reference point, particularly its lists of toxic, reactive, explosive and flammable substances.

Twenty-eight out of 224 chemicals used or produced in the plants visited were identified as hazardous substances according to the World Bank guidelines. Twenty out of 27 factories visited used or produced at least one of the identified hazardous substances.

Arrangement of Premises

The production facilities of the factories visited were generally located next to either living quarters, warehouses, or offices. Spacing between equipment was usually adequate. In smaller plants, loading and unloading of hazardous substances was conducted within the premises without proper accident prevention measures. Large chemical storage tanks were often situated above ground rather than underground. Most of the waste disposal facilities were located inside the premises with the treated effluents directly connected to public sewers. It is difficult to assess the efficiency of the treatment methodologies employed.

Buildings

One of the potential problems was the use of flammable materials such as wooden walls and wooden posts. Most of the factories visited lacked efficient ventilation systems, thus fire could easily spread through openings. In those cases where ventilation systems were installed, they were not functioning properly. This may have been due to poor design or poor maintenance. Emergency exits in buildings were only evident in a few factories.

Firefighting Facilities

A common problem was the lack of an officer designated to look after the fire fighting equipment. In several factories the checking and refilling of fire extinguishers was left to the dealers. Employees in many factories had never participated in an emergency drill. Electrical appliances were not routinely checked and naked electrical switches were common in smaller plants.

Materials, Intermediates and Products

The properties of hazardous chemicals were not well understood by management officers and workers in general. This could lead to negligence and avoidable accidents. The handling of hazardous substances is entirely in the hands of each individual operator.

Processes

Almost none of the factories studied had issued job regulations for the process operation in order to prevent human error and little effort seemed to have been made to minimize the use of dangerous materials. Most of the factories had never examined the possibility of volatile chemical reactions, nor had the potentially affected areas been mapped out in case of an emergency.

Transport, Storage

The transportation of raw materials into the plants was either carried out by company trucks or by outside contractors. Most of the private truck drivers were ignorant about the risk of handling chemicals. The factories themselves rarely issue job regulations for the reception of raw materials. In any case, only a limited number of transport workers understand chemical properties and their health effects.

Process Equipment

There was no systematic inspection of process equipment in most of the factories visited. A systematic approach based on standard practices was found to be lacking in many factories. The survey confirmed that the use of protection devices varied from one extreme to the other in the factories visited – from total compliance with standard practices to complete neglect.

Housekeeping

In general, housekeeping was adequate in large factories due to their more established system. In smaller plants it was observed that good housekeeping is not accorded due priority by the management.

Location

It was observed that all factories were located at a reasonable distance from the support system, that is, fire stations, hospitals. However, in quite a few locations, the accessibility of the factory is a hindrance.

Job Training

Workers were not well informed about the risk associated with the production and use of chemicals in the manufacturing processes. Contingency plans for emergencies were not known to workers. On-the-job training was conducted only in cases of necessity but not as a prerequisite. Training courses on safety and accident prevention were not readily available.

Safety Activities

In the factories visited, safety officers had not fully assumed their responsibilities on accident prevention because of their other workload. Also, safety committees to generate participation of workers on accident prevention had still not been established in most of the factories.

Personal Protective Equipment (PPE)

It was noted that in some cases PPE was wrongly used. For example, in a number of places dust masks were used instead of respirators. Changing PPE wearing habits of workers remains a problem. The heat and humidity of the tropics often make the use of PPEs uncomfortable. Safety campaigns to promote their use was found to be lacking.

Strategies for Major Accident Prevention in the Chemical Industry

Below are the recommendations and proposed actions in formulating strategies for major accident prevention in the chemical industry.

Establishment of a Major Hazard Assessment System

The survey confirmed the lack of an adequate recording system to deal with toxic chemicals. Information on the tracking of pesticides from import, to formulation, to repackaging was not normally available. At the same time the personnel in charge of safety were not fully aware of proper procedures to control accidents. In fact, accident prevention and the associated management of dangerous chemicals was not accorded due priority commensurate with the "explosive" nature of the issue. The potential of major hazards was visible in many of the plants visited, such as the bulk storage of paraquat without proper safety containments and poorly ventilated pesticide-mixing rooms where workers were exposed to the fumes and dust generated from the mixing processes.

It is therefore recommended that a "major hazard assessment system" be developed for Thailand following World Bank guidelines. The system should be based on four premises, namely:

1. For installations with potential major hazards, routine inspection and maintenance are not sufficient. An additional major hazard assessment will be required to reduce the risk of accidents.
2. The responsibility in accident prevention lies with industry not with the government. A factory may use its own safety engineer to conduct a major hazard assessment of its operations, or it may choose to commission outside consultants to undertake such a study. In the case of a large factory with its own scientific or engineering staff, they should perform such an assessment because they are most familiar with industrial processes and understand the complexities of the installation. For the smaller establishments, guidance should be provided by the government.
3. Any new recommendations should be based on the existing foundation already established in various governmental agencies. A new recommendation should refocus direction to face new challenges arising from potential major hazards, not to create another layer of bureaucratic regulation.
4. The issue of major hazard assessment is wide ranging, it involves the work of at least six ministries. With this background in mind, it is recommended that:
   a. All existing national committees dealing with dangerous substances jointly prepare a definitive list of dangerous substances with specific criteria based on both the lethal dose and quantity. A better alternative would be to develop a set of criteria and a list of chemicals based on World Bank guidelines.
   b. The Poisonous Substances Board who executes the Poisonous Substances Act, incorporates the criteria and the master list of dangerous substances as mentioned in a. into PSA. By doing so, all classified dangerous substances will be legally registered before they can be manufactured or imported into the country.
   c. DOL in collaboration with DIW design the guidelines for the preparation of the safety report required in a major hazard assessment. The objective is to enable industry to prepare only one main report, parts of which could then be submitted to the respective government agencies as stipulated in existing laws.
   d. NEB and NSCT jointly review the content of item 10 of its Notification of Types and Sizes of Projects or Activities Requiring an EIA Report to reflect the urgency of accident prevention in industry. Concurrently, the Manual of NEB Guidelines for Preparation of Environmental Impact Evaluations should also be revised using the World Bank guidelines as a reference.
   e. Similarly, DIW review their notifications related to effluent quality and working environment with the objective of ensuring that major hazard assessment will be carried out by the factories.
   f. DOL strengthens its work on classification, identification, and labelling of dangerous substances. The survey conducted in this study reconfirms the need for more stringent precautionary measures and the dissemination of information to both employers and employees alike. The information to be disseminated should be prepared using language which is comprehensive to workers. For large factories, the preparation of major hazard assessment reports can be done by environmental officers

or safety officers which are already required by law in pollution producing industries or in establishments with more than 100 employees.

Accident Prevention Information System

The inadequacy of statistical data on accidents and injuries has been well recognized in Thailand. It is recommended that an Accident Prevention Information System be developed consisting of three elements:

1. Accident reporting system which covers accidents resulting in injuries to the workers and major accident investigation
2. Chronic health hazards
3. Dangerous substances information system

Disposal of Dangerous Substances

Two types of toxic wastes were identified during the survey. The first type consists of chemicals that cannot be sold either because their sell-by date has expired or there is no demand. The second type consists of the sludges resulting from wastewater treatment. Since at present there are no toxic waste dumps in Thailand, the factories visited either bury the wastes within their compounds or have them removed by contractors. Nothing is known about the disposal of these wastes. This is a potentially hazardous situation since the wastes may contaminate ground water or find their way into surface water.

At present, no estimation of the amount of toxic wastes generated annually is available. During the survey, 20 to 30 cartons, each 200 kg in weight, of unsold pesticides were spotted in a number of pesticide manufacturing plants. Assuming that 10% of the total 75 pesticides formulation plants carry the same amount of leftover products, this would work out to be at least 30 tons of unsaleable pesticides to be disposed of.

Plan of the Department of Industrial Works (DIW)

The existence of toxic wastes from other industries, such as metal plating plants, has long been recognized. For example, DIW has estimated that presently there are about 800 cubic metres of wastewater produced daily from textile dyeing plants and about 200 cubic metres of plating wastewaters produced daily that are not properly treated. DIW is currently considering the establishment of three central wastewater treatment plants in the Bangkok Metropolitan area to collect and treat these wastes. The sludges resulting from these wastewater treatment plants will be cast into concrete blocks. These blocks will then be hauled to a burial site in Rajburi. It has been estimated that there will be about 40,000 tons of these concrete blocks per year. When buried at stacks of 5 metres high, an area of 30 hectares will be required. Three of such sites in Rajburi have already been surveyed. Initially, DIW is planning to build these facilities after which they will be contracted to private operators who will recover the costs through charges.

Additional Studies

In the factories surveyed, the treatment methods range from no treatment in small plants to sophisticated treatment in large plants. For example, one plant incinerates all toxic wastes in a high temperature incinerator; while another uses imported Fullers' earth to absorb toxic chemicals and buries the Fullers' earth on site.

It can reasonably be assumed there is a significant number of dumping sites in Thailand. It is therefore encouraging to note that DIW is setting up proper treatment facilities for inorganic wastes, particularly heavy metals from textile mills and electroplating plants. A parallel programme should be launched to take care of hazardous wastes from toxic chemicals. The components of such a programme would include:

1. Surveys to estimate the quantity of toxic wastes
2. Selection of appropriate technologies for the treatment and disposal of toxic wastes
3. Establishment of criteria for the selection of toxic waste dump sites. Special attention should be given to groundwater contamination
4. Establishment of a long-term monitoring programme to ensure that there is no water leaching
5. Determination of the operation of treatment facilities. Should they be operated by a government agency or a private company?
6. Enforcement, ensuring that all toxic wastes are delivered to the treatment facilities

Training and Public Awareness Campaign

From the survey, it was observed that a few factories, mainly affiliates of multinationals, provide safety training for their employees. These companies have the resources, the technical expertise, and the accessibility to technical information compiled at their headquarters. With Bhopal still fresh in mind, large companies are willing to assist the government and share their technical expertise. Enlisting the assistance of large companies to provide training to smaller enterprises is one avenue that can be further explored.

Special courses should be established so that trained inspectors can also offer technical guidance, particularly to small and medium enterprises, on top of their normal regulating functions. In this respect, the on-going training programmes at the National Institute for the Improvement of Working Conditions and Environment (NICE), although still in their beginning period, are steps in the right direction and should be further strengthened to enable NICE to respond effectively to increasing responsibility in accident prevention.

The main objective of an accident prevention campaign programme is to ensure safety of employees and the public. To be successful, the public must be fully informed of the potential hazards from the manufacturing, transportation, use and disposal of dangerous substances. In this connection, NSCT has been successful in its public awareness campaign on television, radio and through other mass media.

A successful public awareness programme does not follow any fixed formula.

It relies on the conscientious effort of the organizer to elicit public participation. The subject of accident prevention should be highlighted by NSCT to educate the public on the potential hazards and to generate public interest and support for this programme.

Acknowledgements

This study was carried out with the funding of the International Labour Regional Office, Bangkok. The authors wish to express their sincere thanks to the President of the Thailand Development Research Institute (TDRI), for granting permission to use the results of this study.

References

1. European Community Directive. 1982. On the Major Accident Hazards of Certain Industrial Activities. 82/501/EEC. *Official Journal of the European Community*, L230, June.
2. UK Health and Safety Executive. 1985. *A Guide to the Control of Industrial Major Accident Hazards Regulations 1984*. HS(R)21. London: HMSO.
3. UK Notification of Installations Handling Hazardous Substances Regulations, 1982. SI No. 1357 UK Health and Safety Executive Guide, HS(R)16.
4. Recommended Procedures for Handling Major Emergencies. London: Chemical Industries Association.
5. USEPA. 1985. Community Preparedness for Chemical Hazards, Part 3: A Guide for Contingency Planning (draft). August. Document 6291H.
6. World Bank. 1985. Manual of Industrial Hazard Assessment Techniques.

30

Environmental Control Measures for Toxic and Hazardous Wastes of the Eastern Seaboard Industrial Complex in Thailand

R.N. Chakrabarty

The Eastern Seaboard Development Programme (ESB) in Thailand is by far the largest industrial and multi-purpose project ever planned and undertaken in the country. This US$4000 million project includes a natural gas-based heavy industrial complex, an industrial estate for light and medium industries, deep sea ports, new urban areas and the associated infrastructures and supporting facilities. Two coastal areas of the Eastern Seaboard, namely, Map Ta Phut in the southeast coast and Laem Chabang in the east coast of the Upper Gulf of Thailand (Figure 30.1) are the sites of the project.

This programme is being conducted under the aegis of the Office of the Eastern Seaboard (OESB – the secretariat of the Eastern Seaboard Development Committee (ESDC) under the Chairmanship of the Prime Minister. For executing the various projects of the ESB, a number of state and private agencies are co-operating with the OESB and are participating in the programme. This paper describes the measures that have been and are being planned and implemented for environmental pollution control, especially for the management of hazardous and toxic industrial wastes of the Map Ta Phut complex.

Features of Map Ta Phut Complex

Map Ta Phut Complex will include a gas-based heavy industrial complex, an industrial estate for supporting industries, a new urban area and a deep sea port. A site plan of the Complex is shown in Figure 30.2.

Industrial Complex

To utilize the natural gas coming ashore from the Gulf of Thailand, the industrial complex at Map Ta Phut is being developed to include a fertilizer complex, a petrochemical complex and other hydrocarbon-based industries, mineral and heavy metal processing industries, for example, a Tantalum and

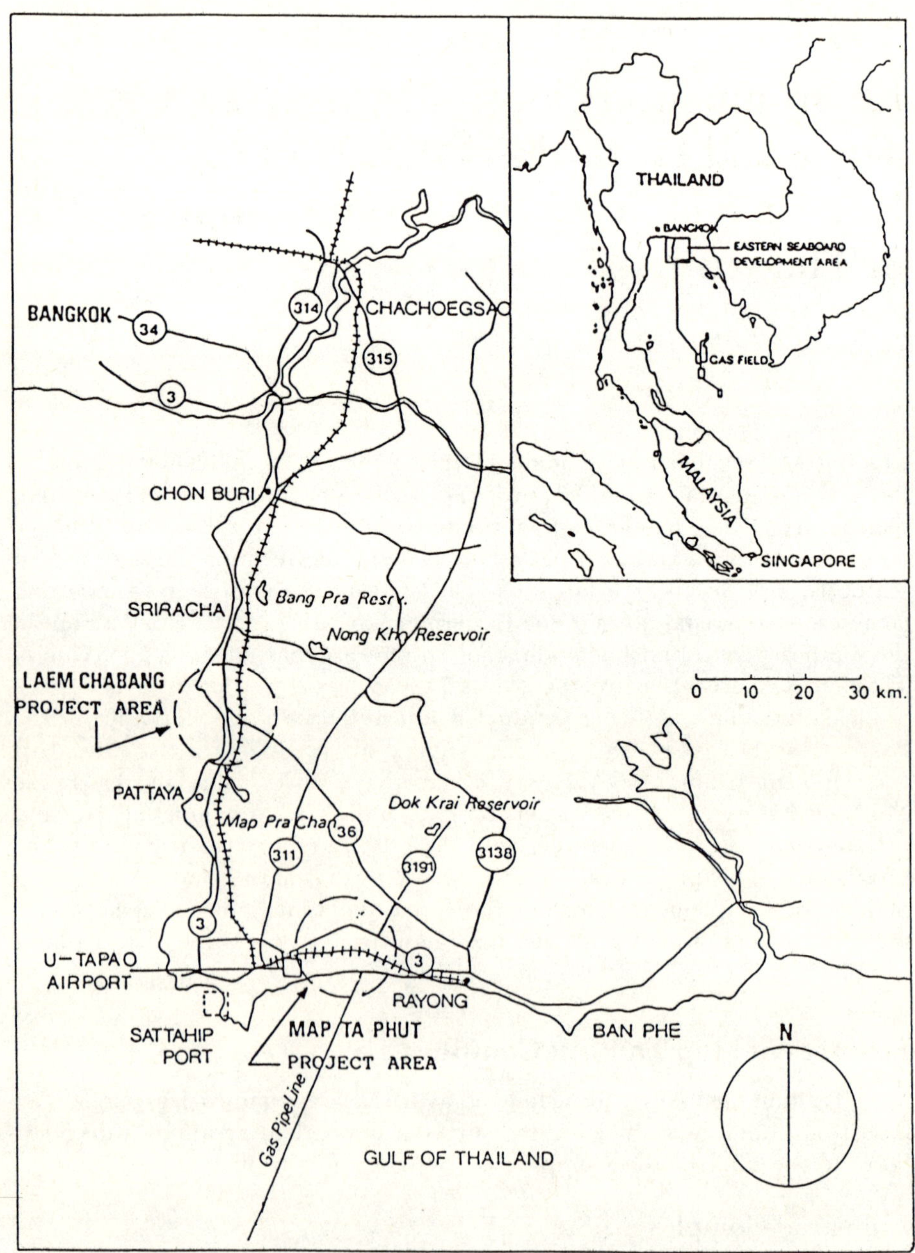

Figure 30.1 Areas of the Eastern Seaboard Development Programme

Niobium Oxide plant and a Hydrofluoric Acid plant, in the first stage of the development programme. A natural gas separation plant has been operating at the Map Ta Phut Complex for over a year.

In the second stage development programme, the establishment of a number of other industries including Caprolactum, Dimethyl Terephthalate (DMT), Acrylonitrile, Styrene, Melamine, Titanium Dioxide and other related industries has been proposed. Figure 30.3 shows the current and proposed industrial projects at the Map Ta Phut industrial complex in the first and second stages of the programme, respectively.

The industries to be established in the Supporting Industry Area of the industrial complex are also expected to be primarily related to petrochemical and fertilizer projects, although they may include other types of industries. However, information about any specific types of industries in the Supporting Industry Area was not available at the time of writing.

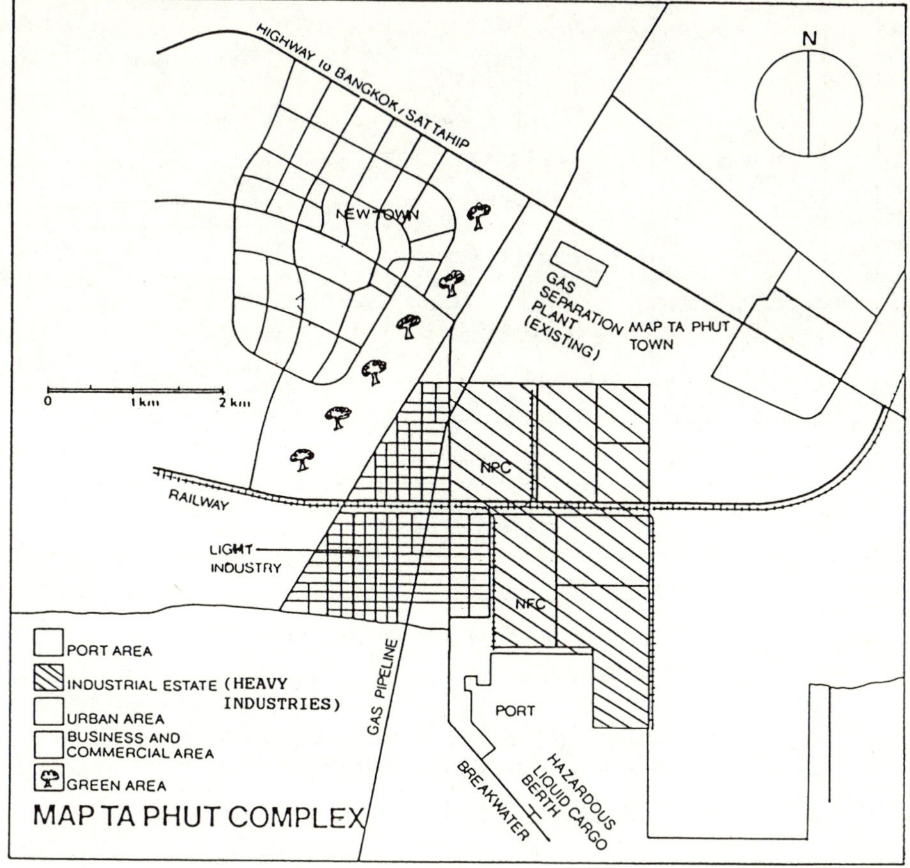

Figure 30.2 Site plan of the Map Ta Phut Complex Comprising an Industrial Estate for Heavy Industries, an Urban Area and a Port

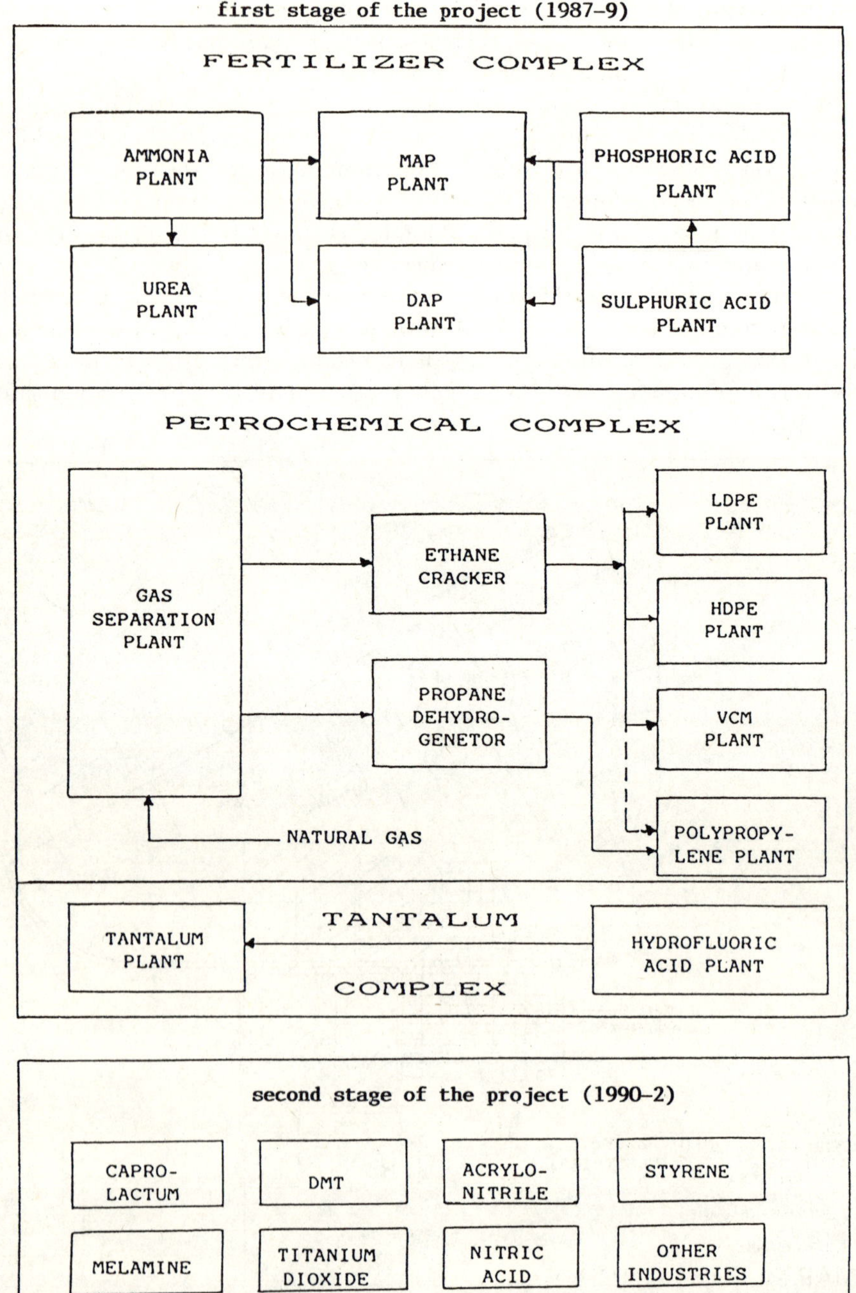

Figure 30.3 Current and Proposed Industrial Projects at the Map Ta Phut Industrial Complex

New Urban Area

A new urban area is being developed at a site northwest of the industrial complex. In the first stage programme, the urban area would be developed for a population of 12,000, and in the second stage, for a population of 40,000.

Industrial Port

The construction of an industrial port south of the fertilizer complex has also been proposed to cater to the needs of the industrial complex as well as other agro-industries and mineral processing industries in the region.

Magnitude of Pollution Potential of Map Ta Phut Industrial Complex

The industries proposed for establishment in the Map Ta Phut industrial complex are well known for their high pollution risk. Besides generating oxygen consuming and refractory chemical pollutants, these industries will release other toxic and hazardous pollutants through the wastewaters, gaseous emissions and solid wastes. It would therefore suffice to list here only those pollutants which are toxic and hazardous, and which would have an adverse impact on the environment. They are:

1. Water-borne Pollutants: Ammonia, amines, fluorides, hydrocarbons, chlorinated hydrocarbons, sulphides, cyanides, methyl isobutyl ketone (MIBK) and heavy metals.
2. Air-borne Pollutants: Hydrocarbons, chlorinated hydrocarbons, hydrogen sulphide, sulphur dioxide, oxides of nitrogen, acid fumes and particulates.
3. Solid Waste Pollutants: Heavy metals, fluoride and phosphate sludge, by-product phosphogypsum and other organic sludges.

Environmental Impact Assessment (EIA) for each of the industrial projects was carried out as required by the National Environment Board (NEB). The Environmental Impact Statements (EIS) thus prepared, identified and estimated, *inter alia*, the possible types and quantum of the chemical pollutants, especially the toxic and hazardous pollutants, to be generated and released through the wastewaters, gaseous emissions and solid wastes from the fertilizer, petrochemical and tantalum plants. A list of these toxic and hazardous chemical pollutants including other significant pollutants from the first stage industrial project at Map Ta Phut is shown in Table 30.1.

Although the marine environment of the inner sector of the Gulf of Thailand, where Laem Chabang is located, has been to some extent affected by pollution from the communities and industries located in its northern coast including Bangkok and other areas, the Map Ta Phut area and its coastal stretch have so far remained unaffected by any man-made pollution.[1,2]

The EIS prepared for these projects therefore emphasized the need for almost complete removal of the toxic and hazardous pollutants and also of adequate removal of other pollutants from the emissions of these plants before their release to the marine and air environment of Map Ta Phut.

However, since the detailed planning and design of any pollution control measures, especially water pollution control measures, is governed by two important factors, such as the methods of disposal and the effluent or emission standards, these factors were evaluated in respect of their applicability and compliance prior to undertaking the planning and design of appropriate control measures.[3] This is discussed below.

Table 30.1 Toxic/Hazardous and other Significant Pollutants of the First Stage Industrial Projects at Map Ta Phut

Type of Project	*Toxic/Hazardous and Other Significant Pollutants in*		
	Wastewaters	*Gaseous Emissions*	*Solid Wastes*
Fertilizer Complex	HN_3–N, NH_4^+–N, Fluoride as F, Phosphorus as P, Radium 226	SO_2,NO_x,SiF_4 Particulates (including urea dust)	Fluoride as F, Phosphorus as P Radium 226
Petrochemical Complex	Hydrocarbons, Chlorinated hydrocarbons, Sulphides, Cyanides	Hydrocarbons, Chlorinated hydrocarbons, SO_2, NO_x, Particulates	
Tantalum Complex	NH_3–N, NH_4^+–N, H_3PO_4, H_2SO_4, MIBK, Heavy Metals	NH_3–N, MIBK, H_2S, Fluorides, Particulates	Heavy metals

Methods of Disposal of Industrial Wastewaters

In view of the coastal location of all the Eastern Seaboard project units, the following alternative systems of wastewater disposal were evaluated:

Alternative 1 Minimum necessary pretreatment of the wastewaters from the individual industrial units including port and urban area, followed by their combined disposal to the Gulf of Thailand through a common submarine outfall.

Alternative 2 Comprehensive treatment of the wastewaters from the individual industrial units, port and urban area, as needed, and their separate disposal either by open discharge to the Gulf or to the nearest water courses leading to the Gulf.

A techno-economic evaluation of the above two alternatives led to the conclusion that to protect the marine environment of Map Ta Phut, especially in respect of the long-term, and possibly irreversible, adverse effects, comprehensive treatment of the wastewaters from the individual industries and their separate or common disposal to the Gulf (Alternative 2) would be cost-effective. An exception in this case, however, was the system of treatment and disposal of the fertilizer complex wastewaters, which is discussed below.

Development and Updating of National Standards

After the policy decision was established for the disposal of treated wastewaters of the Map Ta Phut complex to the Gulf of Thailand, the next important step was to identify the gaps in the existing standards and formulate additional standards, especially for the specific pollutants which are toxic and hazardous and are likely to be discharged by the industries at Map Ta Phut.

In 1979 the Ministry of Industry established the national standards for discharge of sewage and industrial wastewaters into the inland surface waters, which are shown in Table 30.2. The Industrial Estate Authority of Thailand,

Table 30.2 Maximum Standard Effluent Water Quality (Thailand Ministry of Industry Declaration, 1982)

pH	5–9
Permanganate Value, mg/l	60
Total Dissolved Solids, mg/l	5000*
Sulphides (as $-H_2S$), mg/l	1
Cyanide (as HCN), mg/l	0.2
Heavy Metals, mg/l	
Zinc	5
Chromium	0.5
Arsenic	0.25
Copper	1
Mercury	0.005
Cadmium	0.03
Barium	1
Selenium	0.02
Lead	0.2
Nickel	0.2
Manganese	5
Tar	None
Oil and Grease, mg/l	5
Formaldehyde, mg/l	1
Phenol & Cresols, mg/l	1
Free Chlorine, mg/l	1
Insecticides and Radioactivity	None
Total Suspended Solids, mg/l	30, 60, or 150†
BOD_5, mg/l	20
Temperature	40°C

* Represents additional solids over receiving water salinity where receiving water salinity is greater than 2000 mg/l

† If the mixture ratio of wastewater and receiving water is between 1/8 to 1/50, 1/51 to 1/300 and 1/301 to 1/500 total, suspended solids and shall not exceed 30, 60 and 150 mg/l respectively

the executing agency of the Eastern Seaboard Projects, also established its own standards, which were specifically applicable to the industrial estates under its control. However, in the context of the new types of industries proposed for the Map Ta Phut complex and of the types of pollutants to be generated by them, the existing national standards were found to be inadequate for administration of pollution control from the industries.

Additional standards were therefore proposed for free ammonia, ammoniacal nitrogen and dissolved phosphates and fluorides in industrial discharges.These additional standards were necessary to control the emission of pollutants from the new types of industries at Map Ta Phut complex, including the fertilizer complex.

Development of Pollution Control Facilities

In the light of the updated national standards on pollutants discharge and based on the established inputs and the desired outputs, appropriate process flow diagrams, technology for detoxification of specific pollutants and the design criteria for the wastewater treatment and disposal facilities including solid waste disposal facilities for the fertilizer, petrochemical and tantalum industries were developed.

Unlike wastewater treatment technologies and equipment, the toxic and hazardous gaseous emission contrc' technologies and equipment are now largely an integral part of a manufacturing plant. The processes of Double Combustion–Double Absorption (DCDA) to control sulphur dioxide emission from the sulphuric acid plant and a steam stripper and urea hydrolyzer for control of ammonia emission have been employed in the fertilizer plant at Map Ta Phut.

Nevertheless, for the fertilizer and tantalum projects, the use of advanced technologies for removal of specific toxic pollutants present in the wastewaters has been proposed. Air stripping of ammonia, nitrification and denitrification processes have been incorporated in the wastewater treatment systems of some of the aforesaid industrial projects. These advanced waste treatment processes are well proven and have been in use for over a decade now in a number of sewage treatment plants in the United States and a few fertilizer industries in India.[4,5]

Specific Pollution Control Measures Proposed for the Industrial Units at the Map Ta Phut Complex

Fertilizer Complex

The proposed wastewater treatment system for the fertilizer complex at Map Ta Phut is shown in Figure 30.4. The process wastewaters from the phosphoric acid plant and the periodical purges from the cooling water pond, the expected characteristics of which are shown in Table 30.3, are proposed to be equalized in an equalization tank before subjecting them to a two-stage lime treatment

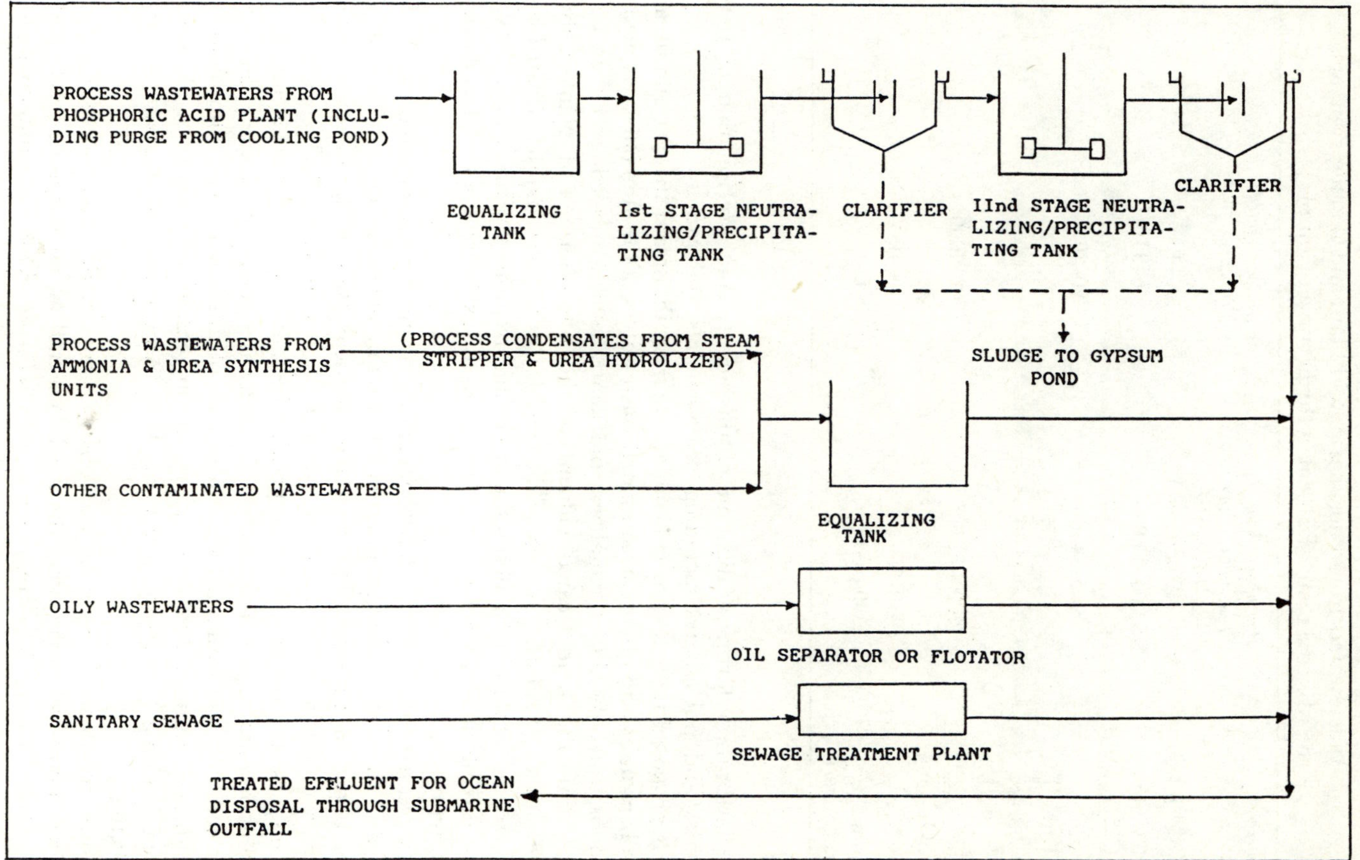

Figure 30.4 Proposed Wastewater Treatment System of the Fertilizer Complex at Map Ta Phut

Table 30.3 Volumes of Flow and Important Characteristics of the Raw Process Wastewaters from Phosphoric Acid Plant and Cooling Pond and the Mixed Treated Effluent of the Fertilizer Complex

Parameters	*Phosphoric Acid Plant and Cooling Pond Wastewaters*	*Mixed Treated Effluent*
Flow, m^3/h	243 (average)	0–400
Total Acidity as $CaCa_3$, mg/l	15,000–50,000	6–9 (pH)
Fluoride, as F, mg/l	900–3400	20–30
Phosphorus as P, mg/l	1000–6000	35–70
Sulphates, as SO_4^-, mg/l	2000–12,000	2700–4000
NH_3-N, as N, mg/l	10–35	3–10
Radium 226, picocuries/l	60–100	0–1

and clarification for neutralization of the acidity and precipitation of the fluorides and phosphates including heavy metals and radium 226.

It is proposed that the process condensate from the steam stripper and urea hydrolizer units and other contaminated wastewaters of the ammonia and urea synthesis units should be equalized and neutralized, if needed, and then be mixed with the lime treated effluent of the phosphoric acid plant. Oily wastewaters and sanitary sewage of the complex would be separately treated in a gravity oil separator or flotator and sewage treatment plant, respectively, before mixing the effluents with the lime treated wastewaters of the phosphoric acid plant.

The treated effluents of the fertilizer complex would be pumped to the Gulf of Thailand through a submarine outfall approximately 2 km offshore and at a water depth of 4 metres. The expected characteristics of the mixed treated effluents of the fertilizer complex are also shown in Table 30.3.

A dispersion modelling study to predict the maximum "near field" increase in contaminant concentrations in the seawater near the outfall diffuser indicated that the estimated concentrations of ammonia, urea, phosphorus, fluoride and radium 226 at the plume impingement area were not likely to exceed 0.45 mg/l, 0.13 mg/l, 0.88 mg/l, 0.38 mg/l and 0.001 picocuries/l, respectively.

However, depending upon the observations made after operating the fertilizer plants for some time, the need to install an air stripping tower and update the two-stage lime treatment system to further reduce the ammonia and phosphorus concentrations in the effluent would be assessed.

The sludges of the two-stage lime treatment unit are to be pumped to the gypsum stack-cum-cooling pond where the by-product phosphogypsum of the phosphoric acid plant would be stored. Incidentally, a study carried out by the UNIDO project in Thailand on the possible recycling of phosphate-rich sludge from the second stage lime treatment unit to the phosphoric acid plant indicated that an additional amount of 4800 tonnes of phosphorus or about 11,000 tonnes of phosphoric acid (as P205) could be recovered per annum, resulting in an annual cost saving of about US$2.0 million.[6] This would also result in an appreciable reduction in the volume of the sludge to be disposed of in the gypsum stack.

Petrochemical Complex

The proposed wastewater treatment system for the petrochemical complex is shown in Figure 30.5. The wastewaters of the upstream unit (ethane cracker and propane dehydrogenerator) of the petrochemical complex, which contain primarily liquid hydrocarbons, sulphides, mercaptans, phenols and other oxygen consuming organic substances, would be equalized in an equalizing basin and then neutralized and treated to reduce the concentration of sulphides. This pretreated wastewater would pass through a Dissolved Air Flotation (DAF) unit in which alum or other coagulants would be used to break the emulsified liquid hydrocarbons and remove them effectively.

The wastewaters from the downstream units (LDPE, HDPE, PP and VCM plants) would be similarly pretreated to remove floating matter and liquid hydrocarbons and be mixed with the pretreated wastewaters of the upstream unit for final biological treatment under the activated sludge process in a common facility.

One of the hazardous chemical pollutants that may be present in the petrochemical plant wastewaters, especially the VCM plant wastewaters, is the chlorinated hydrocarbon in the form of ethylene dichloride (Cl.CH2.CH2.Cl). It is a flammable liquid, a carcinogen and a priority toxic pollutant (according to the U.S. Environmental Protection Agency). Since it is non-biodegradable and is also extremely toxic to the micro-organisms of a biological waste treatment plant, the wastewaters of the VCM plant would be additionally steam stripped to remove the ethylene dichloride before sending the wastewaters to the biological treatment plant.

A guard basin is also to be provided to store the treated effluent and check its quality before allowing the effluent to be disposed of or be recycled to the treatment plant, if it is not found fit for disposal.

Tantalum Complex

Before describing the proposed wastewater treatment system for the Tantalum plant, it may be useful to give the background to the project. The proposed Tantalum project in Thailand, unlike fertilizer, petrochemicals and other traditional types of hydrocarbon processing industries, will probably be the second of its kind in Asia and be one of six or seven such plants currently operating in the United States, the FRG, and other parts of the world.

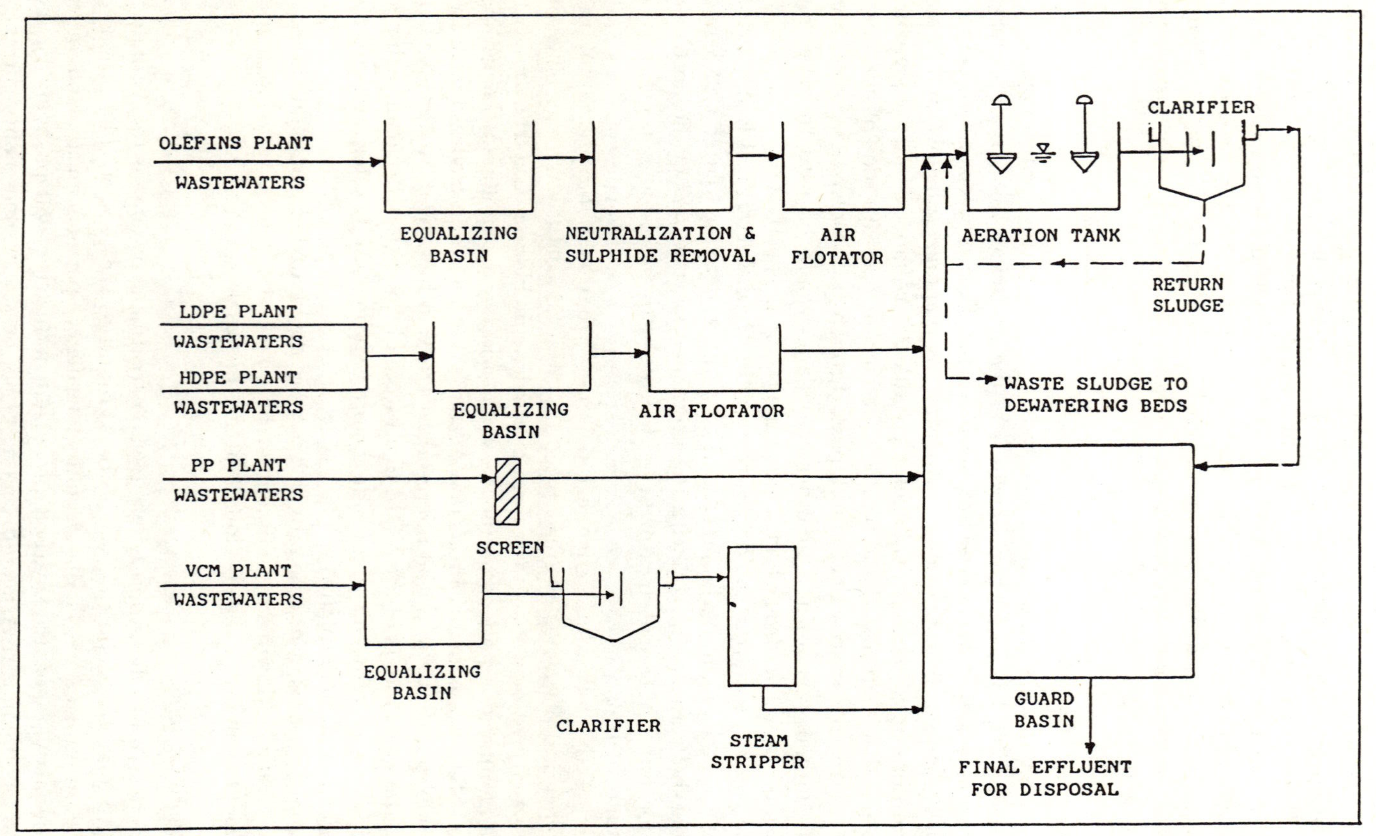

Figure 30.5 Proposed Wastewater Treatment System of the Petrochemical Complex at Map Ta Phut

Information about the processes of extraction and conversion of tantalum and niobium metals to their oxides and other compounds and about the sources and nature of the wastewaters, gaseous emissions and solid wastes including their control measures, is not easily available, except in some special reports or publications, for example, USEPA's "Treatability Manual".[7]

Second, while the methods being presently practised for air pollution control and solid waste disposal in the existing tantalum plants outside Thailand may perhaps be applicable in Thailand also, the methods of treatment and disposal of the wastewaters being followed in other plants may not be applicable in Thailand in the context of the proposed project's location and of the requirements of the local regulatory agency.

In view of this, a joint study[7,8] carried out by the Industrial Estate Authority of Thailand (IEAT) and the UNIDO project in Thailand in co-operation with the Thailand Tantalum Industry Corporation, the owner of the project, led to the identification of the sources, nature and quantum of the hazardous, toxic and other pollutants in the wastewaters, gaseous emissions and solid wastes to be generated in the plant and was helpful in the planning and development of an appropriate system of wastewater treatment to meet the requirements of the local regulatory agency.

Tantalum plant wastewaters are indeed complex in nature, since not only the corrosive and hazardous mineral acids like, hydrofluoric and sulphuric acids but also heavy metals, ammonia, ammonium fluoride, ammonium sulphate and methyl isobutyl ketone (MIBK), an organic solvent, are present in high concentrations in the wastewaters. Although MBIK is slightly soluble in water, it is biodegradable, hence it exerts high BOD. Therefore, segregation and pretreatment of certain wastewater streams followed by combined biological treatment of the pretreated wastewaters has been planned.

Application of advance waste treatment processes had, therefore, to be considered in this case with a view to dealing effectively with the problem. Referring to Figure 30.6, it is proposed that the acidic process wastewaters be equalized and then treated with lime to neutralize the mineral acids and precipitate the heavy metals (tantalum, niobium, zirconium, chromium) and fluorides. After filtration of the treated liquid, it would be subjected to the second-stage process of further precipitation of the residual heavy metals as their sulphides by using sodium sulphide, alum and sulphuric acid, and clarification followed by filtration. The treated liquid would then pass through a cooling tower to reduce its temperature. The sludges from the two-stage treatment would be disposed of by sanitary landfill.

The ammoniacal wastewater from the steam stripper (where the bulk of the free ammonia would be recovered and recycled to the manufacturing plant) containing residual free ammonia and other ammonium salts would be treated in an air stripping tower to remove the free ammonia.

The two pretreated wastewater streams would then be mixed and subjected to an extended aeration activated sludge process for reduction of BOD (due to

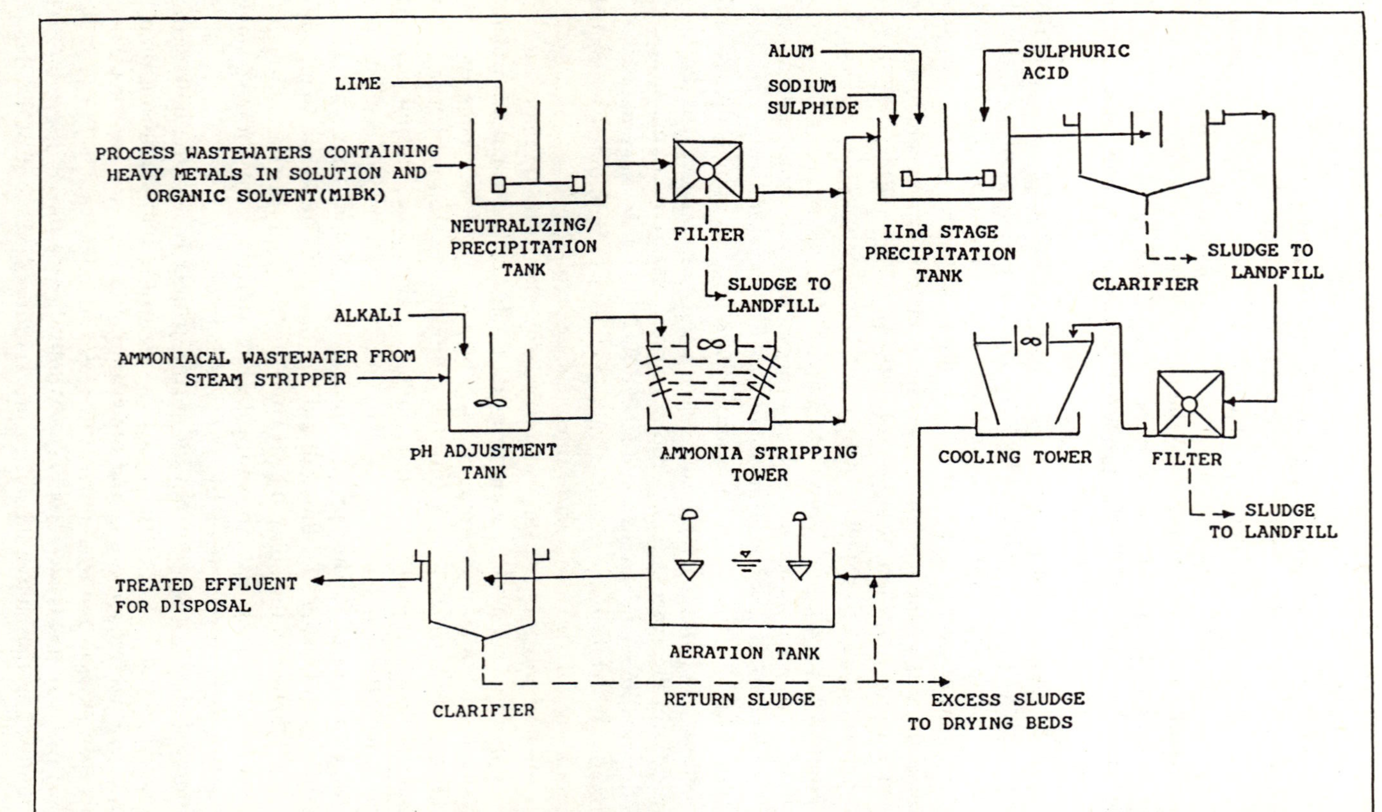

Figure 30.6 Proposed Wastewater Treatment System of the Tantalum Complex at Map Ta Phut

MIBK) and nitrification. The liquid would be settled in a clarifier and be disposed of into the storm drain leading to the sea.

Hazardous Industrial Solid Wastes Management Plan

The hazardous solid wastes that are likely to be generated from the first stage of the Map Ta Phut complex project have been classified as follows:

1. Spent metal catalysts from the manufacturing plants
2. Heavy metals in waste treatment sludges
3. Heavy metals in processed ores and minerals
4. By-product solid wastes

Based on the above classification, the sources and types of hazardous solid wastes and their disposal methods have been determined and are shown in Figure 30.7. While the heavy metal sludges or other hazardous solid wastes from the industries can be disposed of fairly satisfactorily by sanitary landfills, some types of industrial solid wastes, for example, by-product phosphogypsum from the fertilizer industry, still present a disposal problem.

The yearly production of the by-product phosphogypsum from the proposed fertilizer complex at Map Ta Phut has been estimated at about 1.2 million tonnes per annum. About 56 hectares of land would be used as what is called

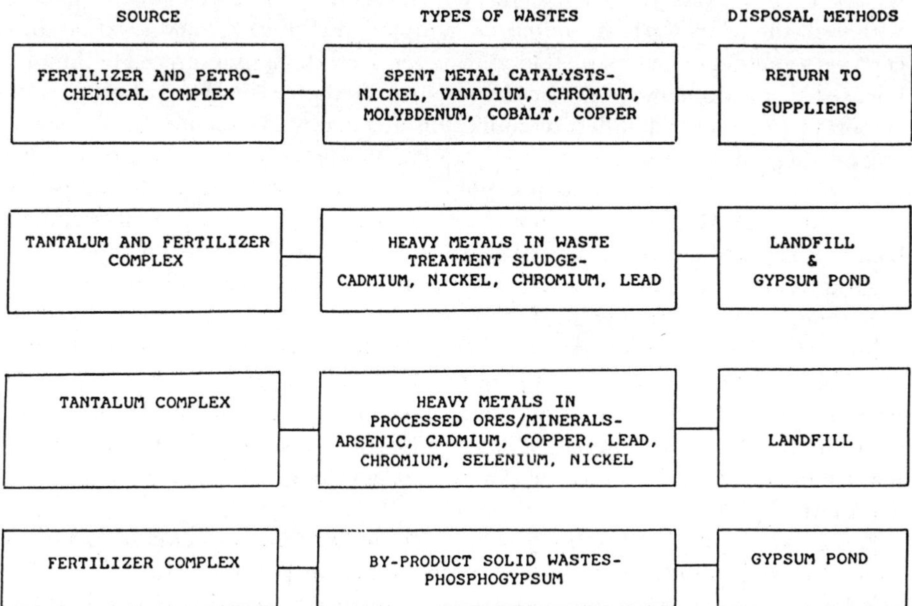

Figure 30.7 Toxic and Hazardous Solid Wastes Management Plan for First Stage Industries of the Map Ta Phut Complex

the "gypsum pond" for storing this solid waste for a period of about five years to a gypsum stack height of 18 metres. As the gypsum pond is to be located within the industrial estate boundary and adjacent to the fertilizer plant, there appears to be no scope of further expansion of the gypsum pond after five years, since the land on all sides of the pond would be occupied by other industries. In view of this, the fertilizer project owner is presently exploring the possibility of export and/or utilization of the phosphogypsum.

Future Hazardous Waste Management Plan for Map Ta Phut Industrial Complex

The second stage of the industrial development project at the Map Ta Phut industrial complex would, like the first stage be largely the gas-based or fertilizer- and petrochemical-based industries and other organic chemical industries. These industries have been planned to include Caprolactum, Dimethyl Terephthalate (DMT), Acrylonitrile, Styrene and Melamine industries. Although the types of industries other than the above-mentioned hydrocarbon-based industries at Map Ta Phut are not precisely known, they are likely to be inorganic chemical industries, including mineral and metal processing. Two such projects, namely, a Titanium Dioxide project and a Nitric Acid project have so far been identified.

The pollution potential of both the organic and inorganic chemical industries is very high, as was revealed in the UNIDO study conducted in Thailand.[9] Although the Map Ta Phut industrial complex, due to its coastal location, has certain advantages in that the many non-hazardous inorganic pollutants released from industry could be disposed of in the sea without the need for elaborate treatment, detailed planning for effective waste management for the second stage of the development programme has nevertheless been carried out. It has been suggested that planning for the second stage of industrial development at Map Ta Phut should take into account the concept of "zoning" based on types of industry. The Map Ta Phut industrial complex will have two separate zones, one for organic and the other for inorganic industries.

Second, as far as practicable, planning for the water pollution control system would also be based on a centralized facility with individual industries providing the necessary pretreatment for their wastewater prior to discharging into the centralized facility. Figure 30.8 shows an outline of the proposed conceptual plan for zoning of the industries and for their wastewater management for the second stage industrial development programme at Map Ta Phut.

A regular monitoring programme for the liquid and gaseous emissions as well as solid wastes from the various sources including the pollution control facilities and for the receiving coastal waters and ambient air quality at Map Ta Phut has also been elaborated by the relevant agencies in co-operation with the owners of the industries.

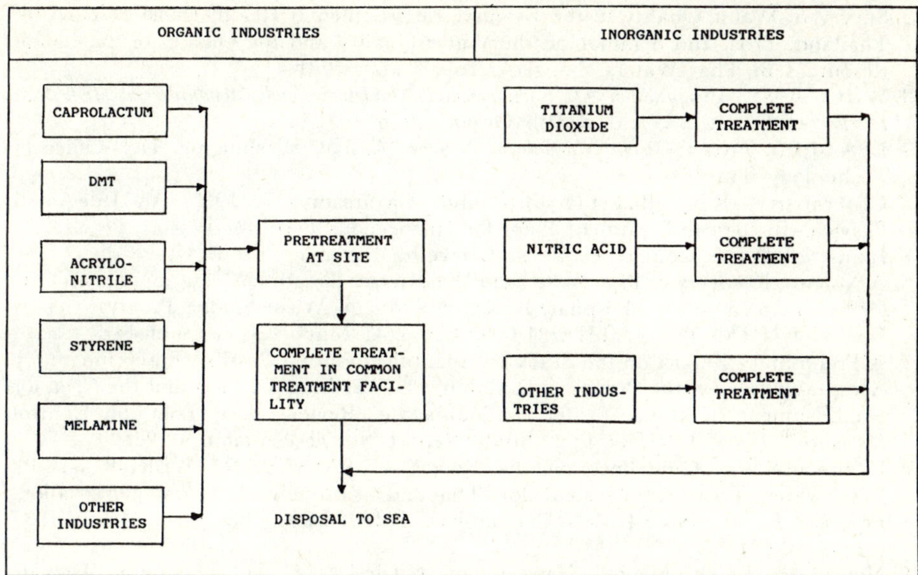

Figure 30.8 Proposed Plan for Wastewater Management of the Second Stage of Industrial Development at the Map Ta Phut Complex

In all the aforesaid activities and programmes relating to environmental pollution control and hazardous wastes management, UNIDO and UNDP are actively co-operating with and providing technical support to the Government of Thailand's Office of the Eastern Seaboard, Industrial Estate Authority of Thailand, National Environment Board and other agencies for the successful implementation of this economic development programme of the country.

Acknowledgements

The author wishes to thank Dr. Savit Bhotivihok, Director, Office of the Eastern Seaboard Development Committee, Mr. Prateeb Chuntaketta, Deputy Governor, Industrial Estate Authority of Thailand, Mr. Y.Y. Kim, Regional Representative, UNDP, Bangkok, and Dr. S. Maltezou, Industrial Development Officer, UNIDO, Vienna, for their valuable co-operation and support in this study. Some of the information and data presented in this report were obtained by courtesy of National Fertilizer Corporation, National Petrochemical Corporation and Thailand Tantalum Industry Corporation, Bangkok.

References

1. Ludwig, Harvey F. 1976. *Environmental Guidelines for Coastal Zone Management in Thailand – Inner Gulf Zone*. Bangkok: National Environment Board.

2. Survey of Water Quality in the Estuary, Environmental Health Division, Govt. of Thailand. Proc. 2nd Seminar on the Water Quality and the Quality of the Living Resources in Thai Waters, Bangkok, 26–28 May 1981.
3. WHO. 1983. *Compendium of Environmental Guidelines and Standards for Industrial Discharges*. Geneva: World Health Organization.
4. EPA. 1975. *Process Design Manual for Nitrogen Control*. Washington, DC: Office of Technology Transfer.
5. Chakrabarty, R.N., Bhalerao, B.B. and Alagarsamy, S. 1977. An Integrated Phsyical–Biological Treatment Plant for Nitrogenous Fertilizer Wastes. Proc. 32nd Industrial Waste Conference. West Lafayette, Indiana: Purdue University.
6. A Note on Recovery of Phosphate from Wastewater Treatment Plant Sludge of NFC for Saving in Import of Phosphate Rock and Reducing Waste Sludge. Progress Report No. 91. UNIDO Project THA/84/009. Bangkok, May 1986 (unpublished).
7. A Preliminary Report on the Study for Identification of a Possible Environmentally Acceptable Site for the Proposed Tantalum Plant at Map Ta Phut and the Need for Establishment of Standards for Pollutants and Review of its Pollution Control Measures. IEAT/UNIDO Joint Study Report No. 92. September 1986.
8. Comments and Guidelines on the Report of PASSAVANT–WERKE on the Wastewater Treatment System for Thailand Tantalum Industry Corporation. Progress Report No. 94. UNIDO Project THA/84/009. Bangkok, March 1987 (unpublished).
9. Notes on Environmental Hazards of Nitric Acid and Titanium Dioxide Manufacturing Plants proposed for Map Ta Phut. Progress Reports No. 83 and 84. UNIDO Project THA/84/009. Bangkok, August 1985 (unpublished).

The Contributors

J. AADAHL, President, Safeware Inc., Santa Clare, California, USA

H. ABELE, University of Economics, Vienna, Austria

S.A.S. ALMEIDA, President, Multiservice Engenharia Ltd., Rio de Janeiro, Brazil

A.K. BISWAS, Vice-President, International Association for Clean Technology, Oxford, UK

L.J.M.J. BLOMEN, Kinetics Technology International, Zoetermeer, the Netherlands

J.D. BRADY, Chairman, Crown Andersen Inc., Peachtree City, Georgia, USA

R.N. CHAKRABARTY, Environmental Pollution Control Adviser (UNIDO), Industrial Estate Authority of Thailand, Bangkok, Thailand

W.B. CLAPHAM, JR., Department of Geological Sciences, Cleveland State University, Cleveland, Ohio, USA

R.L. CUNHA, Consulting Engineer, Rio de Janeiro, Brazil

J.M. DAVE, School of Environmental Sciences, Jawaharlal Nehru University, New Delhi, India

D.P. de ALWIS, National Aquatic Resources Agency, Colombo, Sri Lanka

F. DEFREGGER, Director, Waste Management Division, Bavarian State Ministry for Regional Development and Environmental Affairs, Munich, FRG

A.J. EGGER, Austroplan, Vienna, Austria

K. FEDRA, International Institute for Applied Systems Analysis, Laxenburg, Austria

U. HAUPTMANNS, Gesellschaft für Reactorsicherheit (GRS), Cologne, FRG

D.R. HEDDEN, UOP Inc., Illinois, USA

W.A. IRWIN, Administrative Judge in US Department of Interior, Member of the Commission on Environment, Policy, Law and Administration of the International Union for Conservation of Nature and Natural Resources, (IUCN), Washington DC, USA

R.W. JOHNSON, UOP Inc., Illinois, USA

T.N. KALNES, UOP Inc., Illinois, USA

C.R. KRISHNA MURTI, Cancer Institute (WIA), Adayar, Madras, India

A.S. LAGHATE, Kinetics Technology International, Zoetermeer, the Netherlands

F. LANGEWEG, Head, Waste Products and Emissions, National Institute of Public Health and Environmental Hygiene, Bilthoven, The Netherlands

N.C. LIND, Director, Institute for Risk Research, University of Waterloo, Waterloo, Ontario, Canada

S.P. MAHAJAN, Indian Institute of Technology, Bombay, India

S.P. MALTEZOU, Industrial Operations Technology Division, Department of Industrial Operations, United Nations Industrial Development Organization, Vienna, Austria and Vice-President, International Association for Clean Technology.

J.H. MEYER, Consulting Geologist, Vienna, Austria

C.P. NAISH, Ciba-Geigy, Basle, Switzerland

D. PHANTUMVANIT, Associate Director, Natural Resources and Environment Programme, Thailand Development Research Institute, Bangkok, Thailand

SHI QING, Head, Solid Waste Management Office, National Environmental Protection Agency, Beijing, China

R. SCHAAF, Kinetics Technology International, Zoetermeer, The Netherlands

H. SUTTER, President, International Association for Clean Technology and Department of the Environment, Berlin, FRG

H.J. TEN DOESSCHATE, Kinetics Technology International, Zoetermeer, The Netherlands

N.C. THANH, Director, Department of Environment and Water Resources, International Training Center for Water Resources Management, Sophia Antipolis, France

F.A. URIARTE, Director, Bureau of Science and Technology, ASEAN Secretariat, Jakarta, Indonesia

P.F. VAN DEN OOSTERKAMP, Kinetics Technology International, Zoetermeer, The Netherlands

C.J. VAN KUIJEN, Ministry of Housing, Physical Planning and the Environment, Leidschendam, The Netherlands

J.A.C. VAN STRATUM, Managing Director, Montair Andersen B.V., The Netherlands

N.K. VESTERGAARD, Project Engineer, Chemcontrol A/S, Daqmarhus, Denmark

K. WHITING, Process Engineering Manager, Kaldair Ltd., Sunbury-on-Thames, Middlesex, UK

K.J. YOUTSEY, UOP Inc., Illinois, USA